THE PSYCHOLOGY OF
HUMAN-COMPUTER INTERACTION

The Psychology of Human-Computer Interaction

Stuart K. Card
Thomas P. Moran
Xerox Palo Alto Research Center

Allen Newell
Carnegie-Mellon University

 LAWRENCE ERLBAUM ASSOCIATES, PUBLISHERS

1983   Hillsdale, New Jersey

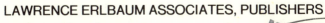

Lawrence Erlbaum Associates, Inc. Publishers
365 Broadway
Hillsdale, New Jersey 07642

Library of Congress Cataloging in Publication Data

Card, Stuart K.
The psychology of human-computer interaction.

Bibliography: p.
Includes index.
1. Interactive computer systems—Psychological aspects.
I. Moran, Thomas P. II. Newell, Allen. III. Title.
QA76.9I58C37 1983   001.64'01'9   82-21045
ISBN 0-89859-243-7
ISBN 0-89859-859-1 (pbk.)
Printed in the United States of America
10 9 8 7 6 5 4

Contents

Preface

Designing interactive computer systems to be efficient and easy to use is important so that people in our society may realize the potential benefits of computer-based tools. Our purpose in this book is to help lay a scientific foundation for an applied psychology concerned with the human users of interactive computer systems. Although modern cognitive psychology contains a wealth of knowledge of human behavior, it is not a simple matter to bring this knowledge to bear on the practical problems of design—to build an applied psychology that includes theory, data, and methodology.

This book is our attempt to span the gap between science and application. We have tackled a small piece of the general problem. With respect to computer science, we have focused on the task domain of text-editing and similar types of highly interactive systems. With respect to psychology, we have focused on the notion of the expert user's cognitive skill in interacting with the system, especially the temporal aspects of the interaction. We have constructed an empirically-based cognitive theory of skilled human-computer interaction in this domain. This theory is our keystone for linking science and application. On one side, we have shown that the theory is a consistent extension of the science of human information-processing. On the other side, we have simplified the theory into practical engineering models, which are the tools for designers to apply the theory. Thus, in addition to putting forth specific psychological models in this book, we have tried to make clear the general framework of an applied psychology, in which these models are but prototypical examples.

THE AUDIENCE FOR THIS BOOK

Interest in the topic of human-computer interaction is shared by people from a range of disciplines. We believe this book makes contact with the specific interests of all of these disciplines. For instance:

vii

(1) Cognitive psychologists will find that theory and empirical methods can be extended to the analysis of a real-world domain and that a practical problem can be a fruitful vehicle for developing basic psychology.

(2) Computer scientists will find that the problem of matching computer power with user abilities may be approached using the theory and methods of the cognitive sciences.

(3) System designers will find that we have derived a number of models and principles of user performance that may be used in design.

(4) Human factors specialists, ergonomists, and human engineers will find that we have synthesized ideas from modern cognitive psychology and artificial intelligence with the old methods of task analysis and brought them to bear on the human-computer interface—which is rapidly becoming the most important domain in human factors practice.

(5) Engineers in several fields concerned with man-machine systems will find that we have extended the notion of work analysis by showing how techniques from cognitive science can be applied to the analysis of procedures that are predominantly mental.

We have used the book as the primary reference in a graduate course on "Applying Cognitive Psychology to Computer Systems," taught (by TM and SC) in the Departments of Psychology and Computer Science at Stanford University (Moran and Card, 1982). Parts of the book, in manuscript, have proven useful to others in teaching similar courses in psychology, computer science, and industrial engineering. The book would be suitable for a variety of courses: (1) a course on human factors in computer systems within a computer science department; (2) a course on human-computer interface design within a computer science department; (3) a course on the psychology of computer users within a psychology department; (4) a course on human-computer interaction within an industrial engineering or human factors department; (5) an advanced research seminar in either computer science, psychology, or industrial engineering; or (6) in a focused short course for industrial professionals. For courses with a design focus, Chapters 1 and 2 can be used to provide psychological background; and Chapters 3, 5, 7, 8, 9, and 12 can be used for analytical and practical content. For courses stressing psychological issues, Chapters 1, 2, 5, 7, 8, 10, and 11 can be used to develop basic concepts and theory.

HISTORY OF THIS RESEARCH

In 1970, Xerox established a new major research center in Palo Alto with the express purpose of exploring digital electronic technologies in support of Xerox's general concern with office information systems. Since that time, the Palo Alto Research Center (PARC) has become well known for its developments in interactive computing, based on personal computers with integral high quality graphic displays (the Alto being the first such computer), connected by a high capacity local network (the Ethernet). It has become known, as well, for being the first living embodiment of this new computational style.

From the start (early 1971) there were discussions between George Pake (then head of PARC), Robert Taylor (now manager of the Computer Sciences Laboratory of PARC), and one of us (AN, as a consultant to PARC) about the possibilities of an active role for psychological research into human interaction with computers. PARC seemed like the perfect place to attempt such an effort. Modern cognitive psychology had come a long way in understanding man as a processor of information, a view that meshed completely with the developments in computer science and artificial intelligence—indeed, derived from them in a number of particulars. The impact of the psychological advances on the human factors of how computers were used was not yet very great, though the potential was clearly there. PARC itself, being both an industrial laboratory with the concomitant underlying emphasis on application and a group engaged in basic research in computer science and artificial intelligence, provided exactly the right environment.

In 1974, opportunity became reality through Jerome Elkind (who had joined PARC to become manager of the Computer Sciences Laboratory). Two of us (TM and SC) joined PARC, and a small unit, called the *Applied Information-Processing Psychology Project* (AIP), was formed. Its charter was to create an applied psychology of human-computer interaction by conducting requisite basic research within a context of application. It was initially located within the Systems Sciences Laboratory, a sister laboratory to the Computer Sciences Laboratory, under William English, who was in charge of a group constructing an experimental interactive office-information system. One reason for its location was the early decision to concentrate on immediate, real-time human-computer interaction, especially as embodied in the use of text-editing systems, rather than on the activities of programming computers. The AIP group has remained intact through many local reorganizations and is presently a part of the Cognitive and Instructional Sciences Group.

The present book, then, presents the results of some of the main strands of the AIP group's research. The group has throughout consisted of just the three of us, in equal collaboration (SC and TM at PARC, with AN as a consultant), supported by research assistants, students, and colleagues in PARC and elsewhere.

ACKNOWLEDGEMENTS

As should be evident from the remarks above, we owe an immense debt to the PARC environment. A few of the people who played a key role in the creation of PARC were mentioned above. It is not possible to enumerate all the individuals who have played a definite role in making our tiny research group viable over the years. We would, however, like to acknowledge a few, both inside and outside of PARC. Harold Hall, Manager of the PARC Science Center, provided support for our studies in his several managerial capacities (the analysis in Chapter 9 is the result of a question he posed to us). Bert Sutherland, as Manager of the Systems Sciences Laboratory, played an important role in supporting us and allowing us the resources to pursue these studies. John Seely Brown, as Area Manager for the Cognitive and Instructional Sciences, has had a major impact on us by creating a stimulating intellectual environment of cognitive scientists around us.

Don Norman and Richard Young provided extensive substantive comments on the research reported in the book. Many productive discussions with colleagues have influenced our thinking and helped us formulate our position. They include: George Baylor, John Black, Danny Bobrow, Ross Bott, Ted Crossman, Jerry Elkind, Austin Henderson, Ron Kaplan, Tom Malone, Jim Morris, William Newman, Beau Sheil, Larry Tesler, and Mike Williams. Several students, working with us at PARC, have kept us on our toes: Terry Roberts, Marilyn Mantei, Jarrett Rosenberg, Allen Sonafrank, Lucy Suchman, Keith Patterson, Kathy Hemenway, Brian Ross, Sally Douglas, Frank Halasz, and Carolyn Foss.

Ralph Kimball, Robin Kinkead, Bill Bewley, and Bill Verplank—in the development divisions of Xerox—gave us valuable advice and helped us test some of the models in this book. Steve Smith and Shmuel Oren provided mathematical consulting. Warren Teitelman and Larry Masinter provided programming help in Interlisp (Teitelman, 1978), the system in which all of our analysis and simulation programs are written. Ron Kaplan and Beau Shiel provided statistical consultation for the analysis of our data and help in using the Interactive Data-analysis Language (Kaplan, Sheil, and Smith, 1978), their statistical analysis system written in Interlisp.

Our experimental work would not have been possible without help and support in building and maintaining our laboratory systems and equipment. Bill Duvall and George Robertson implemented our experiment-running systems; and Jim Mayer, Bill Winfield, and others kept our equipment running. The large amount of experimentation and detailed analysis would not have been possible without the help of several research assistants over the years: Betty Burr, Janet Farness, Steve Locke, Marilyn Mantei, Beverly McHugh, Terry Roberts, Rachel Rutherford, and Betsey Summers.

Many others at PARC have also been of help. Chris Jeffers and Jeanie Treichel provided administrative backing. Barbara Baird, Connie Redell, Malinda Maggiani, and Jackie Guibert provided secretarial support. Giuliana Lavendel and her library staff tracked down many obscure references for us.

A number of people have helped directly with the production of the book. Rachel Rutherford helped edit the text and brought to light numerous errors, inconsistencies, and infelicities of expression. Betsey Summers, Steve Locke, and Leslie Keenan helped manage and proof the text and figures. Bill Bowman gave graphics advice on several figures. Lyle Ramshaw guided us through the intricacies of various document preparation systems. Terri Doughty helped us format the text and tables for galley printing. The galleys for the book were printed on an experimental phototypesetting printer developed at PARC.

In the preparation of this book—much of it about text-editing—we have ourselves been heavy users of the computer text-editors. We have spent several thousands of hours text-editing on BRAVO (one of the systems we describe in the book) at tasks similar to those we have studied on our subjects, performing perhaps a million editing tasks in the process. From this experience and study, we have a great appreciation for the display-based text-editing technology that our colleagues at PARC have been able to fashion.

We have no doubt missed some people who deserve mention. One advantage of writing a book of this kind is that our excuse—that human information-processing systems are limited—is contained herein. As we explain in Chapter 2, searching Long-Term Memory requires considerable effort, and we have not managed to move the full way along the information retrieval curve pictured in Figure 2.27.

SC, TM, AN
Palo Alto
October 1982

Copyright Acknowledgments

Portions of several chapters in this book have been previously published. The authors wish to thank the following publishers for permission to use this material:

Academic Press, Inc. for permission to use in Chapters 5 and 11 portions of Card, S. K., Moran, T. P., and Newell, A.; Computer text-editing: An information-processing analysis of a routine cognitive skill; *Cognitive Psychology 12*, 32–74, (c) 1980.

Taylor & Francis Ltd. for permission to use in Chapter 7 portions of Card, S. K., English, W., and Burr, B.; Evaluation of mouse, rate-controlled isometric joystick, step keys, and text keys for text selection on a CRT; *Ergonomics 21*, 601–613, (c) 1978.

The Association for Computing Machinery for permission to use in Chapter 8 portions of Card, S. K., Moran, T. P., and Newell, A.; The keystroke-level model for user performance time with interactive systems; *Communications of the ACM 23*, 396–410, (c) 1980.

The authors also thank the following for permission to reproduce copyrighted figures:

Figure 2.4 from Fig. 1 (p. 427) of Cheatham, P. G., and White, C. T., *Journal of Experimental Psychology 47*, 425–428, (c) 1954 by the American Psychological Association. Reprinted by permission of the publisher and authors. **Figure 2.9** from Fig. 5 (p. 94) of Michotte, A., *The Perception of Causality*. English translation, (c) 1963 by Methuen and Co, Ltd. Reprinted by permission of Basic Books, Inc. **Figure 2.11** from Fig. 5.3 (p. 146) and Fig. 5.4 (p. 148) of Welford, A. T., *Fundamentals of Skill*, published by Methuen & Co., (c) 1968 by A. T. Welford. Reprinted by permission of the author. **Figure 2.13** from Klemmer, E. T., *Human Factors 4*, 75–79, (c) 1962 by the Human Factors Society. Reprinted by permission of publisher. **Figure 2.21** from Fig. 2 (p. 8) of Posner, M. I., Boies, S. J., Eichelman, W. H., and Taylor, R. L., *Journal of Experimental Psychology 79*, 1–16, (c) 1969 by the American Psychological Association. Reprinted by permission of the publisher and authors. **Figure 2.22** from Fig. 3.1 (p. 62) of Welford, A. T., *Fundamentals of Skill*, published by Methuen & Co., (c) 1968 by A. T. Welford. Reprinted by permission of the author. **Figure 2.23** from Fig. 1 (p. 192) of Hyman, R., *Journal of Experimental Psychology 45*, 188–196, (c) 1953 by the American Psychological Association. Reprinted by permission of the publisher and author. **Figure 2.24** from Glanzer, M., and Cunitz, A. R., *Journal of Verbal Learning and Verbal Behavior 5*, 351–360, (c) 1966 by Academic Press, Inc. Reprinted by permission of the publisher and authors. **Figure 2.25** from Fig. 2 (p. 358) of *Human Experimental Psychology* by Calfee, R. C., (c) 1975 by Holt, Rinehart and Winston, Inc. Reproduced by permission of Holt, Rinehard and Winston, Inc. **Figure 2.26** from Fig. 3 (p. 53) of Underwood, B. J., *Psychological Review 64*, 49–60, (c) 1957 by the American Psychological Association. Reprinted by permission of the publisher and author. **Figure 2.28** from Fig. 3 (p. 122) of Mills, R. G., and Hatfield, S. A., *Human Factors 16*, 117–128, (c) 1974 by the Human Factors Society. Reprinted by permission of the publisher. **Figure 2.30** from Fig. 6.4 (p. 181) of Newell, A., and Simon, H. A., *Human Problem Solving*, (c) 1972. Reprinted by permission of Prentice-Hall, Inc., Englewood Cliffs, New Jersey.

1. An Applied Information-Processing Psychology

A scientific psychology should not only help us to understand our own human nature, it should help us in our practical affairs. In educating our children, it should help us to design environments for learning. In building airplanes, it should help us to design for safety and efficiency. In staffing for complex jobs, it should help us to discover both the special skills required and those who might have them. And on and on. Given the breadth of environments we design for ourselves, there is no limit to the number of domains where we might expect a scientific knowledge of human nature to be of use.

The domain of concern to us, and the subject of this book, is how humans interact with computers. A scientific psychology should help us in arranging this interface so it is easy, efficient, error-free—even enjoyable.

Recent advances in cognitive psychology and related sciences lead us to the conclusion that knowledge of human cognitive behavior is sufficiently advanced to enable its applications in computer science and other practical domains. The years since World War II have been the occasion for an immense wave of new understandings and new techniques in which man has come to be viewed as an active processor of information. In the last decade or so, these understandings and techniques have engulfed the main areas of human experimental psychology[1]: perception,[2] performance,[3] memory,[4] learning,[5] problem solving,[6] psycholinguistics.[7] By now, cognitive psychology has come to be dominated by the information-processing viewpoint.

A major advance in understanding and technique brings with it, after some delay, an associated wave of applications for the new knowledge. Such a wave is about to break in psychology. The information-processing view will lead to a surge of new ways for making psychology relevant to our human needs. Already the concepts of information-processing psychology have been applied to legal eyewitness testimony[8] and to the design of intelligence tests.[9] And in the study of man-machine systems and engineering psychology, it has for some time been common to include a block diagram of the overall human information-processing system in the introductory chapter of textbooks,[10] even though the reach of that block diagram into the text proper is still tenuous. There are already the beginnings of a subfield, for which various names (associating the topic in different ways) have been suggested: user sciences,[11] artificial psycholinguistics,[12] cognitive ergonomics,[13] software psychology,[14] user psychology,[15] and cognitive engineering.[16]

---

[1] For representative examples see Lindsay and Norman's (1977) *Human Information Processing*, Anderson's (1980) *Cognitive Psychology and its Implications*, the *Handbook of Learning and Cognitive Processes* (Estes, ed. 1975-1978), the *Attention and Performance* collections of papers (Kornblum, 1973; Rabbitt and Dornič, 1975; Dornič, 1977; Requin, 1978; Long and Baddeley, 1981), and the journal *Cognitive Psychology*.

[2] Examples: Broadbent (1958), *Perception and Communication*; Green and Swets (1966), *Signal Detection Theory and Psychophysics*; Neisser (1967), *Cognitive Psychology*; Cornsweet (1970) *Visual Perception*.

[3] Examples: Fitts and Posner (1967), *Human Performance*; Welford (1968), *Fundamentals of Skill*; Kintsch (1974), *The Representation of Meaning in Memory*; Tversky (1977), "Feature of similarity"; Posner (1978), *Chronometric Explorations of the Mind*.

[4] Examples: Anderson and Bower (1973), *Human Associative Memory*; Baddeley (1976), *The Psychology of Memory*; Crowder (1976), *Principles of Learning and Memory*; Murdock (1974), *Human Memory, Theory and Data*.

[5] Examples: Fitts (1964), "Perceptual-motor skill learning"; Klahr and Wallace (1976), *Cognitive Development: An Information-Processing View*; Anderson (1981a), *Cognitive Skills and their Acquisition*.

[6] Example: Newell and Simon (1972), *Human Problem Solving*.

Our own goal is to help create this wave of application: to help create an applied information-processing psychology. As with all applied science, this can only be done by working within some specific domain of application. For us, this domain is the human-computer interface. The application is no offhand choice for us, nor is the application dictated solely by its extrinsic importance. There is nothing that drives fundamental theory better than a good applied problem, and the cognitive engineering of the human-computer interface has all the markings of such a problem, both substantively and methodologically. Society is in the midst of transforming itself to use the power of computers throughout its entire fabric—wherever information is used—and that transformation depends critically on the quality of human-computer interaction. Moreover, the problem appears to have the right mixture of industrial application and symbol manipulation to make it a "real-world" problem and yet be within reasonable reach of an extended cognitive psychology. In addition, we have personal disciplinary commitments to computer science as well as to psychology.

This book reports on a program of research directed towards understanding human-computer interaction, with special reference to text-editing systems. The program was undertaken as an initial step towards the applied information-processing psychology we seek. Before outlining individual studies, it is appropriate to sketch how this effort fits in with the larger endeavor.

---

[7] Example: Clark and Clark (1976), *Psychology and Language: An Introduction to Psycholinguistics.*

[8] Loftus (1979).

[9] Hunt, Frost, and Lunneborg (1973).

[10] Sheridan and Ferrell (1974); McCormick (1976).

[11] Vallee (1976).

[12] Sime and Green (1974).

[13] Sime, Fitter, and Green (1975).

[14] Shneiderman (1980).

[15] Moran (1981a).

[16] Norman (1980).

1.1. THE HUMAN-COMPUTER INTERFACE

The human-computer interface is easy to find in a gross way—just follow a data path outward from the computer's central processor until you stumble across a human being (Figure 1.1). Identifying its boundaries is a little more subtle. The key notion, perhaps, is that the user and the computer engage in a communicative dialogue whose purpose is the accomplishment of some task. It can be termed a dialogue because both the computer and the user have access to the stream of symbols flowing back and forth to accomplish the communication; each can interrupt, query, and correct the communication at various points in the process. All the mechanisms used in this dialogue constitute the interface: the physical devices, such as keyboards and displays, as well as computer's programs for controlling the interaction.

At any point in the history of computer technology there seems to be a prototypical user interface. A few years ago it was the teletypewriter; currently it is the alphanumeric video-terminal. But the actual diversity is now much greater. All so-called "remote entry" devices count as interfaces; and a large number of such specialized devices exist in the commercial and industrial world to record sales, maintain inventory records, or control industrial processes. Almost all such devices are fashioned from the same basic sorts of components (keyboards, buttons, video displays, printers) and connect to the same sorts of information-processing mechanisms (disks, channels, interrupt service routines).

The very existence of the direct human-computer interface is itself an emergent event in the development of computers. If we go back twenty years, the dominant scheme for entering information into a computer consisted of a trio of people. First there was the user, someone who wanted to accomplish some task with the aid of the computer. The user encoded what he wanted onto a coding sheet, then sent it to a second person, the keypunch operator, who used an off-line device, the keypunch, to create a deck of punched cards that encoded the same information in a different form. The cards in turn went to a third person, the computer-operator, who entered the cards into the computer via the card reader. The computer then responded by printing messages and data on paper for the operator to gather up and send back to the user. The relationship between the user and the computer was sufficiently remote that it should be likened more to a literary correspondence than to a conversational dialogue. It is the general

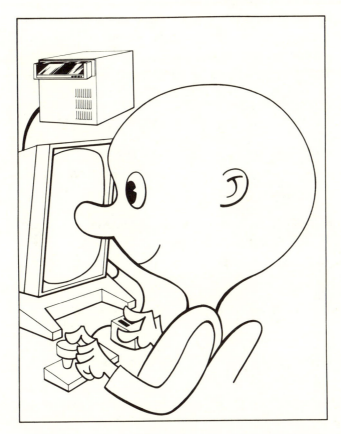

Figure 1.1. The human-computer interface.

demise of such arrangements involving human intermediaries, and the resultant coupling of the user directly to the computer, that has given rise to the contemporary human-computer interface. Whatever continued evolution the interface takes—and it will be substantial—human-computer interaction is unlikely ever to lose this character of a conversational dialogue.

Of course, there is much more to improving computer interfaces than simply making them conversational. Informal evidence from the direct experience of users provides numerous examples of current interface deficiencies:

In one text-editing system, typing the word *edit* while in command mode would cause the system to select _e_very-thing, _d_elete everything, and then _i_nsert the letter _t_ (this last making it impossible to use the system Undo command to recover the deleted text because only the last command could be undone).

In another text-editing system, so many short commands were defined that almost any typing error would cause some disaster to happen. For example, accidentally typing CONTROL-E would cause the printer to be captured by the user. Since no indication of this event was given, no other users would be permitted to print until the other users eventually discovered who had the printer. In an even more spectacular instance, accidentally typing CONTROL-Z would delete all the user's files—permanently.

In one interactive programming system, misspelling a variable name containing hyphens (a common way of marking off parts of a name) would cause the system to rewrite the user's program, inserting code to subtract the parts of the name. In many cases, the user would have to mend his program by hand, laboriously searching for and editing the damaged code.

In a set of different subsystems meant to be used together, the name "List" was given to many different commands, each having a different meaning: (1) send a file to the printer to make a hardcopy, (2) show the directory of files on the display, (3) show the content of a file on the display, (4) copy the workspace to a file, (5) create a particular kind of data structure.

Yet, when one looks at the teletype interfaces of yesterday, it is clear that substantial progress has been made. The emergence of the direct human interface, circumventing the keypuncher and operator, must itself be counted as an improvement of enormous value. We now have interfaces that allow the use of computers for such highly interactive tasks as making engineering drawings and taking airline reservations. But despite considerable advancements, the systems we have are often ragged and in places are sufficiently poor to cripple whole ranges of use.

What strikes one most noticeably about existing interfaces, besides all the little ways they fail, is that their failures appear to be unnecessary. Why, when interaction could be so smooth, even elegant, is it often so rough, even hazardous? Two observations may help explain this perplexing state of affairs.

First, interaction with computers is just emerging as a human activity. Prior styles of interaction between people and machines—such as driver and automobile, secretary and typewriter, or operator and control room—are all extremely lean: there is a limited range of tasks to be accomplished and a narrow range of means (wheels, levers, and knobs) for accomplishing them. The notion of the *operator* of a machine arose out of this context. But the user is not an operator. He does not operate the computer, he communicates with it to accomplish a task. Thus, we are creating a new arena of human action: communication *with* machines rather than operation *of* machines. What the nature of this arena is like we hardly yet know. We must expect the first systems that explore the arena to be fragmentary and uneven.

Second, the radical increase in both the computer's power and its performance/cost ratio has meant that an increasing amount of computational resources have become available to be spent on the human-computer interface itself, rather than on purely computational tasks. This increase of deployable resources exacerbates the novelty of the area, since entirely new styles of interaction become available coincidentally with an increased amount of computational ability available per interaction. These new styles often lead to completely new interfaces, which are then even more ragged than before. At the same time, opportunities for the invention of good interfaces also increase rapidly, accounting for the leaps and bounds we have seen in terms of major improvements in functionality and ease of use.

1.2. THE ROLE OF PSYCHOLOGY

Many in the computer field agree that there is an obvious way to design better human-computer interfaces. Unfortunately, they disagree on what it is. It is obvious to some that psychological knowledge should be applied. Their slogan might be, in the words of Hansen (1971): "Know the user!" It is obvious to others that the interface should simply be designed with more care—that if designers were given the goal of good interfaces, rather than stringent cost limits or tight deadlines, then they would produce good designs. Their slogan might be: "Designers are users too—just give them the time and freedom to design it right!" And it is obvious to others still that one should pour the effort into some new components—flat displays, color graphics, or dynamically codeable micro-processors in the terminal. Their slogan might be: "Make the components good enough and the system will take care of itself!"

Who is to gainsay each of these their point? The technology limits, often severely, what can be done. All the human engineering in the world will not turn a 10-character-per-second teletypewriter into a high-resolution graphics terminal. The history of terminal development so far is writ largely in terms of advances in basic interface components, most notably the resources to allow substantial computational cycles to be devoted to the interface. It is easy to point to current limitations whose lifting will improve the interface by orders of magnitude. Immense gains will occur when the display holds not the common 24 × 80 characters (the typical alphanumeric video terminal, widely available today), but a full page of 64 × 120 characters (the typical 1000 × 800 pixel video terminal, available at a few places today), or even the full drafting board of 512 × 512 characters (not really available anywhere, yet, as far as we know).

Moreover, any accounting will have to credit the majority of the capabilities and advances at the interface to design engineers and only a few of them to psychologists. However many imperfections there remain in the interface, the basic capabilities and inspired creations that do exist came out of an engineering analysis of the functions needed and the fact that the designer, being human, could empathize directly with the user.

And yet, there remain the mini-horror stories—of systems where, after the fact, it became clear that either the nature or the limitations of the user were not appreciated, and some design foolishness was committed. Since it is these stories that come to mind in discussing the role of the human at the interface, it is often assumed that all that one needs are ways of checking to be sure that the obvious is not overlooked; "All we need from psychology is a few good checklists!" might be the slogan here. But as we shall see, there is more to human-computer interaction than can be caught with checklists.

The role psychology might be expected to play in the design of the user-computer interface is suggested by the results it was able to achieve for military equipment during World War II.   At that time, it had become apparent that a strong limiting factor in realizing the potential of man-machine systems, such as radar sets and military aircraft, lay in the difficulty of operating the equipment. Out of a wartime collaboration between natural scientists, engineers, and psychologists came major advances, not only with respect to the man-machine systems being designed, but also with respect to psychological theory itself. Examples of the latter include the theory of signal detection, manual control theory, and a methodology for the design of cockpit instrument displays. That with psychological attention to human performance airplanes became more flyable encourages us to believe that with psychological attention to human performance computers can become more usable.

1.3.   THE FORM OF AN APPLIED PSYCHOLOGY

What might an applied information-processing psychology of human-computer interfaces be like and how might it be used? Imagine the following scenario:

> A system designer, the head of a small team writing the specifications for a desktop calendar-scheduling system, is choosing between having users type a key for each command and having them point to a menu with a lightpen.   On his whiteboard, he lists some representative tasks users of his system must perform. In two columns, he writes the steps needed by the "key-command" and "menu" options. From a handbook, he culls the times for each step, adding the step times to get total task times. The key-command system takes less time, but only slightly.   But, applying the analysis from another section of the handbook, he calculates that the menu system will be faster to learn; in fact, it will be learnable in half the time.   He has estimated previously that an effective menu system will require a more expensive processor: 20% more memory, 100% more microcode memory, and a more expensive display. Is the extra expenditure worthwhile? A few more minutes of calculation and he realizes the startling fact that, for the manufacturing quantities anticipated, training costs

for the key-command system will exceed unit manufacturing costs!   The increase in hardware costs would be
much more than balanced by the decrease in training costs,
even before considering the increase in market that can be
expected for a more easily learned system.   Are there
advantages to the key-command system in other areas,
which need to be balanced?   He proceeds with other
analyses, considering the load on the user's memory, the
potential for user errors, and the likelihood of fatigue.   In
the next room, the Pascal compiler hums idly, unused,
awaiting his decision.

The system designer is engaged in a sort of psychological civil
engineering, trading computed parameters of human performance against
cost and other engineering variables.   The psychological science base
necessary to make possible his design efforts is the sort of applied
psychology that is the topic of this book.   Such a psychology must
necessarily be homogeneous in form with the rest of the engineering
science base to allow tradeoffs between psychological and other design
considerations.   To be useful, we would argue, such a psychology must
be based on task analysis, calculation, and approximation.

Task Analysis.   When psychology is applied in the context of a
specific task,   much of the activity hardly seems like psychology at all,
but rather like an analysis of the task itself.   The reason for this is clear:
humans behave in a goal-oriented way.   Within their limited perceptual
and information-processing abilities, they attempt to adapt to the task
environment to attain their goals.   Once the goals are known or can be
assumed, the structure of the task environment provides a large amount
of the predictive content of psychology.

Calculation.   The ability to do calculations is the heart of useful,
engineering-oriented applied science.   Without it, one is crippled.   Applications are, of course, still possible, as witness mental testing, behavior
modification, assertiveness training, and human-factors investi-gations of
display readability.   But what is needed to support an engineering
analysis are laws of parametric variation, applicable on the basis of a task
analysis.

Psychology is not strong on calculation, though a few useful laws,
such as   Power Law of Practice, exist.   The reason might be thought to
be an inherent characteristic of psychology, or maybe even more
generally, of all human sciences.   Our view is the opposite.   Psychology

is largely non-calculational because it has followed a different drummer. It has been excessively concerned with hypothesis testing—with building techniques to discriminate which of two ideas is right. If one changes what one wants from the science, one will find the requisite techniques. Interestingly, a branch of the human sciences, work-measurement industrial engineering, indeed asked a different question—namely, how long would it take people to do preset physical tasks—and it obtained useful answers.

Approximation. If calculations are going to be made rapidly, they are necessarily going to be over-simplified. Nature—especially human nature—is too complex to be written out on the back of an envelope. But in engineering, approximations are of the essence. It is vital to get an answer good enough to dictate the design choice; additional accuracy is gilding the computational lily.

Again, psychology has in general not asked after approximations, though it has certainly learned to talk in terms of simplified models. The neglect of approximation has been especially encouraged by the emphasis on statistical significance rather than on the magnitude of an effect. A difference of a few percent in performance at two levels of an independent variable is usually of little practical importance and can often be ignored in an approximation, even if the difference is highly significant statistically. But if there is no external criterion—no design decision to be made, for instance—then there is no way to tell which approximations are sufficient.

But, whereas an applied psychology of human-computer interaction should be characterized by task analysis, calculation, and approximation, these are not the only considerations. It is obvious that an applied psychology intended to support cognitive engineering should also be relevant to design. It is less obvious, but nonetheless true, that to be successful, an applied psychology should be theory-based.

RELEVANT TO DESIGN

Design is where the action is in the human-computer interface. It is during design that there are enough degrees of freedom to make a difference. An applied psychology brought to bear at some other point is destined to be half crippled in its impact.

We suspect that many psychologists would tend to pick evaluation as the main focus for application (though some might have picked training). Evaluation is what human factors has done best. Given a real system, one can produce a judgment by experimentation. Thus, the main tool in the human-factors kit has been the methodology of experimental design, supported by concomitant skill in experimental control and in statistics with which to assess the results. The emphasis on evaluation is widespread: There is a whole subfield of psychology whose concern is to evaluate social action programs. The testing movement is fundamentally evaluational in character, whether concerned with intelligence testing or with clinical assessment.

Applying psychology to the evaluation of systems is assuredly easier than applying it to the design of systems. In evaluation, the system is given; all its parts and properties are specified. In design, the system is still largely hypothetical; it is a class of systems. On the other hand, there is much less leverage in system evaluation than in system design. In design, one wants results expressed explicitly as a function of some controllable parameters, in order to explore optimization and sensitivity. In evaluation, this urge is much diminished; experimental evaluation is so expensive as to be prohibitive, permitting exploration of only two or three levels of each independent variable. Most importantly, by the time a system is running well enough to evaluate, it is almost inevitably too late to change it much. Thus, an applied psychology aimed exclusively at evaluation is doomed to have little impact.

There are several choices for how to institutionalize an applied psychology. First, psychologists could be the primary professionals in the field. Though possible in some fields, such as mental health, counseling, or education, we think this arrangement unlikely for computers. The field is already solely in the possession of computer engineers and scientists. Second, psychologists could be specialists, either as members of separate human-factors units within the organizations or as another individual specialty within the primary design team. Our reasons for not favoring separate psychology units reflect the additional separation we believe they imply between the psychology and the development of interfaces. Application of psychology would shift too strongly towards evaluation and away from the main design processes.

We favor a third choice: that the primary professionals—the computer system designers—be the main agents to apply psychology. Much as a civil engineer learns to apply for himself the relevant physics of bridges, the system designer should become the possessor of the relevant applied psychology of human-computer interfaces. Then and only then will it become possible for him to trade human behavioral considerations against the many other technical design considerations of system configuration and implementation. For this to be possible, it is necessary that a psychology of interface design be cast in terms homogeneous with those commonly used in other parts of computer science and that it be packaged in handbooks that make its application easy. Thus, the system designer in our scenario finds the design handbook more efficient to use than plunging blindly into code with his Pascal compiler, although he may still find it profitable to engage in exploratory implementation.

THEORY-BASED

An applied psychology that is theory-based, in the sense of articulating a mechanism underlying the observed phenomena, has advantages of insight and integration over a purely empirical approach. The point can be made by reference to two examples of behavioral science lacking a strong theory in this sense: work-study industrial engineering, referred to earlier, and intelligence testing. Rather than develop the theory of skilled movement, the developers of the several movement time systems chose an empirical approach, tabulating the times to make various classes of movements and ignoring promising theoretical developments such as Fitts's Law (at least until recently). Although their tables of motion times ran to four significant figures, they ignored the variance of the times and interactions between sequential motions, thus rendering the apparent precision illusory. This lack of adequate theoretical development made the work, despite its impressive successes, vulnerable to attacks from outside the field (see Abruzzi, 1956; Schmidtke and Stier, 1961). Similarly, in mental testing, the lack of a psychological theory of the mental mechanisms underlying intelligence (as opposed to a purely statistical theory of test construction) has put the validity of mental tests in doubt despite, again, impressive successes.

It is natural for an applied psychology of human-computer interaction to be based theoretically on information-processing psychology, with the latter's emphasis on mental mechanism. The use of models in which man is viewed as a processor of information also provides a common framework in which models of memory, problem solving, perception, and behavior all can be integrated with one another. Since the system designer also does his work in information-processing terms, the emphasis is doubly appropriate. The lack of this common framework is one reason why it would be difficult to meld in important techniques such as the use of Skinnerian contingent reinforcement. It is not that the techniques are not useful in general, nor that they cannot be applied to the problems of the human-computer interface; but within the framework that underlies this book, they would show up as isolated techniques.

The psychology of the human-computer interface is generally individual psychology: the study of a human behaving within a non-human environment (though, interestingly, interacting with another active agent). But within the study all psychological functioning is included—motor, perceptual, and cognitive. Whereas much psychology tends to focus on small micro-tasks studied in isolation, an applied psychology must dwell on the way in which all the components of the human processor are integrated over time to do useful tasks. For example, it might take into account interactions among the following: the ease with which commands can be remembered, the type font of characters as it affects legibility of the commands, the number of commands in a list, and anything else relevant to the particular interface. The general desirability of such wide coverage has never been in doubt. It appears in our vision of an applied psychology because wide coverage, especially the incorporation of cognition, now seems much more credible than it did twenty years ago. On the other hand, motivational and personality issues are not included. Again, there is hardly any doubt of the desirability of including them in an applied psychology, but it is unclear how to integrate the relevant existing knowledge of these topics.

1.4.   THE YIELD FOR COGNITIVE PSYCHOLOGY

The textbook view is that as a science develops it sprouts applications, that knowledge flows from the pure to the applied, that the backflow is the satisfaction (and support) that comes to a science from benefiting society. We have been reminded often enough that such a view does violence to the realities in several ways. Applied domains have a life and source of their own, so that many ingenious applications do not spring from basic science, but from direct understanding of the task in an applied context—from craft and experience. More importantly in the present context, applied investigations vitalize the basic science; they reveal new phenomena and set forth clearly what it is that needs explanation. The mechanical equivalent for heat, for instance, arose from Count Rumford's applied investigations into the boring of brass cannon; and the bacteriological origin of common infectious diseases eventually arose, in part, out of studies by Pasteur on problems besetting the fermentation of wine. The basic argument was made for psychology by Bryan and Harter (1898); and numerous applied psychological models exist to remind us of what is possible (for example, Bryan and Harter's 1898 and 1899 studies of telegraphy, Book's 1908 studies of typewriting, and Dansereau's 1968 study of mental arithmetic).

These general points certainly hold for an applied cognitive psychology, and on the same general ground that they hold for all sciences. However, it is worth detailing the three main yields for cognitive psychology that can flow from a robust applied cognitive psychology.

The first contribution is to the substance of basic cognitive psychology. The information-processing revolution in cognitive psychology is just beginning. Many domains of cognitive activity have hardly been explored. Such explorations are not peripheral to the basic science. It is a major challenge to the information-processing view to be able to explain how knowledge and skill are organized to cope with all kinds of complex human activities. Each application area in fact becomes an arena in which new problems for the basic science can arise. Each application area successfully mastered offers lessons about the ways in which the basic science can be extended to cover new areas. Ultimately, as a theory becomes solidified, application areas contribute less and less to the basic science. But at the beginning, just the reverse is true.

The domain of human-computer interaction is an example of such an unexplored domain. It has strong skill components. People who interact with computers extensively build up a repertoire of efficient, smooth, learned behaviors for carrying out their routine communicative activities. Yet, the interaction is also intensely cognitive. The skills are wielded within a problem-solving context, and the skills themselves involve the processing of symbolic information. As we shall see in abundance, even the most routine of these activities, such as using a computer text-editing program, requires the interpretation of instructions, the formulation of sequences of commands, and the communication of these commands to the computer.

The second contribution is to the style of cognitive psychology rather than to its substance. We believe that the form of the psychology of human-computer interaction, with its emphasis on task analysis, calculation, and approximation, is also appropriate for basic cognitive psychology. The existing emphasis in psychology on discriminating between theories is certainly understandable as a historical development.

However, it stifles the growth of adequate theory and of the cumulation of knowledge by focusing the attention of the field on the consequences of theories, however uninteresting in themselves, that can be used to tell whether idea A or idea B is correct. Measurements come to have little value in themselves as a continually growing body of useful quantitative knowledge of the phenomena. They are seen instead primarily as indicators fashioned to fit the demands of each experimental test. Since there is no numerical correspondence across paradigms in what is measured, the emphasis on discrimination fosters a tendency towards isolation of phenomena in specific experimental paradigms.

The third contribution is simply that of being a successful application, though it sounds a bit odd to say it that way. Modern cognitive psychology has been developing now for 25 years. If information-processing psychology represents a successful advance of some magnitude, then ultimately it must both affect the areas in which psychology is now applied and generate new areas of application.

1.5. THE YIELD FOR COMPUTER SCIENCE

It is our strong belief that the psychological phenomena surrounding computer systems should be part of computer science. Thus, we see this book not just as a book in applied psychology, but as a book in computer science as well. When university curriculum committees draw up a list of "what every computer scientist should know to call himself a computer scientist," we think models of the human user have a place alongside models of compilers and language interpreters.

The fundamental argument is worth stating: Certain central aspects of computers are as much a function of the nature of human beings as of the nature of the computers themselves. The relevance of both computer science and psychology to the design of programming languages and the interface is easy to argue, but psychological considerations enter into more topics in computer science than is usually realized. The presumption that has governed two generations of operating systems, for instance, that time-sharing systems should degrade response time as the number of users increases, is neither dictated by technology nor independent of the psychology of the user. A sufficiently crisp model of the effects of such a feature on the user could have turned the course of development of operating systems into quite different channels of development (into the logic of guaranteed service, contracted service, or proportionately graded services, for example). The yield for computer science that can flow from an applied psychology of human-computer interaction is engineering methods for taking the properties of users into account during system design.

1.6. PREVIEW

In this book, we report on a series of studies undertaken to understand the performance of users on interactive computing systems. Since new knowledge and insight are often achieved by first focusing on concrete cases and then generalizing, we direct a major portion of our effort towards user performance on computer text-editing systems. From this beginning, we try to generalize to other systems and to cognitive skill generally. We address four basic questions: (1) How can the science base be built up for supporting the design of human-computer interfaces? (2) What are user performance characteristics in a specific human-computer interaction task domain, text-editing? (3) How can our results be cast as practical models to aid in design? (4) What generalizations arise from the specific studies, models, and applications?

SCIENCE BASE

Chapter 2 begins by discussing the existing scientific base on which to erect an applied psychology of the human-computer interface. It does not review all the sources in their own terms—what is available from cognitive psychology, human factors, industrial engineering, manual control, or the classical study of motor skills—rather, it lays out a model of the human information-processor that is suited to an applied psychology and justified by current research.

TEXT-EDITING

Attention then turns to a detailed examination of text-editing as a prototypical example of human-computer interaction. An elementary requirement for understanding behavior at the interface is some gross quantitative information about user behavior, to provide a background picture against which to place more detailed studies in context. The three studies in Chapters 3 and 4 provide such a picture. Two of these (Chapter 3), a benchmark study comparing text-editing systems and a study of the individual user differences, allow one to assess the variability in performance time arising from editing system design and from individual user differences. The third study (Chapter 4) uses the data of Chapter 3 to explore how well a simple model, in which all editing modifications are assumed to take the same time, does at analyzing tradeoffs between using a computer text-editor vs. using a typewriter.

The next three chapters develop an information-processing model for the behavior of users with an editing system. Chapter 5 introduces the basic theory. The user is taken to employ goals, operators, methods, and selection rules for the methods (the GOMS analysis) to accomplish an editing task from a marked-up manuscript. Experimental verification of the analysis is given, and the effect on accuracy due to the detail with which the analysis is applied is also investigated. The routine use of an editing system is discussed as an instance of cognitive skill. Chapter 6 extends the model in three ways. First, the model is reduced to a complete, running computer simulation of user performance. Second, the analysis is extended to user behavior on a display-oriented system. Third, stochastic elements are introduced into the model to predict the distributions of performance times. Chapter 7 examines in detail one suboperation of editing: selecting a piece of text. Four different devices for doing this are tested, and a theoretical account is given for their performance.

ENGINEERING MODELS

Chapters 8 and 9 focus on the ways in which the GOMS analysis can be simplified to provide practical models for predicting the amount of time required by a user to do a task. In Chapter 8, a model at the level of individual keystrokes is presented that is sufficiently simple and accurate to be a design tool. The model is validated over several systems, tasks, and users; and examples are given for ways in which the model could be used in engineering applications. In Chapter 9, a second simplification of the GOMS analysis, this time at a more gross level, is presented This model is suited for cases where, as in the early stages of design, the system to be analyzed is not fully specified.

EXTENSIONS AND GENERALIZATIONS

So far, the studies have focused mostly on manuscript editing and on similar tasks where the user carries out a set of instructions. Chapter 10 extends the same kind of analysis to a particular problem-solving activity:

the use of a computer system to lay out a VLSI electronic circuit. The analysis shows that the user behavior exhibits many of the characteristics of manuscript editing and that the behavior is indeed a routine cognitive skill, partially understandable in terms of the concepts already introduced.

Chapter 11 attempts to place results from the above studies in a larger theoretical context. It continues the discussion of text-editing as an instance of cognitive skill and the relationship between cognitive skill generally and problem solving. Chapter 12 addresses the role of psychological studies in design. It is argued that psychological studies should emphasize the creation of performance models. The several methods of doing this are discussed and provide a framework for summarizing the thrust of the present book. A number of guidelines for systems development that arise from our studies are listed.

SCIENCE BASE

2. The Human Information-Processor

Our purpose in this chapter is to convey a version of the existing psychological science base in a form suitable for analyzing human-computer interaction. To be practical to use and easy to grasp, the description must necessarily be an oversimplification of the complex and untidy state of present knowledge. Many current results are robust, but second-order phenomena are almost always known that reveal an underlying complexity; and alternative explanations usually exist for specific effects. An uncontroversial presentation in these circumstances would consist largely of purely experimental results. Such an approach would not only abandon the possibility of calculating parameters of human performance from the analysis of a task, but would also fail in the primary purpose of giving the reader knowledge in a form relatively easy to assimilate.

Our tack, therefore, is to organize the discussion around a specific, simple model. Though limited, this model allows us to give, insofar as possible, an integrated description of psychological knowledge about human performance as it is relevant to human-computer interaction.

2.1. THE MODEL HUMAN PROCESSOR

A computer engineer describing an information-processing system at the systems level (as opposed, for instance, to the component level) would talk in terms of memories and processors, their parameters and interconnections.[1] By suppressing detail, such a description would help him to envision the system as a whole and to make approximate predictions of gross system behavior.

The human mind is also an information-processing system, and a description in the same spirit can be given for it. The description is approximate when applied to the human, intended to help us remember facts and predict user-computer interaction rather than intended as a statement of what is really in the head. But such a description is useful for making approximate predictions of gross human behavior. We therefore organize our description of the psychological science base around a model of this sort. To distinguish the simplified account of the present model from the fuller psychological theory we would present in other contexts, we call this model the *Model Human Processor.*

The Model Human Processor (see Figures 2.1 and 2.2) can be described by (1) a set of memories and processors together with (2) a set of principles, hereafter called the "principles of operation." Of the two parts, it is easiest to describe the memories and processors first, leaving the description of the principles of operation to arise in context.

The Model Human Processor can be divided into three interacting subsystems: (1) the *perceptual system*, (2) the *motor system,* and (3) the *cognitive system*, each with its own memories and processors. The perceptual system consists of sensors and associated buffer memories, the most important buffer memories being a Visual Image Store and an Auditory Image Store to hold the output of the sensory system while it is being symbolically coded. The cognitive system receives symbolically coded information from the sensory image stores in its Working Memory and uses previously stored information in Long-Term Memory to make decisions about how to respond. The motor system carries out the response. As an approximation, the information processing of the human will be described as if there were a separate processor for each subsystem: a Perceptual Processor, a Cognitive Processor, and a Motor

---

[1] For a survey of computing systems in these terms see Siewiorek, Bell, and Newell (1981).

Processor. For some tasks (pressing a key in response to a light) the human must behave as a serial processor. For other tasks (typing, reading, simultaneous translation) integrated, parallel operation of the three subsystems is possible, in the manner of three pipelined processors: information flows continuously from input to output with a characteristically short time lag showing that all three processors are working simultaneously.

The memories and processors are described by a few parameters. The most important parameters of a memory are

> μ, the storage capacity in items,
> δ, the decay time of an item, and
> κ, the main code type (physical, acoustic, visual, semantic).

The most important parameter of a processor is

> τ, the cycle time.

Whereas computer memories are usually also characterized by their access time, there is no separate parameter for access time in this model since it is included in the processor cycle time.

We now consider each of the subsystems in more detail.

The Perceptual System

The perceptual system carries sensations of the physical world detected by the body's sensory systems into internal representations of the mind by means of integrated sensory systems. An excellent example of the integration of a sensory system is provided by the visual system: The retina is sensitive to light and records its intensity, wave length, and spatial distribution. Although the eye takes in the visual scene over a wide angle, not quite a full half-hemisphere, detail is obtained only over a narrow region (about 2 degrees across), called the *fovea*. The remainder of the retina provides peripheral vision for orientation. The eye is in continual movement in a sequence of saccades, each taking about 30 msec to jump to the new point of regard[2] and dwelling there 60~700 msec for a total duration of

---

[2] Russo (1978).

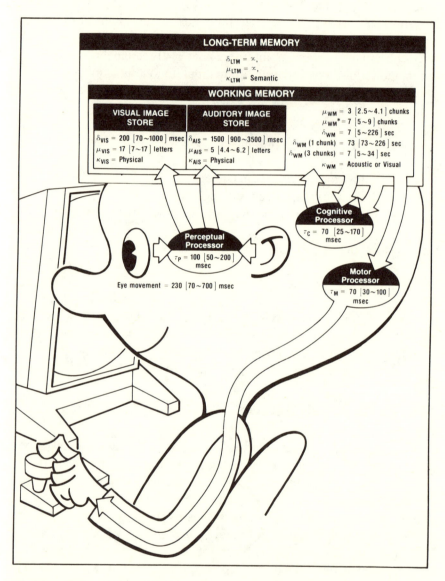

Figure 2.1. The Model Human Processor—memories and processors.

Sensory information flows into Working Memory through the Perceptual Processor. Working Memory consists of activated chunks in Long-Term Memory. The basic principle of operation of the Model Human Processor is the *Recognize-Act Cycle of the Cognitive Processor* (P0 in Figure 2.2). The Motor Processor is set in motion through activation of chunks in Working Memory.

P0. Recognize-Act Cycle of the Cognitive Processor. On each cycle of the Cognitive Processor, the contents of Working Memory initiate actions associatively linked to them in Long-Term Memory; these actions in turn modify the contents of Working Memory.

P1. Variable Perceptual Processor Rate Principle. The Perceptual Processor cycle time τ_P varies inversely with stimulus intensity.

P2. Encoding Specificity Principle. Specific encoding operations performed on what is perceived determine what is stored, and what is stored determines what retrieval cues are effective in providing access to what is stored.

P3. Discrimination Principle. The difficulty of memory retrieval is determined by the candidates that exist in the memory, relative to the retrieval clues.

P4. Variable Cognitive Processor Rate Principle. The Cognitive Processor cycle time τ_c is shorter when greater effort is induced by increased task demands or information loads; it also diminishes with practice.

P5. Fitts's Law. The time T_{pos} to move the hand to a target of size S which lies a distance D away is given by:

$$T_{pos} = I_M \log_2 (D/S + .5),$$ (2.3)

where I_M = 100 [70~120] msec/bit.

P6. Power Law of Practice. The time T_n to perform a task on the nth trial follows a power law:

$$T_n = T_1 n^{-\alpha},$$ (2.4)

where α = .4 [.2~.6].

P7. Uncertainty Principle. Decision time T increases with uncertainty about the judgement or decision to be made:

$$T = I_C H,$$

where H is the information-theoretic entropy of the decision and
I_C = 150 [0~157] msec/bit. For n equally probable alternatives (called Hick's Law),

$$H = \log_2 (n + 1).$$ (2.8)

For n alternatives with different probabilities, p_i, of occurence,

$$H = \Sigma_i p_i \log_2 (1/p_i + 1).$$ (2.9)

P8. Rationality Principle. A person acts so as to attain his goals through rational action, given the structure of the task and his inputs of information and bounded by limitations on his knowledge and processing ability:

Goals + Task + Operators + Inputs
+ Knowledge + Process-limits → Behavior

P9. Problem Space Principle. The rational activity in which people engage to solve a problem can be described in terms of (1) a set of states of knowledge, (2) operators for changing one state into another, (3) constraints on applying operators, and (4) control knowledge for deciding which operator to apply next.

Figure 2.2. The Model Human Processor—principles of operation.

$$\text{Eye-movement} = 230\,[70{\sim}700]\text{ msec}.^3$$

(In this expression, the number 230 msec represents a typical value and the numbers in brackets indicate that values may range from 70 msec to 700 msec depending on conditions of measurement, task variables, or subject variables.) Whenever the target is more than about 30 degrees away from the fovea, head movements occur to reduce the angular distance. These four parts—central vision, peripheral vision, eye movements, and head movements—operate as an integrated system, largely automatically, to provide a continual representation of the visual scene of interest to the perceiver.

PERCEPTUAL MEMORIES

Very shortly after the onset of a visual stimulus, a representation of the stimulus appears in the *Visual Image Store* of the Model Human Processor. For an auditory stimulus, there is a corresponding *Auditory Image Store*. These sensory memories hold information coded *physically*, that is, as an unidentified, non-symbolic analogue to the external stimulus. This code is affected by physical properties of the stimulus, such as intensity. For our purposes we need not enter into the details of the physical codes for the two stores but can instead just write:

$$\kappa_{VIS} = \text{physical},$$
$$\kappa_{AIS} = \text{physical}.$$

For example, the Visual Image Store representation of the number 2 contains features of curvature and length (or equivalent spatial frequency patterns) as opposed to the recognized digit.

The perceptual memories are intimately related to the cognitive Working Memory as Figure 2.1 depicts schematically. Shortly after a physical representation of a stimulus appears in one of the perceptual memories, a recognized, symbolic, acoustically-coded (or visually-coded)

---

[3] Actual saccadic eye-movement times (travel + fixation time) can vary quite considerably depending on the task and the skill of the observer. Russo (1978, Table 2, p. 94) lists 70 msec as the minimum time and 230 msec as a typical time. The largest time given by Busswell (1922, p. 31) for eye-movements in reading is 660 msec (for first-grade children), which we round to 700 msec.

representation of at least part of the perceptual memory contents occurs in Working Memory. If the contents of perceptual memory are complex or numerous (for example, an array of letters) and if the stimulus is presented only fleetingly, the perceptual memory trace fades, and Working Memory is filled to capacity before all the items in the perceptual memory can be transferred to representations in Working Memory (for letters the coding goes at about 10 msec/letter). However, the Cognitive Processor can specify which portion of the perceptual memory is to be so encoded. This specification can only be by physical dimensions, since this is the only information encoded: after being shown a colored list of numbers and letters, a person can select (without first identifying what number or letter it is) the top half of the Visual Image Store or the green items, but not the even digits or the digits rather than the letters.

Figure 2.3 shows the decay of the Visual Image Store and the Auditory Image Store over time. As an index of decay time, we use the half-life, defined as the time after which the probability of retrieval is less than 50%. While exponential decay is not necessarily implied by the use of the half-life, Figure 2.3 shows that it is often a good approximation to the observed curves. The Visual Image Store has a half-life of about

$$\delta_{VIS} = 200 \, [90\!\sim\!1000] \text{ msec },^4$$

but the Auditory Image Store decays more slowly,

---

[4] A least-squares fit to data estimated from figures appearing in Sperling (1960) and Averbach and Coriell (1961) yields the following facts. The half-life of the letters in excess of the memory span that subjects could report in the partial report condition of Sperling's (1960) experiment was 621 msec (9-letter stimulus) and 215 msec (12-letter stimulus). Averbach and Coriell's (1961) experiment gives a half-life of 92 msec (16-letter stimulus). The typical value for δ_{VIS} has been set at 200 msec, representing the middle of these. The lower and upper bounds for δ_{VIS} are set at rounded-off values reflecting the fastest subject in the condition with the shortest half-life and the slowest subject in the condition with the longest half-life. The shortest half-life in these experiments was 93 msec for Averbach and Coriell's Subject GM (16-letter condition); the longest half-life was 940 msec for Sperling's Subject ROR (9-letter condition). It is possible to have the average half-life be 92 msec, shorter than the half-life of any subject, because this average is computed by first taking the mean of each point across subjects, then computing the slope of the best least-square fitting line in semilog coordinates.

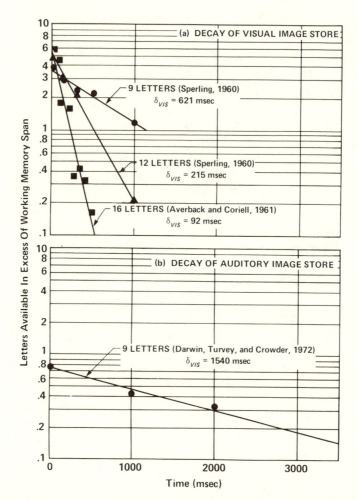

Figure 2.3. Time decay of Visual and Auditory Image Stores.
(a) Decay of the Visual Image Store. In each experiment, a matrix of letters was made observable tachistoscopically for 50 msec. In the case of the Sperling experiments, a tone sounded after the offset of the letters to indicate which row should be recalled. In the case of the Averbach and Coriell experiment, a bar appeared after the offset of the letters next to the letter to be identified. The percentage of indicated letters that could be recalled eventually asymptotes to μ_{WM}^*. The graph plots the percentage of letters reported correctly in excess of μ_{WM}^* as a function of time before the indicator.
(b) Decay of the Auditory Image Store. Nine letters were played to the observers over stereo earphones arranged so that three sequences of letters appear to come from each of three directions. A light lit after the offset of the letters to indicate which sequence should be recalled. The graph plots the percentage of the relevant 3-letter sequence in excess of μ_{WM}^* reported correctly as a function of time before the light was lit.

$$\delta_{AIS} = 1500 \, [900\sim3500] \text{ msec} \,,^5$$

consistent with the fact that auditory information must be interpreted over time. The capacity of the Visual Image Store is hard to fix precisely but for rough working purposes may be taken to be about

$$\mu_{VIS} = 17 \, [7\sim17] \text{ letters} \,.^6$$

The capacity of the Auditory Image Store is even more difficult to fix, but would seem to be around

$$\mu_{AIS} = 5 \, [4.4\sim6.2] \text{ letters} \,.^7$$

PERCEPTUAL PROCESSOR

The cycle time τ_P of the Perceptual Processor is identifiable with the so-called *unit impulse response* (the time response of the visual system to

[5] The half-life of the letters in excess of the memory span that subjects could report in the partial report condition of Darwin, Turvey, and Crowder's (1972) experiment was 1540 msec, which we have rounded to $\delta_{AIS} = 1500$ msec. The difference in decay half-life as a function of letter order in their experiment (963 msec for the third letter, 3466 msec for the first letter) has been rounded to give lower and upper bounds of 900 and 3500. Other techniques have been used to obtain values for the "decay time" of the Auditory Image Store. For example, use of a masking technique gives estimates of around 250 msec full decay (Massaro, 1970), but these experiments have been criticized by Klatzky (1980, p. 42) because they may only measure the time necessary to transmit categorical information to Working Memory. On the other end, experiments that measure the delay at which there is still some facilitation of the identification of a noisy signal (Crossman, 1958; Guttman and Julesz, 1963) give very wide full-decay estimates: from 1000 msec to 15 minutes!

[6] Sperling (1963, p. 22) estimates the capacity of the Visual Image store in terms of the number of letters available at least 17 letters and possibly more. The fewest number of letters available for any subject immediately after stimulus presentation in the 9-letter condition (Sperling, 1960) was 7.4 letters for Subject NJ.

[7] Range is from the number of letters or numbers that could be reported by Darwin, Turvey, and Crowder's (1972) subjects in an experiment in which they had to give the trio of letters coming from one of three directions (indicated by a visual cue shortly after the end of the sounds). Lowest value, 4.4 letters, is for accuracy of recalling second letter of triple when subjects had to name all items coming from a certain direction (Figure 1, p. 259). Highest number, 6.2 letters, is for recall by category when no location was required (Figure 2(B), p. 262).

a very brief pulse of light)[8] and its duration is on the order of

$$\tau_p = 100 \, [50\text{--}200] \text{ msec} .[9]$$

If a stimulus impinges upon the retina at time $t = 0$, at the end of time $t = \tau_p$ the image is available in the Visual Image Store and the human claims to see it. In truth, this is an approximation, since different information in the image becomes available at different times, much as a photograph develops.[10] For example, movement information and low spatial frequency information are available sooner than other information. A person can react before the image is fully developed or can wait for a better image, according to whether speed or accuracy is the more important.

Perceptual events occurring within a single cycle are combined into a single percept if they are sufficiently similar. For example, two lights occurring at different nearby locations within 60~100 msec combine to give the impression of a single light in motion. A brief pulse of light, lasting t msec with intensity I, has the same appearance as a longer pulse of less-intense light, provided both pulses last less than 100 msec, giving rise to Bloch's Law (1885):

$$I \cdot t = k, \quad t < \tau_p .$$

Two brief pulses of light within a cycle combine their intensities in a more complicated way, but still give a single percept.[11] Thus there is a basic quantum of experience; and the present is not an instantaneous dividing line between past and future, but has itself duration.

Figure 2.4 shows the results of an experiment in which subjects were presented with a rapid set of clicks, from 10 to 30 clicks per second, and were asked to report how many they heard. The results show that they heard the correct number when the clicks were presented at 10 clicks/sec, but missed progressively more clicks at 15 and 30 clicks/sec. A simple

[8] See Ganz (1975).

[9] The source of the range is the review by Harter (1967), who also discusses the suggestion that the cycle time can be identified with the 77~125 msec alpha period in the brain.

[10] See Ericksen and Shultz (1978), Ganz (1975).

[11] See Ganz (1975).

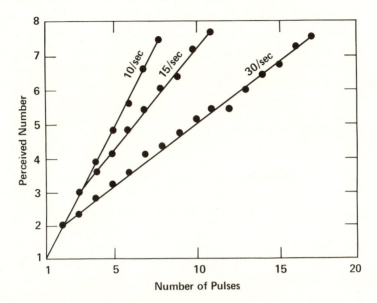

Figure 2.4. Fusion of clicks within 100 msec.

A burst of sound containing an unknown number of auditory clicks at the uniform rate of 10/sec, 15/sec, or 30/sec was presented to the subject. The graph plots the number of clicks/burst reported as a function of the number presented. After Cheatham and White (1954, Figure 1, p. 427).

analysis in terms of the Model Human Processor shows why. When the experimenter plays the clicks at 10 clicks/sec, there is one click for each $\tau_P \simeq 100$ msec interval and the subject hears each click. But when the experimenter plays the clicks at 30 clicks/sec, the three clicks in each 100 msec cycle time are fused into a single percept (perhaps sounding a little louder) and the subject hears only one click instead of three, or 10 clicks/sec. The data in Figure 2.4 show that the number of clicks/sec perceived by the subjects does in fact stay approximately constant in the 10 clicks/sec range (the measured values of the slopes are 9~11 clicks/sec) for the three rates of presentation.

As a second-order phenomenon, the processor time τ_P is not completely constant, but varies somewhat according to conditions. In particular, τ_P is shorter for more intense stimuli, a fact derivable from a more detailed examination of the human information-processor using linear systems theory, but which we simply adopt as one of the principles of operation (Figure 2.2):

> **P1. Variable Perceptual Processor Rate Principle.** *The Perceptual Processor cycle time τ_P varies inversely with stimulus intensity.*

The effect of this principle is such that τ_P can take on values within the 50~200 msec range we have given. Under very extreme conditions of intense, high-contrast stimuli or nearly invisible, low-contrast stimuli, τ_P can take on values even outside these ranges.

The Motor System

Let us now consider the motor system. Thought is finally translated into action by activating patterns of voluntary muscles. These are arranged in pairs of opposing "agonists" and "antagonists," fired one shortly after the other. For computer users, the two most important sets of effectors are the arm-hand-finger system and the head-eye system.

Movement is not continuous, but consists of a series of discrete micromovements, each requiring about

$$\tau_M = 70\,[30\text{~}100]\,\text{msec},[12]$$

which we identify as the cycle time of the Motor Processor. The feedback loop from action to perception is sufficiently long (200~500 msec) that rapid behavioral acts such as typing and speaking must be executed in bursts of preprogrammed motor instructions.

An instructive experiment is to have someone move a pen back and forth between two lines as quickly as possible for 5 sec (see Figure 2.5). Two paths through the processors in Figure 2.1 are clearly visible: (1) The Motor Processor can issue commands ("open loop") about once every $\tau_M = 70$ msec; in Figure 2.5 this path leads to the 68 pen reversals made by the subject in the 5 sec interval, or $\tau_M = 74$ msec/reversal. (2) The subject's perceptual system can perceive whether the strokes are

---

[12] The limit of repetitive movement of the hand, foot, or tongue is about 10 movements/sec (Fitts and Posner, 1967, p. 18). Chapanis, Garner, and Morgan (1949, p. 284) cite tapping rates of 8~13 taps/sec (38~62 movements/sec, assuming 2 movements/tap). Fox and Stansfield (1964) cite figures of 130 msec/tap = 65 msec/movement. Repetition of the same key in Kinkead's data (Figure 2.15*b*) averages to 180 msec/keystroke = 90 msec/movement. The scribbling rate in Figure 2.5 was 74 msec/movement. We summarize these as 70 [30~100] msec/movement.

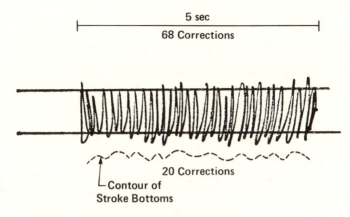

Figure 2.5. Maximum motor output rate.
Marks made by subject moving pen back and forth between two lines as fast as possible for 5 sec.

staying within the lines (the perception process requires τ_P msec) and send this information to the cognitive system, which can then advise (the decision process requires τ_C msec) the motor system to issue a correction (the motor process requires τ_M msec). The total time, therefore, to make a correction using visual feedback ("closed loop") should be on the order of $\tau_P + \tau_C + \tau_M = 240$ msec; in Figure 2.5, this path leads to the roughly 20 corrections about the ruled guidelines as indicated by the dotted line tracing the contours of the bottoms of the strokes, or (5 sec)/(20 movements) = 250 msec/movement.

The Cognitive System

In the simplest tasks, the cognitive system merely serves to connect inputs from the perceptual system to the right outputs of the motor system. But most tasks performed by a person are complex and involve learning, retrieval of facts, or the solution of problems. As would be expected, the memories and the processor for the cognitive system are more complicated than those for the other systems.

COGNITIVE MEMORIES

There are two important memories in the cognitive system: a *Working Memory* to hold the information under current consideration and a *Long-Term Memory* to store knowledge for future use.

Working Memory. Working Memory holds the intermediate products of thinking and the representations produced by the perceptual system. Functionally, Working Memory is where all mental operations obtain their operands and leave their outputs. It constitutes the general registers of the Cognitive Processor. Structurally, Working Memory consists of a subset of the elements in Long-Term Memory that have become *activated;* this intimate association between Working Memory and Long-Term Memory is represented in Figure 2.1 by the placement of Working Memory inside Long-Term Memory. Although Working Memory information can be coded in many ways, the use of symbolic *acoustic* codes is especially common, related, no doubt, to the great importance of verbal materials to the tasks people frequently perform. The user of a telephone, for example, is especially liable to dial numbers mistakenly that sound like the numbers he has just looked up. *Visual* codes, if required by the the task, are also possible (as are some other types of codes). For purposes of the Model Human Processor we consider the predominant code types to be

$$\kappa_{WM} = \text{acoustic or visual .}$$

It is important to distinguish the symbolic, nonphysical acoustic or visual codes of Working Memory, which are unaffected by physical parameters of the stimulus (such as intensity), from the nonsymbolic, physical codes of the sensory image stores, which *are* affected by physical parameters of the stimulus.

The activated elements of Long-Term Memory, which define Working Memory, consist of symbols, called *chunks,* which may themselves be organized into larger units. It is convenient to think of these as nested abstract expressions: CHUNK1 = (CHUNK2 CHUNK3 CHUNK4), with, for instance, CHUNK4 = (CHUNK5 CHUNK6).[13] What constitutes a chunk is as much a function of the user as of the task, for it depends on the contents of the user's Long-Term Memory. The sequence of nine letters below is beyond the ability of most people to repeat back:

B C S B M I C R A

---

[13] It is also possible to think of these as semantic networks, such as those in Anderson (1980) and other recent publications. At the level of our discussion, any of these notations will suffice about equally well. See also Simon (1974) for a technical definition of chunk.

However, consider the list below, which is only slightly different:

CBSIBMRCA

Especially if spoken aloud, this sequence will be chunked into CBS IBM RCA (by the average American college sophomore) and easily remembered, being only three chunks. If the user can perform the recoding rapidly enough, random lists of symbols can be mapped into prepared chunks. A demonstration of this is the mapping of binary digits into hexadecimal digits:

0100001000010011011001101000
0100 0010 0001 0011 0110 0110 1000
4213668

This last can be easily remembered. The coding must be done in both directions, binary to hexadecimal and hexadecimal to binary, and takes substantial practice before it can be carried out as part of a regular memory-span test, but it can be done. Indeed, with extended effort, the digit span can be increased enormously. A Carnegie-Mellon University student holds the current record at 81 decimal digits, presented at a uniform rate of 1 digit per second.[14] This particular event occurred as part of a psychological study, where it could be verified that all the gain was due to elaborate recoding and immense practice in its use and development, rather than any physiological endowment.

Chunks can be related to other chunks. The chunk ROBIN, for example, sounds like the chunk ROBERT. It is a subset of the chunk BIRD, it has chunk WINGS, it can chunk FLY. When a chunk in Long-Term Memory is activated, the activation spreads to related chunks and to chunks related to those. As the activation spreads to new chunks, the previously activated chunks become less accessible, because there is a limited amount of activation resource. The new chunks are said to *interfere* with the old ones. The effect of this interference is that the chunk appears to fade from Working Memory with time (unless reactivated), as the decay curves in Figure 2.6 show. The curves are significantly affected by other variables, including the number of other chunks the user is trying to remember, retrieval interference with similar

---

[14] Ericsson, Chase, and Faloon (1980); Chase and Ericsson (1981).

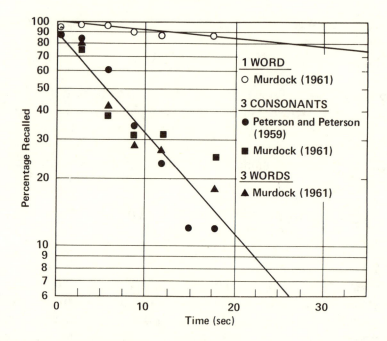

Figure 2.6. Working Memory decay rate.
Subject is given either one or three words or consonants to remember. He counts backwards (preventing rehearsal) for a time and then recalls stimulus. Graph plots proportion of items correctly recalled as a function of the time elapsed until recall began.

chunks in Working Memory, and input and retrieval memory strategies of the user. As a working value we take the half-life of 7 sec from the curve in Figure 2.6, which together with other data gives

$$\delta_{WM} = 7 \,[5{\sim}226] \,\sec.^{15}$$

The decay parameter δ_{WM} has a wide range, because most of the apparent decay comes about from the details of interference, as we have noted above. But these details are difficult to analyze, so it is most convenient to accept the range and talk in terms of decay. Since the

[15] For three chunks, Peterson and Peterson's (1959) data (Figure 2.6) give a half-life of about 5 sec. Murdock's data (Murdock, 1961) in Figure 2.6 give a half-life of about 7 sec for 3 words and also 9 sec for 3 consonants. On the other hand, Melton's (1963) data give a much longer half-life of 34 sec. For one chunk, Murdock's data in Figure 2.6 and Melton's (1963) give half-lives of 73 sec and 226 sec, respectively.

decay rate is particularly sensitive to the number of chunks in the recalled item, it is useful to record the decay rate of representative item sizes:

$$\delta_{WM}(\text{1 chunk}) = 73 \,[73\sim226] \sec .^{15}$$
$$\delta_{WM}(\text{3 chunks}) = 7 \,[5\sim34] \sec .^{15}$$

When people are asked to recall information a few seconds after hearing it, they use both Working Memory and Long-Term Memory to do so. Experimentally, these two systems have been teased apart showing that there is a *pure capacity of Working Memory* (example: number of immediately preceding digits recallable from a long series when the series unexpectedly stops),

$$\mu_{WM} = 3 \,[2.5\sim4.1] \text{ chunks} .^{16}$$

When this pure capacity is augmented by the use of Long-Term Memory, the *effective capacity of Working Memory* μ_{WM}^{*} (example: longest number that can be repeated back) extends to the familiar 7 ± 2 chunks,

$$\mu_{WM}^{*} = 7 \,[5\sim9] \text{ chunks} .^{17}$$

Long-Term Memory. Long-Term Memory holds the user's mass of available knowledge. It consists of a network of related chunks, accessed associatively from the contents of the Working Memory. Its contents comprise not only facts, but procedures and history as well.

Apparently, there is no erasure from Long-Term Memory,

$$\delta_{LTM} = \infty.$$

However, successful retrieval of a chunk depends on whether associations to it can be found. There are two reasons the attempt to retrieve a chunk might fail: (1) effective retrieval associations cannot be found, or

---

[16] Crowder (1976) reviews several methods. Estimates are Waugh and Norman (1965) method, 2.5 items; Raymond (1969) method, 2.5 items; Murdock (1960b, 1967) method, 3.2~4.1 items; Tulving and Colatla (1970) method, 3.3~3.6 items. See also Glanzer and Razel (1974).

[17] Miller (1956).

(2) similar associations to several chunks interfere with the retrieval of the target chunk. The great importance of these links between particular chunks in Long-Term Memory, that is, the *semantic* coding of information, leads us to list it as the predominant code type,

$$\kappa_{LTM} = \text{semantic} .$$

To be stored in Long-Term Memory, information from the sensory memories must ultimately be encoded into symbolic form: a pattern of light and dark might be coded as the letter A, an extended pattern coded as a system error message.   When the information from Working Memory becomes part of Long-Term Memory, the precise way in which it and the coincident Working Memory contents were encoded determines what cues will be effective in retrieving the item later. Suppose a user names a computer-imaging file LIGHT (as opposed to DARK). If he later scans a directory listing of file names to identify which ones were the ones he created and thinks of LIGHT (as opposed to HEAVY), he will not be able to recognize the file, because he will be using a different set of retrieval cues. As a principle of operation,

> *P2. Encoding Specificity Principle.*[18]   *Specific encoding operations performed on what is perceived determine what is stored, and what is stored determines what retrieval cues are effective in providing access to what is stored.*

Because of interference with other chunks in memory that are more strongly activated by the associations used as retrieval cues, information, despite being physically present, can become functionally lost. Stated as a principle,

> *P3. Discrimination Principle.* *The difficulty of memory retrieval is determined by the candidates that exist in the memory, relative to the retrieval cues.*

Items cannot be added to Long-Term Memory directly (accordingly, Figure 2.1 shows no arrow in this direction); rather, items in Working

---

[18] Tulving and Thompson (1973).

Memory (possibly consisting of several chunks) have a certain probability of being retrievable later from Long-Term Memory. The more associations the item has, the greater its probability of being retrieved. If a user wants to remember something later, his best strategy is to attempt to associate it with items already in Long-Term Memory, especially in novel ways so there is unlikely to be interference with other items. Of course this activity, by definition, activates more items in Long-Term Memory, causing new items to appear in Working Memory, and use capacity. On a paced task, where a user is given items to remember at a constant rate, the percentage of the items recalled later increases as the time/item increases (the probability the item will be stored in Long-Term Memory and linked so it can be retrieved increases with residence time in Working Memory), until the time allowed per item is of the same magnitude as the decay time of Working Memory (after which, more time available for study does not increase the time the item is in Working Memory), around δ_{WM} sec/chunk = 7 sec/chunk.[19]

Storing new chunks in Long-Term Memory thus requires a fair amount of time and several Long-Term Memory retrievals. On the other hand, Long-Term Memory is accessed on every 70 msec cognitive-processing cycle. Thus the system operates as a fast-read, slow-write system. This asymmetry puts great importance on the limited capacity of Working Memory, since it is not possible in tasks of short duration to transfer very much knowledge to Long-Term Memory as a working convenience.

COGNITIVE PROCESSOR

The *recognize-act cycle*, analogous to the fetch-execute cycle of standard computers, is the basic quantum of cognitive processing. On each cycle, the contents of Working Memory initiate associatively-linked actions in Long-Term Memory ("recognize"), which in turn modify the contents of Working Memory ("act"), setting the stage for the next cycle. Plans, procedures, and other forms of extended organized behavior are built up out of an organized set of recognize-act cycles.

Like the other processors, the Cognitive Processor seems to have a cycle time of around a tenth of a second:

[19] Newell and Simon (1972, p. 793) reviews experiments that gives times of 8~13 sec/chunk.

$$\tau_C = 70\ [25\sim 170]\ \text{msec} .^{20}$$

The cycle times for several types of tasks are given in Figure 2.7. The times vary in the 25~170 msec/cycle range, depending on the specific experimental phenomenon and experimental circumstances with which one wishes to identify the cycle. We have chosen as a nominal value 70 msec, about at the median of those in Figure 2.7, but have included within the upper and lower limits all the estimates from the figure. As with the Perceptual Processor, the cycle time is not constant, but can be shortened by practice, task pacing, greater effort, or reduced accuracy.

> *P4.* *Variable Cognitive Processor Rate Principle.* *The Cognitive Processor cycle time τ_C is shorter when greater effort is induced by increased task demands or information loads; it also diminishes with practice.*

The cognitive system is fundamentally parallel in its recognizing phase and fundamentally serial in its action phase. Thus the cognitive system can be aware of many things, but cannot do more than one deliberate thing at a time. This seriality occurs on top of the parallel activities of the perceptual and motor systems. Driving a car, reading roadside advertisements, and talking can all be kept going by skilled intermittent allocation of control actions to each task, along the lines of familiar interrupt-driven time-sharing systems.

Summary. This completes our initial description of the Model Human Processor. To recapitulate, the Model Human Processor consists of (1) a set of interconnected memories and processors and (2) a set of

[20] On the fast end, memory scanning rates go down to 25 msec/item (Sternberg, 1975, p. 225, Figures 8 and 9, lower error bar for LETTERS). Michon (1978, p. 93) summarizes the search for the "time quantum" as converging on 20~30 msec. On the slow end, silent counting, which takes about 167 msec/item (Landauer, 1962), has sometimes been taken as a minimum cognitive task. It has sometimes been argued (Hick 1952) that the subject in a choice reaction time experiment makes one choice for each bit in the set of alternatives, in which case a typical value would be 153 msec/bit (Figure 2.22). Welford (1973, in Kornblum) has proposed a theory of choice reaction in which the subject makes a series of choices, each taking 92 msec. Blumenthal (1977) reviews an impressively large number of cognitive phenomena with time constraints in the tenth of a second range.

Rate at which an item can be matched against Working Memory:

Digits	33 [27~39] msec/item	Cavanaugh (1972)
Colors	38 msec/item	Cavanaugh (1972)
Letters	40 [24~65] msec/item	Cavanaugh (1972)
Words	47 [36~52] msec/item	Cavanaugh (1972)
Geometrical shapes	50 msec/item	Cavanaugh (1972)
Random forms	68 [42~93] msec/item	Cavanaugh (1972)
Nonsense syllables	73 msec/item	Cavanaugh (1972)

Range = 27~93 msec/item

Rate at which four or fewer objects can be counted:

Dot patterns	46 msec/item	Chi & Klahr (1975)
3-D shapes	94 [40~172] msec/item	Akin and Chase (1978)

Range = 40~172 msec/item

Perceptual judgement:

92 msec/inspection	Welford (1973)

Choice reaction time:

92 msec/inspection	Welford (1973)
153 msec/bit	Hyman (1953)

Silent counting rate:

167 msec/digit	Landauer (1962)

Figure 2.7. Cognitive processing rates.
Selected cycle times (msec/cycle) that might be identified with the Cognitive Processor cycle time.

principles of operation. The memories and processors are grouped into three main subsystems: a perceptual system, a cognitive system, and a motor system. The most salient characteristics of the memories and processors can be summarized by the values of a few parameters: processor cycle time τ, memory capacity μ, memory decay rate δ, and memory code type κ. Each of the processors has a cycle time on the order of a tenth of a second.

A model so simple does not, of course, do justice to the richness and subtlety of the human mind. But it does help us to understand, predict, and even to calculate human performance relevant to human-computer interaction. To pursue this point, and to continue our development of the Model Human Processor, we now turn to an examination of sample phenomena of human performance.

2.2. HUMAN PERFORMANCE

We have said that in order to support cognitive engineering of the human-computer interface, an applied information-processing psychology should be based on task analysis, calculation, and approximation. These qualities are important for the Model Human Processor to possess if we are to address the practical prediction of human performance. Although it might be argued that the primitive state of development in psychological science effectively prevents its employment for practical engineering purposes, such an argument overlooks the often large amounts of uncertainty also encountered in fields of engineering based on the physical sciences. The parameters of soil composition under a hill, the wind forces during a storm, the effects of sea life and corrosion on underwater machinery, the accelerations during an earthquake—all are cases where the engineer must proceed in the face of considerable uncertainty in parameters relevant to the success of his design.

A common engineering technique for addressing such uncertainty is to settle on nominal values for the uncertain parameters representing low, high, and typical values, and to design to these. Thus a heating engineer might calculate heating load for a building at design temperatures of 10°F. for winter, 105°F. for summer, and a more common 70°F. day.

A similar technique helps us to address the uncertainties in the parameters of the Model Human Processor. We can define three versions of the model: one in which all the parameters listed are set to give the worst performance (*Slowman*), one in which they are set to give the best performance (*Fastman*), and one set for a nominal performance (*Middleman*).

The difference between the results of the Middleman (nominal) and the Fastman-Slowman (range) calculations must be kept clearly in mind. Secondary effects, outside the scope of the model, may mean that the appropriate parameter value for a particular calculation lies at a place in the range other than that given as the nominal value: the real predictions of the Model Human Processor are that a calculated quantity will lie somewhere within the Slowman~Fastman range. On the other hand, because these ranges are set by extreme and not particularly typical values, the range is pessimistically wide. The nominal value for each parameter allows a complement to the range calculations based on a typical value for the parameter at some increased risk of inaccuracy due to secondary effects. The two types of calculation, range and nominal, can be used together in a number of ways depending on whether we are more interested, say, in assessing the sensitivity of a nominal calculation to secondary effects or in identifying the upper or lower boundary at which some user performance will occur.

We turn now to examples of human performance bearing potential relevance to human-computer interaction, relating these, where possible, to the Model Human Processor. The performances are drawn from the areas of perception, motor skill, simple decisions, learning and retrieval, and problem solving.

Perception

Many interesting perceptual phenomena derive from the fact that similar visual stimuli that occur within one Perceptual Processor cycle tend to fuse into a single coherent percept. As an example, consider the problem of the rate at which frames of a moving picture need to be changed to create the illusion of motion.

MOVING PICTURE RATE

> **Example 1.** Compute the frame rate at which an animated image on a video display must be refreshed to give the illusion of movement.

Solution. Closely related images nearer together in time than τ_p, the cycle time of the Perceptual Processor, will be fused into a single image. The frame rate must therefore be such that:

$$\text{Frame rate} > 1/\tau_p = 1/(100 \text{ msec/frame})$$
$$= 10 \text{ frames/sec} . \blacksquare$$

This solution can be augmented by realizing that in order to be certain that the animation will not break down, the frame rate should, of course, be faster than this number. How much faster? A reasonable upper bound for how fast the rate needs to be can be found by redoing the above calculation for the Fastman version of the model ($\tau_P = 50$ msec):

$$\text{Max frame rate for fusion} = 1/(50 \text{ msec/frame})$$
$$= 20 \text{ frames/sec}.$$

This calculation is in general accord with the frame rates commonly employed for motion picture cameras (18 frames/sec for silent and 24 frames/sec for sound).

The Model Human Processor also warns us of secondary phenomena that might affect these calculations. By the Variable Perceptual Processor Rate Principle, τ_P will be faster for the brighter screen of a cinema projector and slower for the fainter screen of a video display terminal.

MORSE CODE LISTENING RATE

Because stimuli within τ_P fuse into the same percept, the cycle time of the Perceptual Processor sets fundamental limits on the speed with which the user can attend to auditory or visual input.

> **Example 2.** In the old type of Morse Code device, dots and dashes were made by the clicks of the armature of an electromagnet, dots being distinguished from dashes by a shorter interval between armature clicks. Subsequently, oscillators came into use which allowed the dots and dashes to be done by bleeps of different lengths. Should there be any difference between the two devices in the maximum rate at which code can be received?

Solution. With the older device, a dot requires the perception of two events (two clicks of the armatures). According to the model, this requires $2\tau_P$ msec, if each of these events is to be separately perceived. Officially a dash is defined as 3 dots in length, leading to an estimate of $6\tau_P$. However, high speed code often differs from the standard; and an expert should be able to perceive a dash as different than a dot if it is at least τ_P longer, giving $2\tau_P + \tau_P = 3\tau_P$ msec as the minimum time for a dash. Assuming a minimum $1\tau_p$ space between letters and $2\tau_p$ space between words, we can calculate the reception rate for random text by first computing the minimum reception time per letter and then weighting that by English letter frequencies, with an appropriate adjustment for word spacing. This calculation should underestimate somewhat the reception rates for each system, since it is only based on a first-order approximation to English below the word level; but it will allow a relative comparison. The probabilities for the letters in English are given in Figure 2.8 together with their Morse Code representation and the time/letter computed by the rates given above, assuming $\tau_p = 100$ [50~200] msec. Weighting the time/code by the frequency of its occurrence gives a mean time of 709 [354~1417] msec/letter (including spacing between letters). Assuming 4.8 char/word (the value for Bryan and Harter's 1898 telegraphic speed test) gives:

$$
\begin{aligned}
\text{Max reception rate} &= (.709\,[.354{\sim}1.417]\,\text{sec/letter} \\
&\quad \times\ 4.8\ \text{letters/word}) \\
&\quad +\ .200\,[.100{\sim}.400]\,\text{sec/word-space} \\
&= 3.6\ [1.9{\sim}7.0]\ \text{sec/word} \\
&= 17\ [9{\sim}32]\ \text{words/min}\,.
\end{aligned}
$$

For the oscillator-based telegraph, on the other hand, a dot requires the perception of only one event. This should require τ_p. Assuming that a dash can be distinguished from a dot if the dash is $2\tau_p$ long, the time per letter would be 453 [227~907] msec and the calculation is:

$$
\begin{aligned}
\text{Max reception rate} &= (.453\,[.227{\sim}.907]\,\text{sec/letter} \\
&\quad \times\ 4.8\ \text{letters/word}) \\
&\quad +\ .200\,[.100{\sim}.400]\,\text{sec/word-space} \\
&= 2.4\ [1.3{\sim}4.6]\ \text{sec/word} \\
&= 25\ [13{\sim}47]\ \text{words/min}\,.\ \blacksquare
\end{aligned}
$$

So it would be expected that operators could receive code faster with the newer oscillator-based system than with the older system. Informal evidence suggests that this is true and that the oscillator-based rates are at least in the right vicinity. Current reception rates are faster than the rates of turn-of-the-century telegraphers, although this comparison may be confounded with the effect of sending equipment. Whereas 20~25 words/min with the old telegraph was regarded as the range for very

Letter	p	Morse Code	Calculated Minimum Reception Time	
			Armature System (msec)	Oscillator System (msec)
E	.1332	•	300 [150~600]	200 [100~400]
T	.0978	–	400 [200~800]	300 [150~600]
A	.0810	• –	600 [300~1200]	400 [200~800]
H	.0772	• • • •	900 [450~1800]	500 [250~1000]
O	.0663	– – –	1000 [500~2000]	700 [350~1400]
S	.0607	• • •	700 [350~1400]	400 [200~800]
N	.0601	– •	600 [300~1200]	400 [200~800]
R	.0589	• – •	800 [400~1600]	500 [250~1000]
I	.0515	• •	500 [250~1000]	300 [150~600]
L	.0447	• – • •	1000 [500~2000]	600 [300~1200]
D	.0432	– • •	800 [400~1600]	500 [250~1000]
M	.0248	– –	700 [350~1400]	500 [250~1000]
C	.0236	– • – •	1100 [550~2200]	700 [350~1400]
U	.0309	• • –	800 [400~1600]	500 [250~1000]
W	.0287	• – –	900 [450~1800]	600 [300~1200]
G	.0218	– – •	900 [450~1800]	600 [300~1200]
Y	.0212	– • – –	1200 [600~2400]	800 [400~1600]
F	.0179	• • – •	1000 [500~2000]	600 [300~1200]
B	.0163	– • • •	1000 [500~2000]	600 [300~1200]
P	.0153	• – – •	1100 [550~2200]	700 [350~1400]
K	.0107	– • –	900 [450~1800]	600 [300~1200]
V	.0099	• • • –	1000 [500~2000]	600 [300~1200]
J	.0015	• – – –	1200 [600~2400]	800 [400~1600]
X	.0014	– • • –	1100 [550~2200]	700 [350~1400]
Q	.0008	– – • –	1200 [600~2400]	800 [400~1600]
Z	.0006	– – • •	1100 [550~2200]	700 [350~1400]

Figure 2.8. Morse codes arranged in order of frequency of individual letters.

Frequencies (as a proportion of total letters) in column p are based on Mayzner and Tresselt (1965).

good, experienced railroad telegraphers by Bryan and Harter (1898), reception rates of 45~50 words/minute are seen with the oscillator-based code (and the world record is over 75 words/minute!). This comparison is in the predicted order and, as expected, somewhat faster than our calculation based on a first-order approximation to English. A better approximation to the first-order assumptions of our calculation (but, alas, for Russian) is the set of rates achieved by a set of non-Russian-speaking telegraphers whose job it was to transliterate Russian Morse Code: 30 words/minute average, 38~40 words/minute maximum, and 45 words/minute top (Robin Kinkead, personal communication)—rates consonant with our oscillator-based calculation.

PERCEPTUAL CAUSALITY

One way for two distinct stimuli to fuse is for the first event to appear to *cause* the other.

> **Example 3.** In a graphic computer simulation of a pool game, there are many occasions upon which one ball appears to bump into another ball, causing the second one to move. What is the time available, after the collision, to compute the initial move of the second ball, before the illusion of causality breaks down?

Solution. The movements of the first and second balls must appear to be part of the same event in order for the collision to appear to cause the movement of the second ball, if the movement occurs within one cycle of 100 msec. Since the illusion will break down in the neighborhood of 100 msec, the program should try to have the computation done well before this time. The designer can be sure the illusion will hold if designed for Fastman, with the computation done in 50 msec. ∎

Figure 2.9 shows the results of an experiment analogous to Example 3 in which subjects had to classify collisions between objects (immediate causality, delayed causality, or independent events) as a function of the delay before the movement of the second object. The perception of immediate causality ends in the neighborhood of 100 msec; some degradation of immediate causality begins for some subjects as early as 50 msec.

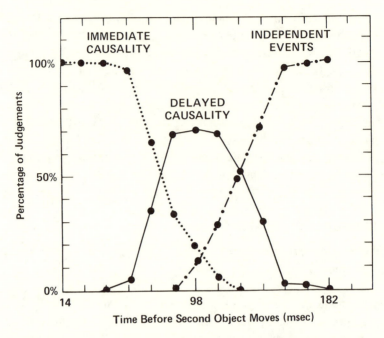

Figure 2.9. Perceived causality as a function of inter-event time between the motion of two objects.
Three types of perceived causality are shown as a function of the interval separating the end of Object A's motion and the beginning of the second object's motion. Average over three subjects. From Michotte (1963, Figure 5, p. 94).

READING RATE

Many perceptual phenomena concern a visual area large enough that the fovea of the eye must be moved to see them. When eye movements are involved, they can dominate the time required for the task.

Example 4. How fast can a person read text?

Solution. Assuming 230 msec/saccade (from Figure 2.1), a reading rate can be calculated from assumptions about how much the reader sees with each fixation. If he were to make one saccade/letter (5 letters/word), the reading rate would be:

$$(60 \text{ sec/min})/(.230 \text{ sec/saccade} \times 5 \text{ saccade/word})$$
$$= 52 \text{ words/min} .$$

For one saccade/word, the rate would be:

$$(60 \text{ sec/min})/(.230 \text{ sec/saccade} \times 1 \text{ saccade/word})$$
$$= 261 \text{ words/min} .$$

For one saccade/phrase (containing the number of characters/fixation found for good readers, 13 chars = 2.5 words), the rate would be:

$$(60 \text{ sec/min})/(.230 \text{ sec/saccade} \times 1/2.5 \text{ saccade/word})$$
$$= 652 \text{ words/min} . ∎[21]$$

How much the reader takes in with each fixation is a function of the skill of the reader and the perceptual difficulty of the material. If the material is conceptually difficult, then the limiting factor for reading rate will not be in the eye-movement rate, but in the cognitive processing. The calculation implies that readers who claim to read much more than 600 words/min do not actually see each phrase of the text. In other words, speed readers skim.

Motor Skill

Just as fundamental limits on the rate of user perceptual performance were set by the cycle time of the Perceptual Processor, limits on movement are set by the rates of the Perceptual and Motor Processors. Two basic kinds of movement occur in human-computer interaction: (1) movement of the hand towards a target and (2) keystrokes.

FITTS'S LAW

The first kind of movement, moving the hand towards a target, can be understood, and an expression for movement time derived, using the Model Human Processor plus some assumptions.[22] Suppose a person wishes to move his hand D cm to reach an S cm wide target (see Figure 2.10). The movement of the hand, as we have said, is not continuous, but consists of a series of microcorrections, each with a certain accuracy.

[21] This calculation is discussed in Hochberg (1976, p. 409).

[22] This derivation is similar to that of Crossman and Goodeve (1963) and Keele (1968).

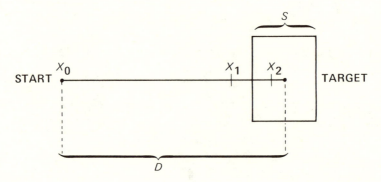

Figure 2.10. Analysis of the movement of a user's hand to a target.

The hand starts from the point labeled START and is to move to anywhere inside the TARGET as fast as possible. D is the distance to the target and S is the width of the target.

To make a correction takes at minimum one cycle of the Perceptual Processor to observe the hand, one cycle of the Cognitive Processor to decide on the correction, and one cycle of the Motor Processor to perform the correction, or $\tau_P + \tau_C + \tau_M$. The time to move the hand to the target is then the time to perform n of these corrections or $n(\tau_P + \tau_C + \tau_M)$. Since $\tau_P + \tau_C + \tau_M \simeq 240$ msec, n is the number of roughly 240-msec intervals it takes to point to the target.

Let X_i be the distance remaining to the target after the ith corrective move and X_0 $(= D)$ be the starting point. Assume that the relative accuracy of movement is constant, that is, that $X_i / X_{i-1} = \varepsilon$, where $\varepsilon < 1$ is the constant error. On the first cycle the hand moves to

$$X_1 = \varepsilon X_0 = \varepsilon D.$$

On the second cycle, the hand moves to

$$X_2 = \varepsilon X_1 = \varepsilon(\varepsilon D) = \varepsilon^2 D.$$

On the nth cycle it moves to

$$X_n = \varepsilon^n D. \tag{2.1}$$

The hand stops moving when it is within the target area, that is when

$$\varepsilon^n D \leq \frac{1}{2}S.$$

Solving for n gives

$$n = -\log_2(2D/S) / \log_2 \varepsilon.$$

Hence the total movement time T_{pos} is given by

$$T_{pos} = n(\tau_P + \tau_C + \tau_M)$$

$$T_{pos} = I_M \log_2(2D/S), \qquad (2.2)$$
$$\text{where } I_M = -(\tau_P + \tau_C + \tau_M) / \log_2 \varepsilon.$$

Equation 2.2 is called Fitts's Law. It says that the time to move the hand to a target depends only on the relative precision required, that is, the ratio between the target's distance and its size. Figure 2.11a plots movement time according to Equation 2.2 for an experiment in which subjects had to alternate tapping between two targets S in. wide, D in. apart. The points fall along a straight line as predicted, except for points at low values of $\log_2(2D/S)$.

The constant ε has been found to be about .07 (see Keele, 1968; Vince, 1948), so I_M can be evaluated:

$$I_M = -240 \text{ msec} / \log_2(.07) \text{ bits}$$
$$= 63 \text{ msec/bit}.$$

A Fastman~Slowman calculation gives a range of $I_M = 27 \sim 122$ msec/bit. Several methods have been used to measure the correction time. One is to turn out the lights shortly after a subject starts moving his hand to a target and note the minimum light-on time that affects accuracy.[23] Another is to detect the onset of correction from trajectory acceleration changes.[24] These methods have given cycle time values in the range

[23] For a discussion, see Welford (1968).

[24] Carlton (1980); Langolf (1973); Langolf, Chaffin, and Foulke (1976).

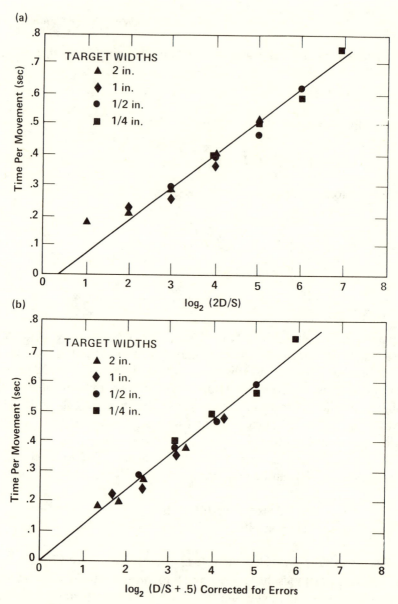

Figure 2.11. Movement time as a function of two versions of Fitts's Law.

From Welford (1968, Figures 5.3 and 5.4).

(a) Times for reciprocal tapping with a 1 oz. stylus plotted in terms of Equation 2.2. Data from an experiment by Fitts (1954). Each point is based on a total of 613~2669 movements obtained from 16 subjects.

(b) The same data as in (a) plotted in terms of Equation 2.3, corrected for errors by Crossman's method (see Welford, 1968).

$\tau_P + \tau_C + \tau_M$ = 190~260 msec/cycle (we calculated $\tau_P + \tau_C + \tau_M$ = 240 msec). The measured correction times correspond to I_M = 50~68 msec/bit (we calculated 63 [27~122] msec/bit).

Measurements of I_M determined directly by plotting observations according to Equation 2.2 give somewhat higher values centering around I_M = 100 msec/bit. The slope of the line drawn through the points in Figure 2.11a is about I_M = 104 msec/bit. Slopes from other experiments are in the I_M = 70~120 msec/bit range. Since I_M will be useful for later calculations, we set here a value based on several experiments:

$$I_M = 100 \, [50{\sim}120] \, \text{msec/bit}.^{[25]}$$

This value is a refinement of the value calculated from the Model Human Processor.

The problem of the points that wander off the line for low values of $\log_2(D/S)$ and the slight curvature evident in Figure 2.11a can be straightened by adopting a variant of Fitts's Law developed by Welford (1968):

$$T = I_M \log_2 (D/S + .5). \tag{2.3}$$

In Figure 2.11b the same data are plotted using Equation 2.3 (and a method of correcting for errors). All the points now lie on the line and the slight bowing has been straightened. This equation gives a somewhat higher estimate for I_M in Figure 2.11b, I_M = 118 msec/bit.

---

[25] For single, discrete, subject-paced movements, the constant is a little less than I_M = 100 msec/bit and closer to the 50~68 msec/bit value cited above for other experimental methods and for our nominal calculation. Fitts and Peterson (1964) get 70~75 msec/bit. Fitts and Radford (1966) get a value of 78 msec/bit (12.8 bits/sec). Pierce and Karlin (1957) get maximum rates of 85 msec/bit (11.7 bits/sec) in a pointing experiment. For continuous movement, repetitive, experimenter-paced tasks, such as alternately touching two targets with a stylus or pursuit tracking, the constant is a little above I_M = 100 msec/bit. Elkind and Sprague (1961) get maximum rates of 135 msec/bit (7.4 bits/sec) for a pursuit tracking task. Fitts's original dotting experiment (Figure 2.11) gives 118 msec/bit using Equation 2.3. Welford's (1968) study using Equation 2.3 and the actual distance between the dots gives 120 msec/bit.

Example 5. On a certain pocket calculator, the heavily used gold **f** button employed to shift the meaning of the keys is located on the top row (see Figure 2.12). How much time would be saved if it were located in a more convenient position just above the numbers?

Solution. Assume that the position of the **5** button is a fair representation of where the hand is just before pressing the **f** button. From the diagram, the distance from the **5** button to the present **f** button is 2 in., to the proposed location, 1 in. The button is 1/4 in. wide. By the Equation 2.3 version of Fitts's Law, movement time is $I_M \log_2 (D/S + .5)$, where I_M is expected to be about 100 msec/bit. So the difference in times required by the two locations is

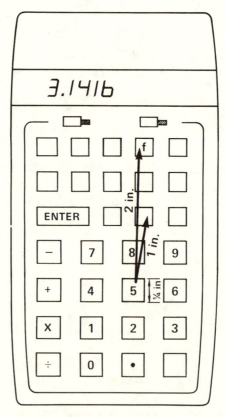

Figure 2.12. Location of keys on the pocket calculator in Example 5.

$$\Delta T = 100 \, [\log_2 (2/.25 + .5) - \log_2 (1/.25 + .5)]$$
$$= 100 \, (3.09 - 2.17)$$
$$= 90 \, \text{msec} . \blacksquare$$

A test of this calculation by an informal experiment is in agreement with the predicted result. The time to press the **f** button was measured by counting the number of times the hand could alternate between the **f** and **5** button in 15 sec at both the old and the proposed location. By this method, the mean time/movement is just 15 sec/number of movements. The experiment was repeated three times:

	Old Time	*New Time*
Trial 1:	290 msec	200 msec/button-press
Trial 2:	240 msec	170 msec/button-press
Trial 3:	230 msec	180 msec/button-press
Mean:	250 msec	180 msec/button-press
Observed difference:		70 msec/button-press
Calculated difference:		90 msec/button-press

Notice that the time to press the **f** button is greater than what it could be in a more favorable location by over 1/3 (70 msec difference in a 180 msec operation). Of course, it is important to keep in mind that the design of the entire calculator will entail some trade-offs in individual key locations.

POWER LAW OF PRACTICE

Before considering the second type of motion, keystrokes, it is useful to digress to consider a learning principle applicable to perceptual-motor learning generally: The time to do a task decreases with practice. It was Snoddy (1926) who first noticed that the rate at which time improves is approximately proportional to a power of the amount of practice as given by the following relationship.

P6. Power Law of Practice. The time T_n to perform a task on the nth trial follows a power law:

$$T_n = T_1 n^{-\alpha} \tag{2.4}$$

or

$$\log T_n = C - \alpha \log n ,\tag{2.5}$$

*where T_1 is the time to do the task on the first trial.
$C = \log T_1$, and α is a constant.*

It can be seen in Equation 2.5 that performance time declines linearly with practice when plotted in log-log coordinates. Typical values for α are in the .2~.6 range.

> **Example 6.** A control panel has ten keys located under ten lights. The user is to press a subset of the keys in direct response to whatever subset of lights is illuminated. If the user's response time was 1.48 sec for the 1000th trial and 1.15 sec for the 2000th trial, what is the expected response time for the 50,000th trial?

Solution. Using Equation 2.5, we can solve for T_1 in order to eliminate it.

$$T_1 = T_n n^\alpha$$
$$(T_{1000})1000^\alpha = (T_{2000})2000^\alpha$$
$$\alpha = \log (T_{1000}/T_{2000}) / \log (2000/1000) = .36 .\tag{2.6}$$

Solving for T_1 using Equation 2.6,

$$T_1 = (T_{1000}) 1000^{.36} = 18 \text{ sec} .$$

The entire equation is

$$T_n = 18 n^{-.36} .\tag{2.7}$$

Thus, the expected time on the 50,000th trial is

$$T_{50,000} = (18)(50,000^{-.36}) = .37 \text{ sec} . \blacksquare$$

Figure 2.13 shows the results of an experimental study of this situation carried out to 75,000 trials. The response time on the 50,000th trial was .40 sec compared to the .37 sec calculated. Characteristically,

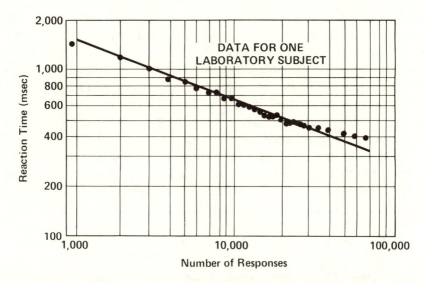

Figure 2.13. An example of the Power Law of Practice.
Improvement of reaction time with practice on a 1023-choice task.   Subjects pressed keys on a ten-finger chordset according to pattern of lights directly above the keys.   After Klemmer (1962).

the data here are well fit by Equation 2.5, except at the ends. Estimating by eye, the best-fitting straight line in the linear portion of the curve gives $T = 21n^{-.38}$, comparable to Equation 2.7.

The Power Law of Practice applies to all skilled behavior, both cognitive and sensory-motor.[26]   However, practice does not cover all aspects of learning. It does not describe the acquisition of knowledge into Long-Term Memory or apply to changes in the quality of performance. Quality does improve with practice, but it is measured on a variety of different scales, such as percentage of errors, total number of errors, and preference ratings, that admit of no uniform treatment.

KEYING RATES

The Power Law of Practice plays an important role in understanding user keystroking performance.   Keying data into a system is a highly repetitive task: in a day's time, a keypuncher might strike 100,000 keys. The Power Law of Practice has three practical consequences here. First, there is a wide spread of individual differences based primarily on the

[26] See Newell and Rosenbloom (1981).

amount of previous typing practice. Typing speed ranges from 1000 msec/keystroke for an absolute novice to 60 msec/keystroke for a champion typist, more than a factor of 15 difference. Second, the power function form for the practice curve (Equation 2.4) has a very steep initial slope (linear in the log means it drops through the first factor of 10 in one hundredth the time it takes to drop through the second factor of ten—consult Figure 2.13). Thus typists pass through an initial unpracticed state to one of moderate skill rather rapidly. Third, the practice curve becomes relatively flat after a short time (though it never entirely ceases to improve, according to the Power Law). This means that, for users of moderate skill, performance is relatively stable and one can indeed talk about constant rates for typing and keying.

> **Example 7.** How fast can a user repetitively push with one finger a key on the typewriter keyboard? How fast can he push two keys using alternate hands?

Solution. In the case of a repeated keystroke, the finger must first be cocked back, then brought forward. Each half of the stroke, according to the Model Human Processor, will take $\tau_M = 70$ msec and the whole stroke will take $\tau_M + \tau_M = 140$ msec. In the case of keystrokes between alternate hands, it should be possible for one hand to stroke while the other is cocking if the strokes are coordinated, so in these cases strokes could follow each other within 70 msec. ∎

These two are the fastest and slowest cases, hence the typing rate for a skilled typist might be expected to lie somewhere within 70~140 msec/keystroke for a mixture of same-hand and different-hand stroke combinations (if the typist is given sufficient look-ahead so that perceptual and cognitive processing overlaps motor processing).

Figure 2.14 gives data-entry rates for some keystroke-operated devices. For typewriter-like devices, expert typing rates hover in the 100~300 msec range, as expected. Champion keypunch and typing performance is in the 60~80 msec range, faster than the Middleman calculation above, but slower than the 30 msec lower bound set by a Fastman calculation. As Figure 2.14 shows, difficult text or lack of expertise exact perceptual and cognitive costs that slow the rate.

More detailed calculations of user performance can be made using data for individual interkeystroke times such as those collected by Kinkead (1975) and reproduced in Figure 2.15, which breaks down

Typewriters	(msec/stroke)	
Best keying	60	Dresslar (1892)
Typing text	158~231	Hershman and Hillix (1965)
Typing random words	200~273	Hershman and Hillix (1965)
Typing random letters	462~500	Hershman and Hillix (1965)
Typing (1 char look-ahead)	750~1500	Hershman and Hillix (1965)
Unskilled typing of text	1154	Devoe (1967)
10-Key Pads	(msec/stroke)	
Numeric keypunching	112~400	Neal(1977)
Keypunching	300~444	Klemmer and Lockhead (1962)
10-key telephone	789~952	Pollock and Gildner (1963), Deininger (1960)
10-key adding machine	1091	Minor and Revesman (1962)
Other Keyboards	(msec/stroke)	
Simple pushbuttons	570~690	Munger, Smith, and Payne (1962)
5×5 adding machine	600~800	Pollock and Gildner (1963)
Coded physician's order	779~2222	Minor and Pittman (1965)
10×10 adding machine	1200	Minor and Revesman (1962)
Chord Sets	(msec/chord)	
Stenotypists	333	Seibel (1964)
8-key chordset	508~1017	Pollock and Gildner (1963)
Mail sorting	517~882	Cornog and Craig (1965)
Hand Entry	(msec/char)	
Hand printing	545~952	Devoe (1967)
Handwriting	732	Devoe (1967)
Mark sensing	800~3750	Devoe (1967), Kolesnick and Teel (1965)
Hand punching	3093	Kolesnick and Teel (1965)

Figure 2.14. Keying times for selected input techniques.

interkeystroke times by key and by whether the preceding keystroke was on the same hand, finger, or key as the current keystroke. These times can be used to make approximate comparisons between keyboard layouts.

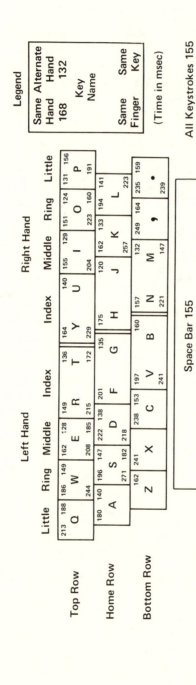

Figure 2.15. Interkeystroke typing times.
Based on 155,000 keystrokes from 22 typists (from Kinkead, 1975).

Example 8. A manufacturer is considering whether to use an alphabetic keyboard (see Figure 2.16) on his small business computer system. Among several factors influencing his decision is the question of whether experienced users will find the keyboard slower for touch-typing than the standard Sholes (QWERTY) keyboard arrangement. What is the relative typing speed for expert users on the two keyboards?

Solution. Figure 2.15 gives the time/keystroke t_i for all but the most infrequent letter keys, broken down by whether the previous key was the same key, the same finger, the same hand, or the other hand. Figure 2.17 gives the frequencies f_i with which two-letter combinations appear in English (punctuation and space digraphs are, unfortunately, not available in the table). The expected typing rate is just the weighted average,

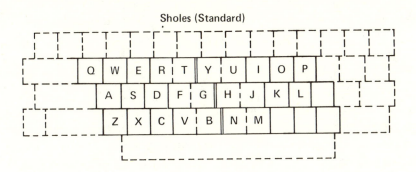

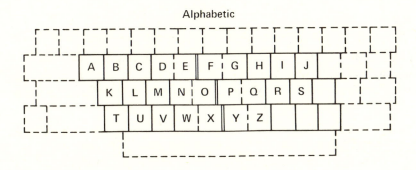

Figure 2.16. Arrangement of letter keys on Sholes and on one possible alphabetic typewriter.

First Letter	Second Letter												
	A	B	C	D	E	F	G	H	I	J	K	L	M
A	2	229	354	242	9	115	214	13	375	19	142	842	335
B	182	15	—	2	547	—	—	—	121	13	—	227	—
C	562	—	49	—	496	—	4	543	248	—	168	125	—
D	172	—	—	36	660	8	34	6	403	—	—	51	11
E	880	13	337	1213	433	112	110	19	165	2	38	583	310
F	174	2	—	—	233	127	—	—	290	—	—	66	—
G	136	—	—	—	380	2	53	312	170	—	—	61	2
H	1056	9	—	4	3139	8	2	—	848	—	—	8	6
I	210	66	589	310	329	218	265	—	—	—	59	543	339
J	32	—	—	—	44	—	—	—	4	—	—	—	—
K	8	4	—	2	293	4	2	4	138	—	—	17	—
L	452	13	6	337	937	61	4	2	655	—	25	740	34
M	547	106	—	—	757	9	—	—	325	—	—	6	76
N	250	—	254	1476	846	36	1190	19	288	15	70	79	28
O	64	68	132	208	45	942	62	11	74	6	87	365	553
P	343	—	—	—	435	—	—	61	142	—	2	295	6
Q	—	—	—	—	—	—	—	—	—	—	—	—	—
R	577	32	108	167	1730	19	76	15	615	—	112	129	117
S	252	34	131	2	797	11	2	473	464	—	74	72	102
T	456	9	62	4	1103	8	—	3397	971	2	—	138	42
U	98	55	161	55	131	15	182	—	91	—	4	352	297
V	78	—	—	—	929	—	—	—	229	—	—	—	—
W	571	—	4	6	507	—	—	490	231	—	2	23	2
X	23	—	34	—	28	4	—	6	25	—	—	2	—
Y	25	9	15	4	140	—	—	4	38	—	—	13	28
Z	17	—	—	—	61	—	—	—	8	—	—	6	—

Figure 2.17. Frequencies of English digraphs.

Probability of digraph occurrence x 10^5. Computed from data of Underwood and Schulz (1960, Appendix D).

$$\text{Typing rate} = \Sigma_i f_i t_i .$$

Applying this formula to both the Sholes keyboard (the conventional one) and the alphabetic keyboard of Figure 2.16 (and dividing the result by $\Sigma_i f_i$ to compensate for the fact that only about 90% of the digraph times are given in Figure 2.17) gives

Typing rate (Sholes) $= 152$ msec/keystroke (72 words/min)
Typing rate (alphabetic) $= 164$ msec/keystroke (66.5 words/min).

The alphabetic arrangement is calculated to be about 8% slower than the Sholes arrangement. ∎

First Letter	Second Letter												
	N	O	P	Q	R	S	T	U	V	W	X	Y	Z
A	2146	2	193	2	1128	1028	1362	115	252	70	13	272	25
B	2	293	—	—	140	15	4	246	4	—	—	127	—
C	—	653	—	2	333	9	333	81	—	—	—	32	—
D	34	257	4	—	108	161	2	131	21	8	—	70	—
E	1355	72	149	25	2106	1285	431	13	288	170	185	204	4
F	6	431	—	—	210	—	127	123	—	—	—	4	—
G	32	184	6	—	176	81	19	87	—	—	—	13	—
H	13	—	2	—	98	23	197	127	2	11	—	19	—
I	2394	471	68	2	386	1105	1238	8	288	—	26	—	62
J	—	89	—	—	—	—	—	57	—	—	—	—	—
K	97	—	2	—	2	59	2	—	—	—	—	15	—
L	11	378	28	—	9	112	106	100	26	25	—	481	—
M	2	386	206	—	19	78	2	142	—	—	—	114	—
N	64	486	4	8	6	384	967	87	34	—	2	134	9
O	1487	390	225	2	1239	284	466	1306	138	435	21	42	8
P	2	252	174	—	343	49	62	91	—	—	—	13	—
Q	—	—	—	—	—	—	—	115	—	—	—	—	—
R	202	819	17	—	114	458	299	134	62	8	—	252	—
S	25	331	157	23	2	386	1151	242	—	47	—	61	—
T	8	694	2	—	413	363	263	216	—	78	—	202	—
U	460	—	142	—	541	481	524	—	9	—	2	8	2
V	—	55	—	—	—	—	—	2	—	—	—	6	—
W	89	274	—	—	25	28	6	—	—	—	—	11	—
X	—	2	61	—	—	—	34	4	—	—	—	—	—
Y	11	352	17	—	6	104	30	—	2	9	—	—	—
Z	—	6	—	—	—	—	—	—	—	—	—	8	13

Kinkead (1975) used a similar calculation to show that the Dvorak keyboard would be expected to be only 2.6% faster than the Sholes keyboard. This calculation makes two strong assumptions. The first is that the frequencies of the digraphs will not seriously affect the digraph times, a reasonable assumption by the Power Law argument above. A more difficult assumption is that there are no substantial leveling effects, in which slow digraphs slow down faster ones. This last assumption has been disputed by Yamada (1980a, 1980b).

Simple Decisions

We have discussed how simple calculations are possible for perceptual and motor performance; now we can consider how the perceptual and motor systems, together with central cognitive mechanisms, combine in simple acts of behavior.

SIMPLE REACTION TIME

The basic reaction time for simple decisions can be derived from Figure 2.1.

> **Example 9.** A user sits before a computer display terminal. Whenever any symbol appears, he is to press the space bar. What is the time between signal and response?

Solution. Let us follow the course of processing through the Model Human Processor in Figure 2.1. The user is in some state of attention to the display (Figure 2.18a). When some physical depiction of the letter A (we denote it α) appears, it is processed by the Perceptual Processor, giving rise to a physically-coded representation of the symbol (we write it α') in the Visual Image Store and very shortly thereafter to a visually coded symbol (we write it α'') in Working Memory (Figure 2.18b). This process requires one Perceptual Processor cycle τ_P. The occurrence of the stimulus is connected with a response (Figure 2.18c), requiring one Cognitive Processor cycle, τ_C. The motor system then carries out the actual physical movement to push the key (Figure 2.18d), requiring one Motor Processor cycle, τ_M. Total time required is $\tau_P + \tau_C + \tau_M$. Using Middleman values, the total time required is $100 + 70 + 70 = 240$ msec. Using Fastman and Slowman values gives a range 105~470 msec. ∎

In practice, measured times for a simple reaction under laboratory conditions range anywhere from 100 to 400 msec.

PHYSICAL MATCHES

If the user has to compare the stimulus to some code contained in memory, the processing will take more steps.

> **Example 10.** The user is presented with two symbols, one at a time. If the second symbol is identical to the first, he is to push the key labeled YES, otherwise he is to push NO. What is the time between signal and response for the YES case?

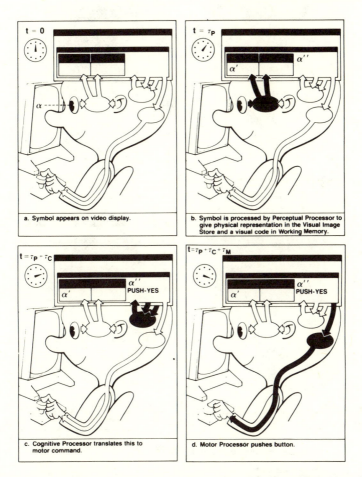

Figure 2.18. Simple reaction-time analysis using the Model Human Processor.

Solution. The first symbol is presented on the screen where it is observed by the user and processed by his Perceptual Processor, giving rise to associated representations in his Visual Image Store and Working Memory. The second symbol is now flashed on the screen and is similarly processed (Figure 2.19*a*). Since we are interested in how long it takes to respond to the second symbol, we now start the clock at 0. The Perceptual Processor processes the second symbol to get an iconic representation in Visual Image Store and then a visual representation in

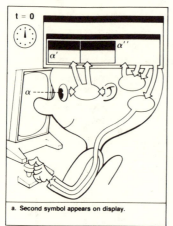

a. Second symbol appears on display.

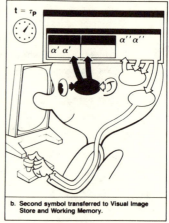

b. Second symbol transferred to Visual Image Store and Working Memory.

c. Symbols matched on basis of Visual Codes.

d. Cognitive Processor translates successful match to motor command.

e. Motor Processor pushes button.

Figure 2.19. Physical name-match analysis using the Model Human Processor.

68

Working Memory (Figure 2.19b), requiring one cycle, τ_P. If not too much time has passed since the first symbol was presented, its visual code is still in Working Memory and the Cognitive Processor can match the visual codes of the first and second symbols against each other to see if they are the same (Figure 2.19c). This match requires one Cognitive Processor cycle, τ_C. If they match, the Cognitive Processor decides to push the YES button (Figure 2.19d), requiring another cycle, τ_C. Finally, the Motor Processor processes the request to push the YES button (Figure 2.19e), requiring one Motor Processor cycle, τ_M. The total elapsed reaction time, according to the Model Human Processor, is

$$
\begin{aligned}
\text{Reaction time} &= \tau_P + 2\tau_C + \tau_M \\
&= 100\,[50{\sim}200] + 2{\times}(70\,[25{\sim}170]) + 70\,[30{\sim}100] \\
&= 310\,[130{\sim}640]\ \text{msec}\,.
\end{aligned}
$$

As our analyses become more complex, it becomes convenient to use a more concise notation. Such a notation can be had by writing symbolically what the contents of the memories are after each step. This has been done for the last two examples, Examples 9 and 10, in Figure 2.20.

NAME MATCHES

If the user has to access a chunk from Long-Term Memory, the response will take longer.

> **Example 11.** Suppose in Example 10 the user was to press YES if the symbols had the same name (as do the letters A and a), regardless of appearance and NO if they did not. What is the time between signal and response for the YES response?

The analysis is similar to the previous example except that instead of performing the match on the visual codes, the user must now wait (see Figure 2.20 Step 2.01) until the visual code has been recognized and an abstract code representing the name of the letter is available. The consequence of adding the new step is the addition of one more Cognitive Processor cycle,

$$
\begin{aligned}
\text{Reaction time} &= \tau_P + 3\tau_C + \tau_M \\
&= 100\,[50{\sim}200] + 3{\times}(70\,[25{\sim}170]) + 70\,[30{\sim}100] \\
&= 380\,[155{\sim}810]\ \text{msec}\,.
\end{aligned}
$$

Step		Display	VIS	WM	Hand	Elapsed Time
Example 9. Simple reaction						
	State at start of clock:					
1.	Symbol appears	α				0
2.	Transmitted to VIS		α'	α''		τ_P
3.	Initiate response		α'	α''. PUSH-YES		$\tau_P + \tau_C$
4.	Process motor command		α'	α''. PUSH-YES	PUSH-YES	$\tau_P + \tau_C + \tau_M$
Example 10. Physical match						
	State at start of clock:		α'	α''		
1.	Second symbol appears	α	α'	α''		0
2.	Transmitted to VIS		α'. α'	α''. α''		τ_P
2.1.	Match		α'. α'	α''. α'', MATCH = TRUE		$\tau_P + \tau_C$
3.	Initiate response		α'	α''. α''. PUSH-YES		$\tau_P + 2\tau_C$
4.	Process motor command			α''. α''. PUSH-YES	PUSH-YES	$\tau_P + 2\tau_C + \tau_M$
Example 11. Name match						
	State at start of clock:		α_1'	α_1'':A		
1.	Second symbol appears	α_2	α_1'	α_1'':A		0
2.	Transmitted to VIS		α_1'. α_2'	α_2''. α_1'':A		τ_P
2.01.	Recognize		α_1'. α_2'	α_2'':A. α_1'':A		$\tau_P + \tau_C$
2.1.	Match		α_1'. α_2'	MATCH = TRUE		$\tau_P + 2\tau_C$
3.	Initiate response		α_2'	PUSH-YES		$\tau_P + 3\tau_C$
4.	Process motor command			PUSH-YES	PUSH-YES	$\tau_P + 3\tau_C + \tau_M$
Example 12. Class match						
	State at start of clock:		α'	α'':A:LETTER		
1.	Second symbol appears	β	α'	α'':A:LETTER		0
2.	Transmitted to WM		α'. β'	β''. α'':A:LETTER		τ_P
2.01.	Recognize		α'. β'	β'':B. α'':A:LETTER		$\tau_P + \tau_C$
2.02.	Classify		α'. β'	β'':B:LETTER. α'':A:LETTER		$\tau_P + 2\tau_C$
2.1.	Match		β'	MATCH = TRUE		$\tau_P + 3\tau_C$
3.	Initiate response			PUSH-YES		$\tau_P + 4\tau_C$
4.	Process motor command			PUSH-YES	PUSH-YES	$\tau_P + 4\tau_C + \tau_M$

Figure 2.20. Trace of the Model Human Processor's memory contents for simple decision tasks.

The symbols α and β stand for the unrecognized visual representation of the input; the symbols α' and β' stand for the physical representation of the input in the Visual Image Store (VIS); the symbols α'' and β'' stand for the visual code of the input in Working Memory (WM); and the symbols **A** and **LETTER**, stand for the abstract representation. The notation α'':**A** means that both visual and abstract codes exist in Working Memory and are associated with one another.

CLASS MATCHES

It might happen that the user has to make multiple references to Long-Term Memory.

> **Example 12.** Suppose in Example 11 the user was to press YES if both symbols were letters, as opposed to numbers. What would be the time between signal and response?

The analysis is similar (see Figure 2.20) to the previous example except that a new step, Classify, is required to convert both versions of the symbol to the same representation.

$$\begin{aligned} \text{Reaction time} &= \tau_P + 4\tau_C + \tau_M \\ &= 100\,[50{\sim}200] + 4{\times}(70\,[25{\sim}170]) + 70\,[30{\sim}100] \\ &= 450\,[180{\sim}980]\ \text{msec} . \blacksquare \end{aligned}$$

Experiments have been performed by many researchers to collect empirical data on the questions presented in these examples. The results are that name matches take about 70 msec longer than physical matches and that class matches take about 70 msec longer yet. (70 msec is the nominal value we have used for τ_C.) Figure 2.21 shows one such experimental result. Name matches are about 85 msec slower than physical matches when there is very little time between the first and second symbol. By the time 2 sec have elapsed, the visual code in Working Memory has decayed so that the extra step of getting the name must occur and, in fact, performance is close to that required for a name match. For these predictions, the relative, nominal value calculation gives good agreement with the data, but the absolute values of the reaction times are low (data: 525 msec, calculation: 380 [155~810] msec), reflecting some systematic, second-order effect adding a constant time to all the data points. The absolute values remain within the Fastman~ Slowman range however.

CHOICE REACTION TIME

If the user has to make a choice between two responses, we can analyze the task as in Example 10 where the choices were YES and NO. If there are a larger number of choices, the situation is more complicated, but still the task can be analyzed as a sequential set of decisions made by the Cognitive Processor, each adding a nominal $\tau_C = 70$ msec to the response.[27] Regardless of the detailed analysis of the mental steps involved in choosing between alternatives, more alternatives require more steps and, hence, more time. The relationship between time required and number of alternatives is not linear because people apparently can arrange the processing hierarchically (for example, dividing the responses into groups, then on the first cycle deciding which group should get

[27] See Welford (1973) and Smith (1977).

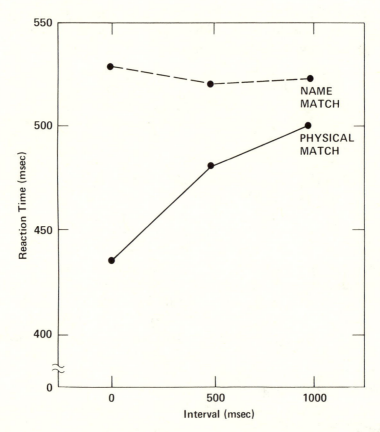

Figure 2.21. Reaction times for matching successively presented letters as a function of the inter-stimulus interval.
From Posner, Boies, Eichelman, and Taylor (1969, Figure 2, p. 8).

further consideration).   The minimum number of steps necessary to process the alternatives can be derived from information theory and, *to a first order of approximation,* the response time of people is proportional to the information-theoretic entropy of the decision.

> *P7. Uncertainty Principle: Decision time T increases with uncertainty about the judgment or decision to be made:*
> $T = I_C H$, *where H is the information-theoretic entropy of the decision and I_C is a constant.*

For the case where a person observes *n* alternative stimuli, which are associated one-to-one with *n* responses (example: sorting multiple-part business forms by color), this principle can be given a simple mathematical formulation:

$$H = \log_2 (n + 1).\tag{2.8}$$

The equation, a variant of Hick's Law, may be taken as an empirical relationship that simply fits many measured situations, in that no particular mechanism is proposed. However, the equation is clearly related to rational ways of processing that minimize expected time. H is a function of $n+1$ rather than just n because there is uncertainty about whether to respond or not, as well as about which response to make. As an illustration, Figure 2.22 shows the reaction time required between the onset of one of n equally probable signals and the pressing of the appropriate button. The figure plots the reaction time against the

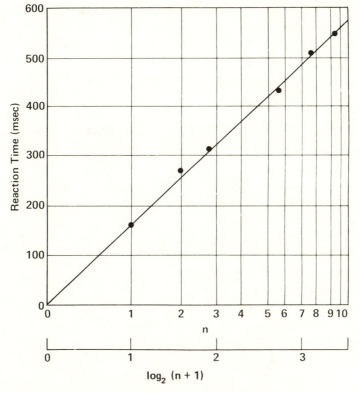

Figure 2.22. Hick's Law of choice reaction time.
After Welford (1968, p. 62). At the onset of one of n lights, arranged in a row, the subject is to press the key located below the light.

number of alternatives (1 to 10), on a log scale showing that the measurements form the straight line predicted from the equation.

Equation 2.8 can be generalized to the case where the n alternatives have different probabilities of occurring,

$$H = \Sigma_{i=1}^{n} p_i \log_2 (1/p_i + 1). \qquad (2.9)$$

Although the probability in the formula is the person's subjective probability, it often can be estimated from the task. When all of the probabilities are equal ($= 1/n$), $p_i \log (1/p_i + 1) = (1/n) \log_2 (n+1)$ and Equation 2.9 reduces to Equation 2.8.

> **Example 13.** A telephone call director has 10 buttons. When the light behind one of the buttons comes on, the secretary is to push the button and answer the phone. What is the percentage difference in reaction time required between the cases where (1) each one of the telephones receives an equal number of calls and (2) two of the telephones are used heavily, receiving 50% and 40% of the calls, with the remaining 10% uniformly distributed among the remaining phones?

Solution. By the Uncertainty Principle and Equation 2.9, the reaction time to signals of unequal probability is

$$T = I_C H,$$

where
$$H = \Sigma_{i=1}^{n} p_i \log_2 (1/p_i + 1).$$

For case (1), $p_i = .1$ and

$$H = 10 (.1 \log_2 (1/.1 + 1)) = 3.46 \text{ bits}.$$

For case (2), $p_1 = .5$, $p_2 = .4$, and $p_i = .0125$ (where $3 \leq i \leq 10$),

$$H = .5 \log_2 (1/.5 + 1) + .4 \log_2 (1/.4 + 1)$$
$$+ (8)(.0125)(\log_2 (1/.0125) + 1)$$
$$= 2.14 \text{ bits}.$$

The difference is $\Delta H = 3.46 - 2.14 = 1.32$ bits. So the response time for case (2) is calculated to be $2.14/3.46 = 62\%$ of the reaction time for case (1). ∎

Example 13 discussed one form of weighted occurrence probability. Another way of creating uncertainty is not to have signals occurring with fixed frequencies, but to have sequential dependencies of the signals. For instance, suppose at each trial either the signal for response #1 or response #2 can occur. However, the signal for response #1 occurs with .8 probability after a previous signal for response #1, but only with .2 probability after a signal for response #2. One can apply the same information-theoretic formula to compute the uncertainty. Hyman (1953) tried these different ways of inducing uncertainty, with the results shown in Figure 2.23. As can be seen, all the different ways of inducing uncertainty fit the same curve.

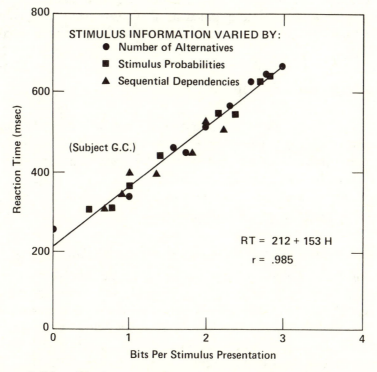

Figure 2.23. Choice reaction time for three different ways of manipulating the stimulus information H.
Data for a single subject. Hyman (1953, Figure 1, p. 192, subject G.C.).

Figure 2.23 shows that it takes about $I_C = 150$ msec/bit of uncertainty, above a base of about $C = 200$ ms, which we could identify as $C = \tau_P + \tau_M$. Using these values we can estimate the actual reaction times in Example 13: (1) Where each of the telephones receives an equal number of calls, the reaction time would be 200 msec + (150 msec/bit)(3.46 bits) = 719 msec. (2) Where two of the telephones are heavily used, the reaction time would be 521 msec. When the 200 msec intercept is taken into account, case (2) is 72% of case (1).

There are also situations in which we do not know how to compute H, but in which we do know that relatively more mental steps must be involved in one case than in another. For example, if the lights and keys in Example 13 were paired randomly with each other, the user would require more mental steps, I_C would be increased, and the response could be expected to take more time. The relative number of mental steps required as a function of the features of a particular set of inputs and outputs of an interface is called its stimulus-response compatibility. As the result of practice, fewer mental steps are required and I_C becomes smaller.

Learning and Retrieval

Most user behavior is, of course, more complex than the simple decisions we have just been discussing for the fundamental reason that most user behavior is performed in complex system environments and depends on the user's knowledge and understanding of those environments. How knowledge about systems and procedures is stored and retrieved is, therefore, of some importance.

FORGETTING JUST-ACQUIRED INFORMATION

Recall again the flow of information in Figure 2.1 from perceptual memory to Working Memory to Long-Term Memory. The ratio between the decay times of these stores is large, on the order of 200 msec : 7000 msec : ∞, which reduces to $1:35:\infty$. The characteristics of retrieval will depend on the elapsed time since the information was stored, because that will determine which memories, if any, preserve the item. For retrievals done a few seconds after input, items may be stored in either Working Memory or Long-Term Memory, or in both. For retrievals done a few minutes after input, items are retrievable only from Long-

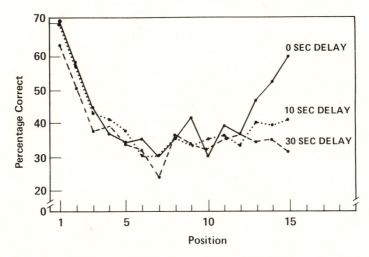

Figure 2.24. Probability of recalling a word from a list as a function of the position of the word in the list and of the delay before starting recall.

From Glanzer and Cunitz (1966, Figure 2, p. 358). Each point represents the mean for five lists and 46 subjects.

Term Memory. This fact is illustrated by Figure 2.24, which shows the results of an experiment in which people were given a list of words to learn and later to recall (in any order). Between presentation of the list and recall they were prevented from rehearsal (that is, from physically or mentally saying the list over and over) by the introduction of a different task.

The curves show the probability of recall at each position of the studied items (position 1 is the earliest one presented). The top curve shows that both the initial and the final words in the list are remembered better than the ones in the middle. The bottom curve shows what happens if a delay of 30 seconds occurs before recall is started, allowing new items to be activated in Working Memory, interfering with those to be remembered. As can be seen, the difference is that the final words lose all their extra memorability. The middle curve simply confirms the analysis by showing that a delay of 10 sec is intermediate in its effect.

> **Example 14.** A programmer is told verbally the one-syllable file names of a dozen files to load into his programming system. Assuming the names are all arbitrary, in

which order should the programmer write down the names so that he remembers the greatest number of them (has to ask for the fewest number to be repeated)?

Solution. Twelve arbitrary file names means the programmer has to remember 12 chunks (assuming one chunk/name), which is larger than μ_{WM}, so some file names will be forgotten. The act of trying to recall the file names will add new items to Working Memory, interfering with the previous names. The items likely to be in Working Memory but not yet in Long-Term Memory are those from the end of the list. If the user tries to recall the names from the end of the list first, he can snatch some of these from Working Memory before they are displaced. The probability of recalling the first names will not be affected since they are in Long-Term Memory. Thus, the programmer should recall the last names first, then the others. ∎

> **Example 15.** Suppose that in Example 14, the 12 files did not have arbitrary names, but rather names such as INIT1, INIT2, INIT3, INIT4, PERF1, PERF2, PERF3, PERF4, SYSTEMS1, SYSTEMS2, SYSTEMS3, SYSTEMS4. In which order should the programmer write down the file names so that he remembers the largest number of them?

Solution. Unlike the case in Example 14 where each file was a separate chunk, here there are only 4 chunks: INIT#, PERF#, SYSTEMS#, and the rule for #. The number of chunks is within the user's Working Memory span and hence the order of recalling the files should make little difference. ∎

> **Example 16.** Show that the amount of time a programmer can delay typing the name of the file before forgetting it (with probability > .5) is much longer if the file name is CAT than if it is TXD. (Assume the work involved does not permit the user to rehearse the file name.)

Solution. The file name TXD is assumed to be a nonsense word and therefore must be coded in three chunks. From Figure 2.1, $\delta_{WM}(3$ chunks$) = 7$ [5~34] sec, but the file name CAT is one chunk, $\delta_{WM}(1$ chunk$) = 73$ [73~226] sec. Nominally, the user can remember the meaningful name on the order of 73 sec / 7 sec = 10 times longer. ∎

Actually, the advantage of meaningful names is likely to be even greater than this calculation shows, since meaningful names are easier to transfer to Long-Term Memory and have more associates to get them back.

Two more comments are in order. First, we have treated chunks as if they were all alike. Experimental confirmation of the *approximate* equivalence of chunks for memory decay appeared in Figure 2.6. The figure thus shows that a list of three consonants like TXD is forgotten at the same rate as a list of three words like (CAT PIG MAN). Second, we have assumed intervening demands on the user that prevented him from rehearsing the chunks in Working Memory. If rehearsal is possible, a small number of chunks can be kept in Working Memory indefinitely, at the cost of not being able to perform many other mental tasks.

INTERFERENCE IN WORKING MEMORY

According to the Discrimination Principle, it is more difficult to recall an item if there are other similar items in memory. The similarity between two items in memory depends on the mental representation of each item, which depends in turn on the memory in which the item resides. The two most important dimensions of interference are acoustic interference and semantic interference. Items in Working Memory are usually more sensitive to acoustic interference (they are confused with other items that sound alike) because they usually (but not necessarily) use κ = acoustic coding (Conrad, 1964). Items in Long-Term Memory are more sensitive to semantic interference (they are confused with other items with similar meaning) because they use κ = semantic coding.

> **Example 17.** A set of error indicators in a system have been assigned meaningful three-letter words as mnemonics. The idea is that, since each word is a single chunk, more codes can be remembered and written down at a glance, and since each code is only three letters the codes will be fast to write. When the system crashes, the operator is to write down a set of up to five code words that appear in a special alphanumeric display. Which is more important to avoid (in order to minimize transcription errors), codes that are similar in sound or codes that are similar in meaning?

Because the codes are to be written down immediately, the codes will be held largely in Working Memory during transcription. Because

	Experiment I (Spoken)				Experiment III (Visual)	
	Group A (N = 20)		Group S (N = 21)		Group AV (N = 10)	
	Acoustically Similar	Control	Semantically Similar	Control	Acoustically Similar	Control
Word Set	mad, man, mat, map, cad, can, cat, cap	cow, day, far, few, hot, pen, sup, pit	big, long broad, great, high, tall, large, wide	old, deep, foul, late, safe, hot, strong, thin	Same as Expt. I plus cab, max	Same as Expt. I plus rig, day
Percentage Correctly Recalled	10%	82%	65%	71%	2%	58%

Figure 2.25. Acoustic vs. semantic interference in Working Memory.
Subjects studied 25 five-word lists. The words in the lists were either acoustically similar, semantically similar, or unrelated (control condition). The numbers in the table are the proportion of lists recalled entirely correctly and in the proper order. Data of Baddeley (1966) as presented in Calfee (1975, Figure 17.6).

Working Memory uses largely acoustic coding, transcription errors will occur mainly from interference between acoustic codes. Similar sounding codes should therefore be avoided. ∎

Figure 2.25 shows the result of a similar experiment in which subjects had to remember lists of five words, then recall them twenty seconds later. They made many errors with the acoustically similar lists (only 1~2% of the lists were recalled error-free), but substantially fewer with the semantically similar lists (13% of the lists were recalled error-free), and this was true regardless of whether they were given the lists aurally or visually.

INTERFERENCE IN LONG-TERM MEMORY

The Discrimination Principle P4 says that the difficulty of recall depends on what other items can be retrieved by the same cues. Thus, as the user accumulates new chunks in Long-Term Memory, old chunks that are semantically similar to the new chunks become more difficult to remember.

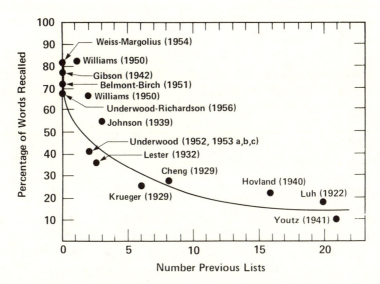

Figure 2.26. Interference of previously learned material with later learning.

Recall of serial lists 24 hours later as a function of number of previous lists learned. Revised version of Underwood (1957, Figure 3, p. 53).

A demonstration of this fact is shown in Figure 2.26. When people learn lists of words in the laboratory, they forget a large fraction of them within 24 hours. Underwood (1957) managed to find 16 separate published studies that both recorded the amount of forgetting after 24 hours and gave enough detail to determine the number of previous lists that had been learned prior to the one tested. Even though these lists differed in length, time per list item, and details of experimental procedure, it is clear that learning more prior lists results in more forgetting and that this accounts for a very large fraction of the forgetting that occurs. The size of the interference effect shows that much of what passes for forgetting is failure to retrieve, not actual loss from the memory.

> **Example 18.** A user is about to learn how to use a new, line-oriented text-editor, identical to one he already knows except for the command names (such as ERASE instead of DELETE). Will his learning of the new editor interfere with his ability to remember the command names of the old one?

Solution. Yes. When the user learns the new editor, there will be new chunks in memory similar to those of the old editor and, by the Discrimination Principle, these may interfere with retrievals about the old editor. Indeed, it is a common experience for programmers to be unable to recall how to use an old system on which they have spent hundreds of hours after learning a similar new one. ∎

Not only does just-acquired knowledge interfere with previous knowledge in Long-Term Memory, it also interferes with subsequent knowledge, although usually with smaller effect.[28]

SEARCHING LONG-TERM MEMORY

Information is retrieved from Long-Term Memory with each basic cycle of the cognitive processor, but retrieval of the desired item is not always successful. When sufficiently long times are available for search, strategies can be used to probe Long-Term Memory repeatedly. Retrieving the name for a known but rarely used command is a typical example.

It is worth emphasizing the difficulty faced by the user attempting to retrieve an item from his Long-Term Memory, as given by the Encoding Specificity Principle. When he learned the item, it was encoded in some way. This encoding included various possible cues for recalling the item. At retrieval time, the user knows neither the desired item nor its recall cues. He must therefore guess, placing cues in Working Memory where they will serve as calls on Long-Term Memory on the next cycle. The guesses may be good and succeed immediately or, even if they fail, may retrieve some information that can help on a subsequent try.

A graphic example of Long-Term Memory search, emphasizing its capacity, the requirement for interactive strategic search, and the fact that Long-Term Memory is in many ways an *external* body of knowledge, like a phone book or an encyclopedia, is shown in Figure 2.27. The subject was asked, seven years after being graduated, to remember the names of all 600 members of her high school graduating class. (The experimenter had the year book.) As the graph shows, even after ten hours of trying, the subject was still retrieving new information from Long-Term Memory. Her strategy was an elaborate version of the interactive retrieval strategy above: In her mind, the subject scanned for faces, attended old parties, worked the alphabet, wandered down familiar streets asking for the house occupants. The process also produced fabrications

---

[28] Murdock (1963).

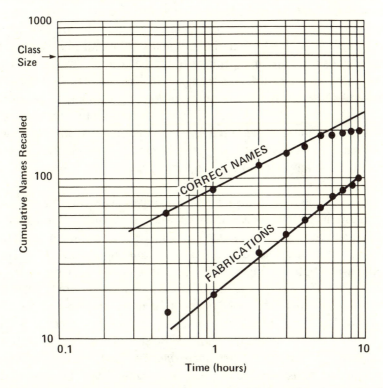

Figure 2.27. Recall of the names of high-school graduating class, seven years after being graduated.
Replotted data from Subject S1 in Williams and Hollan (1981).

where non-classmate names were recalled somewhat uncertainly during early sessions and were later misrecalled as classmate names.

Complex Information-Processing

The psychological phenomena we have discussed so far comprise the building blocks out of which more complex user behavior is composed. This more complex behavior spans longer times and is rationally organized.

OPERATOR SEQUENCES

More complex activities must ultimately be composed of the sorts of elementary actions we have been discussing. These rudimentary actions operate to cause physical changes in the state of the world or mental changes in the state of the user, and to emphasize this property we call them *operators*. It has been realized, in an insight into the structure of behavior dating at least from the Gilbreths (Gilbreth, 1911), that the operators are sufficiently independent of the behavioral situation in which they are observed that different segments of behavior can be seen to be composed of the same few operators differently combined. It further turns out that it is possible to define operators sufficiently independent of each other that the time required by an operator in isolation is a good approximation to the time it requires as part of a sequence (although there are generally second-order interactions that set limits to this additivity).

Figure 2.28 shows a direct attempt to investigate whether the time required by an operator was the same when it occurred in isolation as when it occurred as part of a sequence. The tasks were simple operations of reading analogue and digital dials, looking up values in a table, computing a simple arithmetic formula, and entering data by keying it.

As the figure shows, the mean operator time required when the operator is combined with other operators is about the same as the time required in isolation, but the variability in the operator times is greater when the operator is combined, with coefficients of variation roughly 15-20% higher.[29]   Thus, to a *first approximation* (and when careful task definitions and measurements are made), integrated task behavior could be decomposed, in this case, into component operators, which could be defined and measured in independent contexts.

> **Example 19.**   In the experiment reported in Figure 2.28*b*, the total time to do the combined task was 51.56 sec (*SD* = 18.85).   How close is this result to the times predicted from Figure 2.28*a*?

The total time to do the combined task should be the sum of the mean times for the individual tasks:

$$T = 6.24 + 3.45 + 9.26 + 34.20$$
$$= 53.15 \text{ sec} .$$

---

[29] It is convenient to express variability in terms of the coefficient of variation *CV* = Standard Deviation / Mean, because it makes variability from distributions with different means more easily comparable; we often use this statistic in preference to the standard deviation.

(a) INDEPENDENT TASKS

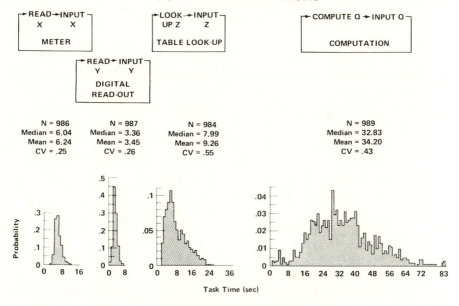

```
┌─READ─►INPUT─┐        ┌─LOOK-►INPUT─┐              ┌─►COMPUTE Q ─►INPUT Q─┐
│   X      X  │        │ UP Z     Z  │              │                      │
│             │        │             │              │                      │
│   METER     │        │ TABLE LOOK-UP│              │      COMPUTATION     │
└─────────────┘        └─────────────┘              └──────────────────────┘
          ┌─READ─►INPUT─┐
          │   Y      Y  │
          │             │
          │  DIGITAL    │
          │  READ-OUT   │
          └─────────────┘
```

N = 986	N = 987	N = 984	N = 989
Median = 6.04	Median = 3.36	Median = 7.99	Median = 32.83
Mean = 6.24	Mean = 3.45	Mean = 9.26	Mean = 34.20
CV = .25	CV = .26	CV = .55	CV = .43

(b) INTEGRATED TASK

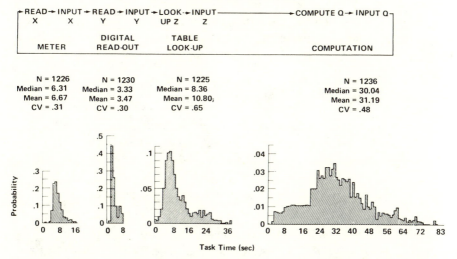

```
┌─READ─►INPUT─►READ─►INPUT─►LOOK-►INPUT──────────────►COMPUTE Q─►INPUT Q─┐
│   X      X     Y      Y    UP Z     Z                                  │
│              DIGITAL       TABLE                                       │
│   METER      READ-OUT      LOOK-UP                    COMPUTATION      │
└───────────────────────────────────────────────────────────────────────┘
```

N = 1226	N = 1230	N = 1225	N = 1236
Median = 6.31	Median = 3.33	Median = 8.36	Median = 30.04
Mean = 6.67	Mean = 3.47	Mean = 10.80;	Mean = 31.19
CV = .31	CV = .30	CV = .65	CV = .48

Task Time (sec)

Figure 2.28. Time distributions for four operators (a) when measured in isolation and (b) when measured as part of an integrated task.

Five university students performed each of the following operators: READ-METER-AND-TYPE-INPUT, READ-DIGITAL-DISPLAY-AND-TYPE-INPUT, READ-X-Y-AND-LOOKUP-Z, READ-X-Y-Z-AND-COMPUTE-Q. They performed the operators both in isolation and as part of a larger integrated task. From Mills and Hatfield (1974, Figures 3 and 4).

85

The measured task time was $(53.15 - 51.56)/53.15 = 3\%$ higher than calculated. ∎

The variance of the combined task should be the sum of the variance for the individual tasks, assuming independence among the tasks:

$$SD = \sqrt{[1.53^2 + .90^2 + 5.10^2 + 14.77^2]}$$
$$= 15.73 \text{ sec}$$
$$CV = SD/Mean = 15.73/53.15 = .30 \, .$$

The measured coefficient of variation is $18.85/51.56 = .37$, which is $(.37 - .30)/.30 = 23\%$ higher than calculated.

THE RATIONALITY PRINCIPLE

A person attempts to achieve his goals by doing those things the task itself requires to be done. Much of the complexity of human behavior derives not from the complexity of the human himself (he is simply trying to achieve his goals), but from the complexity of the task environment in which the goal-seeking is taking place.[30] It follows that, to understand and predict the course of human behavior, one should analyze a task to discover the paths of rational behavior. We come, therefore, to what might be called the fundamental principle of task analysis:

> **P8. Rationality Principle.** *A person acts so as to attain his goals through rational action, given the structure of the task and his inputs of information, and bounded by limitations on his knowledge and processing ability:*
> *Goals + Task + Operators + Inputs*
> *+ Knowledge + Process-limits → Behavior.*

The principle really offers a nested set of formulations that can be used in order to predict a person's behavior. The first version, *Goals + Task + Operators,* takes into account only the objective situation; the other factors reflect hidden constraints, namely what the person can perceive, what he knows, and, finally, how he can compute. The additional factors offer successive approximations to how he will behave,

---

[30] See Simon (1947, 1969), Newell and Simon (1972).

with the shorter equations being easier to use, but giving cruder approximations.

THE PROBLEM SPACE PRINCIPLE

Rational behavior can often be given a more precise description. Suppose a person has the goal to prove a theorem using the rules of symbolic logic. There is a set of mental states through which he passes (describable in terms of symbolic expressions) and a number of operators for changing one state into another (operations in symbolic logic). This set of states and operators is called a *problem space*. In general:

> *P9. Problem Space Principle. The rational activity in which
> people engage to solve a problem can be described in terms of
> (1) a set of states of knowledge, (2) operators for changing
> one state into another, (3) constraints on applying operators,
> and (4) control knowledge for deciding which operator to
> apply next.*

There are different problem spaces for different tasks, and there may well be changes in problem spaces over time, as the user acquires more knowledge about the structure of the task.

An example of a short problem-solving task, and one that has been examined in detail, is the cryptarithmetic puzzle. As shown below, each letter is to be assigned a different digit so that replacing the letters by their digits forms a correct addition. For example:

$$
\begin{array}{ccccccl}
  & D & O & N & A & L & D \\
+ & G & E & R & A & L & D \\
\hline
  & R & O & B & E & R & T & \quad D=5
\end{array}
$$

A typical way in which a person goes about solving such a problem is a combination of elementary reasoning and trial-and-error. For example:

> ...I can, looking at the two D's (pause) each D is 5; therefore T is 0.
> So I think I'll start by writing that problem here. I'll write 5, 5 is O.
> Now do I have any other T's? No. But I have another D. That
> means I have a 5 over the other side. Now I have 2 A's and 2 L's that
> are each somewhere and this R, 3 R's, 2L's equal and R. Of course I'm
> carrying a 1. Which will mean that R has to be an odd number.

```
      D   O   N   A   L   D

  +   G   E   R   A   L   D        D = 5
      _____

      R   O   B   E   R   T
```

Informal Description: Letters in the above array are to be replaced by numerals from zero though nine, so that all instances of the same letter are replaced by the same numeral. Different letters are to be replaced by different numbers. The resulting array is to be a correctly worked problem in arithmetic. The assignment for the letter D is already given to be 5.

States: Assignments of numbers to letters.

Operators: (ASSIGN Letter Number)
                       (PROCESS-COLUMN Column)
                       (GENERATE-DIGITS Letter)
                       (TEST-DIGIT Number)

Path Constraint: D + D = T, etc.

Figure 2.29.   External problem space for a cryptarithmetic task.

Because the 2 L's, any two numbers added together has to be an even number and 1 will be an odd number. So R can be 1... [Excerpt from protocol for Subject S3, Newell and Simon, 1972, p. 230].

The problem space for this subject (see Figure 2.29) consists of assignments of numbers to letters (R = 3), and various relations that can be known about the letters and digits (R > 5, R odd, R unassigned). The mental operators used by this subject can be identified:

ASSIGN   Assign a number to a letter.

PROCESS-COLUMN   Infer other assignments and
                        constraints from a column.

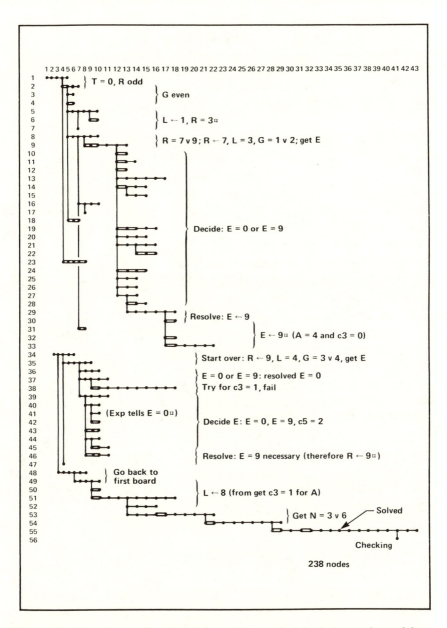

Figure 2.30. Search of subject through his internal problem space for the cryptarithmetic task.

Subject S3, Newell and Simon (1972, Figure 6.4, P. 181) for DONALD + GERALD = ROBERT. Each dot in the diagram represents a state of knowledge of the subject. Each link is the result of applying an operator.

GENERATE-DIGITS	Determine what numbers are possible for a letter.
TEST-DIGIT	Determine if a digit can be assigned to a letter.

There is also a more general operator:

SET-UP-GOAL	Set up goal to obtain a certain result or to check that a knowledge expression is true.

These operators embody the limitations of human information-processing in various ways. For example, with only ten digits to be assigned and with the assignments just having been made, one might think that an intelligent problem solver would always know what digits were available. Not directly. Unless the **TEST-DIGIT** operator is applied, the problem-solver will not know whether a digit has been assigned to another letter.

Figure 2.30 gives a graphic presentation of the behavior of the subject whose protocol was excerpted above. Each state of knowledge of the subject is represented by a point and the operation of an operator by a connecting line. The double lines are places where the person repeats a path previously trod. This repeating of a path is a reflection of Working Memory limitations, it being easier to drop back repeatedly to an anchor state than to remember the intermediate states. The graph can be summarized by saying that: (1) the subject is involved in heuristic search; and (2) upon close examination the apparently complex behavior resolves into a small number of elements (the parts of a state and the operators) interacting with the complex constraints of the task, an illustration of how complexity in behavior arises from the environment.

2.3.   CAVEATS AND COMPLEXITIES

We have attempted to convey a version of existing psychological knowledge in a form suitable for analyzing human-computer interaction. We have summarized this knowledge in a simple model of the human processor and have suggested, through examples, how it might be used with task analysis, calculation, and approximation to support engineering calculations of cognitive behavior. Although it is hoped that the model itself will be useful, the real point is in the spirit of the enterprise: that knowledge in cognitive psychology and related sciences is sufficiently advanced to allow the analysis and improvement of common mental tasks, provided there is an understanding of how knowledge must be structured to be useful. The present chapter is an illustration of one possible way for structuring this knowledge.

In the foregoing description, we have chosen to concentrate on a picture of basic human information-processing capabilities relevant for human-computer interaction rather than to detail human engineering studies of particular systems or techniques. Human-engineering studies relevant to our particular concerns are referenced in context in later chapters. For general reviews of behavioral studies of human-computer interaction, the reader is directed to Moran (1981b), Ramsey, Atwood, and Kirshbaum (1978), Ramsey and Atwood (1979), Rouse (1977), Miller and Thomas (1977), and Bennett (1972). For reviews of the general "man-machine" literature, the reader is directed to Rouse (1980), Pew, Baron, Feehrer, and Miller (1977), Meister (1976), Sheridan and Ferrell (1974), and Parsons (1972).

There are also many papers that either review, or for other reasons provide convenient entry into, specialized portions of the human-computer interaction literature. Perceptual issues of video displays are treated in Cakir, Hart, and Stewart (1980), Shurtleff (1980), and Gould (1968). Reviews of the large literature on devices for data entry can be found in Sperandio and Bisseret (1974), Seibel (1972), Alden, Daniels, and Kanarick (1972), and Devoe (1967). The design of command languages is treated in Barnard, Hammond, Morton, Long, and Clark (1981); Moran (1981a); Boies (1974); Fitter and Green (1979); Reisner (1981); and Martin (1973). Programming has received considerable attention: Sheil (1981); Shneiderman (1980); Brooks (1977); Shepard, Curtis, Milliman, and Love (1979); and Smith and Green (1980). And finally, a number of systems have been proposed as frameworks for the human operation of machines; for example, Lane, Streib, Glenn, and Wherry (1980); Siegal and Wolf (1969); and Quick (1962).

The model of human information-processing that we have presented is our own synthesis of the current state of knowledge. In many respects (though not all) it corresponds to the dominant model of the seventies (Fitts and Posner, 1967; Neisser, 1967; Atkinson and Shiffrin, 1968; Welford, 1968; Newell and Simon, 1972; Lindsay and Norman, 1977; Anderson, 1980). But beyond any general model, a large amount of detailed knowledge is available in the literature on all the phenomena we have examined. In order to make the reader aware in some general way of the limits of our model, we mention briefly a number of the complexities documented in the literature and some of the alternative theoretical views.

BOXES VS. DEPTH OF PROCESSING

The dominant model of the seventies had as an underlying heuristic the assumption that there was an elaborate logic-level structure of many separate registers (the "boxes"), each with its own distinct memory parameters and connected by a distinct set of transfer paths. There was a Short-Term Memory consisting of seven chunks, brought into prominence by Miller (Miller, 1956; cf. Blankenship, 1938); forgetting was accomplished by displacement from fixed slots in the registers. Short-Term Memory was separate from Long-Term Memory, in contradistinction to the earlier theory, which simply posited a single structure of stimulus-response connections. The discovery by Sperling (1960) of the Visual Image Store, which was clearly distinct from the Short-Term Memory, provided impressive support for the "box" view.

A number of difficulties have beset this model, mostly in increased complexities and muddying-up of initially clean distinctions, as experimental evidence has accumulated. Initially it appeared that all information in the Short-Term Memory was coded acoustically (Conrad, 1964) and all information in Long-Term Memory coded semantically, but this has proved not to be the case. For instance, in some of the examples in this chapter, the use of visual codes in Working Memory is evident. Initially, rehearsal seemed to play the key role in the transfer of information from the Short-Term Memory to the Long-Term Memory—the more an item was rehearsed, the better chance it had of being stored away permanently. It has since seemed necessary to distinguish maintenance rehearsal, which has no implications for permanent memory, from elaborative rehearsal, which does. This distinction proved to be the crack in the edifice. It resulted in a new general view, called *depth of processing*, which attempts to do away with the structural boxes entirely and substitute a continuum of processing depth to determine how well material is remembered. "Depth" is defined somewhat intuitively: examining the letters of words is shallow, finding rhymes a little deeper, and creating stories using the words deeper still. This view is now itself under serious attack (Wicklegren, 1981) for lack of precision in its theory and for its unsuccessful predictions.

WORKING MEMORY SPAN

The original view of Working Memory, following Miller (1956), was that it had a capacity of 7±2 items, coinciding with the immediate memory span. Gradually, much of the support for the existence of an independent Working Memory came from the recency effect in free recall (the fading ability to remember the last few items heard that we examined in Figure 2.24). Various ways of calculating Working Memory size from the recency effect all give answers in the range 2.5~4.1 items for the capacity. This implies that the immediate memory is a compound effect of more than one process, which is the way we have described it.

At the opposite end of the spectrum from sizes of 2.5~4.1 vs. 7±2 is the notion of Working Memory as an activation of Long-Term Memory, hence, of essentially unlimited instantaneous extent, but of limited access. The model presented here couples such a view with that of decay to get the limited access. This view, though not widely stated explicitly, is represented in a few places in the literature (Shiffrin and Schneider, 1977).

The Model Human Processor has moved some distance from the model of the early seventies in replacing separate memory registers with registers that are subregisters of each other: Working Memory is the subset of activated nodes in Long-Term Memory, and the Visual and Auditory Image Stores are not completely separate from Working Memory. Baddeley (1976, 1981) and his co-workers have used the term Working Memory functionally to include additional components of the human limited-capacity short-term storage system, which combine for skilled tasks such as reading to provide a capacity somewhat larger than our μ_{WM}. Chase and Ericsson (1982) have used the term Working Memory to include rapid accessing mechanisms in Long-Term Memory, what we have termed Effective Working Memory. They showed in a series of ingenious experiments that, through extensive practice, people can enormously increase their Effective Working Memory beyond our $\mu_{WM}{}^*$. The upshot of the Baddeley and Chase and Ericsson results is to emphasize the intimate connection between Working Memory, Long-

Term Memory, and attention. For the sake of simplicity, we have not attempted to incorporate these ideas into the Model Human Processor, pending their further development.

MEMORY STRENGTH VS. CHUNKS

The notion that memories have strengths, and can be made stronger by repetition, has been a central assumption of much psychological theorizing. Wicklegren (1977) gives a good account of this view for the whole of memory. The notion that memories come in discrete chunks, which either exist or do not exist in Long-Term Memory, provides an alternative conception that has risen to prominence with the information-processing view of man. It is this view we have presented.

It is difficult to determine in a simple, experimental way which of these two positions holds in general. Each type of theory can mimic and be mimicked by the other. One basic difficulty is that memory phenomena, being inherently errorful and varying, always lead to data samples that show considerable variation. One can never tell easily whether the variation arose from corresponding variation of strength or from discrete probabilistic events. The same effects producible by gradation in strengths also flow from multiple copies of chunks (Bernbach, 1970). Such multiplicity, far from being contrived, might be expected if a system manufactured chunks continually from whatever was being attended to.

WHAT IS LIMITING?

That humans are limited in their abilities to cope with tasks is clear beyond doubt. Where to locate the constraint is less clear. One general position has focused on memory as the limiting agent, as in the notion of the register containing a fixed set of slots. Another general position has focused on processing. A more sophisticated notion is that processing and memory may each be limiting but in different regions of performance (Norman and Bobrow, 1975). The processing position has usually taken the form of some sort of homogeneous quantity called *processing capacity*, which is allocated to different tasks or components of a task, usually within a *parallel* system. Another form of processing limit is to posit a *serial* system and permit it only one operation at a time.

Again, it is not possible to formulate experimental ways of distinguishing these alternatives in general. Serial processing systems can mimic parallel ones by rapid switching, and parallel systems of limited capacity can show the most obvious sign of serial processing, linear time effects.

INTERFERENCE VS. DECAY

The Model Human Processor incorporates spontaneous decay over time and interference as mechanisms that produce memory-retrieval failure. Typically these are held to be alternative mechanisms and much effort has gone into trying to determine to which one forgetting is attributable. Actually, with the advent of information-processing models, a third alternative occurred: *displacement* of old items by new ones. This is clearly a version of interference, though one that involves total loss at storage time (of the interfered-with item), not of interaction at retrieval time.

The strong role of interference in long-term forgetting has been well-established. However, no one has ever accounted for the losses in very long term memory (weeks, months, or years) in a way that excludes genuine forgetting, although at least one investigator (Wickelgren, 1977) believes he can separate true forgetting from interference in the long term.

EXPANSIONS OF THE MODEL HUMAN PROCESSOR

There are at least three areas where the description of the Model Human Processor might be significantly expanded at some cost in simplicity. The first area is the semantic description of Long-Term Memory. As the study of Long-Term Memory proceeded, it became evident to psychologists that, in order to understand human performance, the semantic organization of Long-Term Memory would have to be taken into account. We have not described semantic memory in any depth here, since the details of such an account would carry us beyond the bounds set for this chapter. For surveys of the relevant literature, the reader is referred to Anderson (1980), Lindsay and Norman (1977), Norman and Rumelhart (1975), and Anderson and Bower (1973).

The second area is the description of the Perceptual Processor. In the simplified description we have given of perceptual processing, we have skipped over considerable detail that is appropriate at a more refined level of analysis. A description based on Fourier analysis could be used to replace various parts of the model for describing the interactions of visual stimuli with intensity and distance (Cornsweet, 1970; Ganz, 1975; Breitmeyer and Ganz, 1976).

The third area is the description of the Cognitive Processor. We have not said much in detail about the control structure of the Cognitive Processor; but it is necessary to consider the processors's control discipline if interruptability, errors, multiple-tasking, automaticity, and other phenomena are to be thoroughly understood. A more detailed description of the recognize-act cycle, and how the characteristics of simple decisions arise from it, might be given in terms of a set of recognize-act rules, called *productions* (Newell, 1973). According to this description, the productions themselves reside in Long-Term Memory. On each cycle, the recognition conditions of the rules are compared with the contents of Working Memory (or said another way, some of the recognition conditions of the rules are activated through spreading activation in Long-Term Memory). The rule with the best match (the highest state of activation) fires and causes its associated action to occur, altering the contents of Working Memory (activating other chunks in Long-Term Memory). Perceptual input whose recognition activates previously non-activated chunks in memory may, through this mechanism, interrupt and redirect the previous course of processing. The description might be elaborated to give both an account of skilled behavior that requires little conscious attention and an account of unskilled behavior. A production system description has also been used to give a description of complex information-processing where each action might involve several dozen recognize-act cycles (for examples, see Newell and Simon, 1972; Young, 1976; Anderson 1976).

THE EXISTENCE OF ALTERNATIVES

Does the existence of alternatives to various features of the Model Human Processor, like those we have just mentioned, and the fact that agreement on them is very difficult to obtain, rob the model of its usefulness or show that it is impossible to settle things in psychology? Not at all, and for two reasons.

The first reason is a technical issue about making progress in psychology. Many of the difficulties arise because classes of quite different mechanisms can mimic each other rather closely, as in the case of interference and decay. However, this mimicking works only over narrow ranges of behavior. For instance, if only one specific task is considered—say, the immediate memory distractor task (Figure 2.6) in which a single item is given, then counting backward by sevens, then attempting to recall the item—it is easy to generate several explanations (decay, interference, displacement) that are indistinguishable, even in principle, by unlimited precision in the data.   But if these same mechanisms are required to provide the explanation in many diverse tasks, it becomes much harder for the mimicking to succeed.   Thus, the comments we have made apply locally—mechanism X competes with mechanism Y to explain a given phenomenon, but only when that phenomenon is considered in relative isolation.

The current style in psychology is to have a highly elaborated base of quantitative data over many diverse phenomena, with many local theories.   The science has not yet succeeded in putting together general theories that are tight enough quantitatively so that the same posited mechanism (for example, Working Memory decay) is forced to show itself in action in a large diversity of tasks.   Such comprehensive theories may soon emerge—the groundwork seems well-laid for them—but there has not yet been enough of this theorizing to settle the issues reflected in this section.

The second reason that the existence of alternatives does not rob the model of its usefulness concerns the use to which our model is to be put. The model's purpose is to provide a sufficiently good approximation to be useful.   Its function is synthesis, not discrimination of alternative underlying mechanisms.   If basic mechanisms are not distinguishable in a domain where there has been extensive empirical investigation, there is some assurance that working with either will provide a reasonable first approximation.   Then it is important to obtain a single overall picture based on one set of mechanisms that works globally and fits in with an appropriate unified theoretical perspective.   This we have done.

Our purpose in this chapter has been to prepare the way for the specific set of studies of human-computer interaction that is to follow. Though these studies do not take the details we have been presenting for granted, they do presume the basic orientation laid out here.

TEXT-EDITING

3. System and User Variability

The use of a computer for editing text is a paradigmatic example of human-computer interaction, and for several reasons. (1) The interaction is commonly rapid: A user completes several transactions a minute for sustained periods. (2) The interaction is intimate: A text-editor, like all well-designed tools, becomes an unconscious extension of its user, a device to operate *with* rather than operate *on*. (3) Text-editors are probably the single most heavily-used programs: There is currently a massive effort to introduce text-editing systems into offices and clerical operations. Even in a systems programming environment, one study (Boies, 1974) found that 75% of the system commands issued were text-editor commands. And (4) computer text-editors are similar to, and can therefore be representative of, other systems for human-computer interaction: Like most other systems, they have a discrete command language and provide ways to input, modify, and search for data. The physical details of their interfaces are not particularly unique. Because of these similarities, progress in understanding user interaction with text-editors should help us to understand interaction with other systems as well.

The study of text-editors is a task that is reasonably within the range of the analytic tools we have available from cognitive psychology and computer science. It is a symbolic task of substantial, but manageable, complexity. Because of the intrinsic importance of the task itself, the similarities with other tasks, and the task's tractable complexity, studies of computer text-editing are a natural starting point in the study of human-computer interaction.

3.1.   THE STUDY OF TEXT-EDITING

Before proceeding to the description of our studies on text editing, it is useful to set the stage by describing what is known from previous studies, the details of the physical environment for the systems we shall study, and a sample of typical text-editor dialogue.

STUDIES OF TEXT-EDITING

Despite its practical application and its apparent fruitfulness as a research problem, there have been few studies of computer text-editing (other than reports on specific editors). Previous work on editors falls into two groups, analytical studies of editor design and behavioral studies of users.

Analytical studies of editors have focused on editing time and comparative functionality. By making idealized assumptions, Oren (1972, 1974, 1975) was able to derive equations for editing time as a function of several system properties of "word-processing" systems; but he did not report empirical validation of his models. Van Dam and Rice (1971) compared several types of editors informally. Riddle (1976) and Roberts (1979) both derived taxonomies of editing features and used these to compare the functionality of widely used systems.

Behavioral studies have focused on editing time and to a lesser extent, on the methods actually used by users, users' errors, and learning. Embley, Lan, Leinbaugh, and Nagy (1978) analyzed editors in terms of the number of commands and number of keystrokes users required to perform benchmark tasks. They also tried to predict the commands and keystrokes required by deriving the editing commands from a comparison of the file before and after editing (Anandan, Embley, and Nagy, 1980). Hammer (1981) derived the minimum number of keystrokes required to make an edit and compared that with human performance. We (Card, Moran, and Newell, 1976, 1980a, 1980b; Card, 1978) videotaped users of text-editors to determine their methods and predict, using cognitively-oriented models, their editing time. These studies are elaborated in the present book. DeLaurentiis (1981) used keyboard protocols to determine how users' methods change as they move from novice to expert. Hammer and Rouse (1979) tried to summarize users methods as a Markov transition matrix. Roberts (1979) constructed a method for evaluating editors from behavioral tests of editing time, learning time, and errors, and also investigated mental loading. The behavioral studies have recently been reviewed by Embley and Nagy (1981).

PHYSICAL TEXT-EDITING ENVIRONMENT OF THE STUDIES

The physical arrangement of the user, his computer terminal, and a text manuscript, though particular to our experiments, is entirely typical of the arrangements commonly encountered in offices where computer-assisted document preparation systems are in use. This arrangement may be assumed in the experiments we describe unless contradicted.

A person (the "user") sits before a computer terminal with a keyboard for input and a video display terminal for output (see Figure 3.1). In the

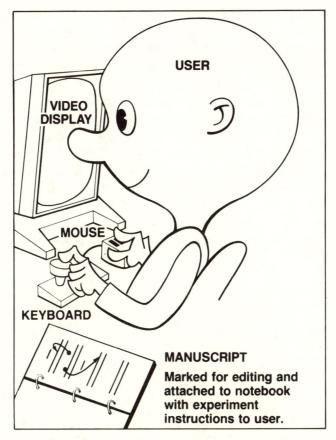

Figure 3.1. Physical layout for the manuscript-editing task.

computer is a text file. To the user's left is a manuscript, consisting of a printout of the text file on which modifications have been marked with red ink. The user, working via a computer program for text-editing, is to effect each of the marked modifications in the text file, producing an updated file. Variations on the task occur with variations in the nature of the computer, the editor program, the terminal, the size of the manuscript, the kind and number of corrections, the physical layout, and the familiarity of the user with the manuscript and the editor.

The keyboard the user employs is similar to that of a standard American electric typewriter save for the addition of a few special keys, such as ESC ("escape"), CONTROL (a type of shift key), LINEFEED, BACKSPACE, and DELETE. The extra keys are used for special system-dependent functions.

Many systems we discuss also employ a special "mouse" pointing device to select items on the video display. The mouse is a small box set atop wheels or ball bearings and attached to the keyboard by a flexible wire. The user can roll it about the table causing analogous movements of the cursor on the video display. He can push one of the three buttons protruding from the top of the mouse to indicate selection of the letter, word, or text fragment indicated on the screen by the cursor.

For the most part, users whose behavior we observe are in daily interaction with the systems on which they are recorded, as part of their job duties. Overall, the experimental arrangement is very similar to the user's natural setting when working with the system. Experiments are run in a room much like the user's own office. The only difference is the presence of a television camera in the room, but the user is typically not much aware of the camera once he becomes absorbed in his task.

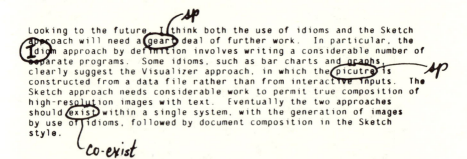

Figure 3.2. Sample fragment of a marked-up manuscript.
Four modifications are indicated by the markings on this fragment. The marks on the manuscripts given to users in the experimental sessions are in red ink.

SAMPLE DIALOGUE WITH A TEXT-EDITOR

To make the details of editing concrete, let us consider the task of making the modifications indicated on the manuscript fragment shown in Figure 3.2 using a typical teletypewriter-oriented editor, POET. The first instruction on the manuscript indicates that the word **great** has been mistyped in the manuscript as **geart**. A typical dialogue between the system and the user is as follows (the inputs by the user and the outputs by the system are shown on the left, comments are shown on the right):

SYSTEM:	**#**	Prompts the user for a command.
USER:	**"geart"**	The user indicates the line he wants by typing between quotation marks a string of characters, **geart**, that identify the line. The system responds by finding the next line in the file that contains the characters and locating itself at that line, called the *Current Line*.
SYSTEM:	**#**	Prompts for a command.
USER:	**/**	This command tells the system to print the Current Line on the display. The user's purpose is to ensure that the system is located at the intended line.
SYSTEM:	**approach will need a geart deal of ...**	(In the dialogue, an elipsis "..." is used to shorten long typeouts by the system.)
SYSTEM:	**#**	The system displays the Current Line on the bottom line of the display screen and prompts for the next command. The user decides to make the change by using the Substitute command to substitute the characters **grea** for the characters **gear**.
USER:	**s**	Invokes the Substitute command.
SYSTEM:	**ubstitute**	Completes the command name and waits for the first argument to the Substitute command.
USER:	**grea RETURN**	Types the new text to be substituted and terminates it with RETURN.
SYSTEM:	**(for)**	Prompts the user for second argument of Substitute.
USER:	**gear RETURN**	Types the old text to be replaced and terminates it with RETURN.
SYSTEM:	**[OK]**	Asks the user to confirm that the command is stated correctly before executing it.

USER:	**RETURN**	Types RETURN to confirm. The system then makes the substitution every time it can on the Current Line.
SYSTEM:	**1**	Responds by printing the number of substitutions it made—in this case, only one was made.
SYSTEM:	**#**	Prompts for a command.
USER:	**/**	Again, the user wishes to look at the Current Line to verify that the modification was made as he intended.
SYSTEM:	**approach will need a great deal of ...**	
SYSTEM:	**#**	The system prompts for a command. The modification was correct. The user now proceeds to the next task, in which the word "idiom" is to be capitalized. The user decides he will do this by substituting the characters **Idi** for **idi.** Because the task is located on the line immediately following, this time the user decides to use the LINEFEED command (instead of the "quoted string method") to locate the line..
USER:	**LINEFEED**	Invokes command to increase Current Line by 1 (move to the next line on the file) and display it.
SYSTEM:	**idiom approach by definition involves ...**	
SYSTEM:	**#**	Prompts for a command.
USER:	**s**	Invokes Substitute command, etc.
SYSTEM:	**ubstitute**	
USER:	**Idi RETURN**	
SYSTEM:	**(for)**	
USER:	**idi RETURN**	
SYSTEM:	**[OK]**	
USER:	**RETURN**	
SYSTEM:	**1**	
SYSTEM:	**#**	
USER:	**/**	
SYSTEM:	**Idiom approach by definition involves ...**	
SYSTEM:	**#**	

The user proceeds in this manner through the rest of the manuscript, making the indicated modifications as he encounters them on the marked-up manuscript.

The POET editor, discussed above, is typical of a large class of editors designed to be usable on a teletypewriter terminal. An example of a rather different sort of editor is the "display-oriented" editor, RCG (see Englebart and English, 1968). This editor uses a five-key chordset for entering commands and uses the mouse for pointing. With RCG, the user could perform the task as follows:

USER:	**rc**	Typed on chordset.
SYSTEM:	**Replace Character**	
		Displayed at top of screen.
USER:		Points to first **i** in **idiom** with mouse.
	MOUSE1	Presses button 1 on mouse.
SYSTEM:		Underlines character.
USER:		Moves hands to keyboard.
	I	Capital "I" typed on keyboard.
SYSTEM:		The word **idiom** instantly changes to **Idiom**. User moves left hand to chordset, right hand to mouse.
	MOUSE1	Presses button 1 on mouse to indicate termination of command.
SYSTEM:		Redisplays entire screen of text with change made.

The description is shorter because the more complex operations required by POET to indicate the target text are replaced in RCG by a simpler pointing and select operation.

There are many other schemes for designing an editor. Some will have effects on user performance. The twin questions naturally arise, just how much effect does the design of an editor have on the time to edit a manuscript, and how do differences between editors compare with differences between people? Before embarking on more detailed investigations, it is important to get an approximate answer to these questions. If the design of the editor makes little difference in editing time, then there is little point to investigating editing rates for different designs unless they are radically different from current ones. If differences in editing time between users are much larger than those between editors, then more leverage is gained by studying individual differences. Consequently, we describe two exploratory experiments that bear on these points.

3.2.   TIME DIFFERENCES AMONG EDITORS (EXPERIMENT 3A)

In order to discover how much difference the design of an editor makes to the speed with which text can be edited, the obvious thing to do is compare the speed of several editors on benchmark tasks.

METHOD FOR EXPERIMENT 3A

Editing Systems. Five editing systems of substantially different design (see Figure 3.3) were chosen for study: POET, SOS, TECO, BRAVO, and RCG. Three of the systems (POET, SOS, and TECO) are teletypewriter-oriented; they assume a discipline imposed by a typewriter with a long scroll of paper (although they were actually tested with a video display on which the last 40 lines could be seen). One line at a time is typed on the scroll, with both the system's output and the user's input intermixed. The two remaining editors (BRAVO and RCG) are display-oriented. They operate by showing the user a picture of a page of text and updating the picture after each editing modification.

Benchmark Tasks.   The editors were compared by testing user performances on four benchmarks (see Figure 3.4): (1) a Letter Typing benchmark, in which the user typed a letter from scratch; (2) a Manuscript Modification benchmark, in which the user made corrections to a text file; (3) a Text Assembly benchmark, in which the user assembled a document from stored paragraphs; and (4) a Table Typing benchmark, in which the user typed a table of numbers and labels into the system.

Users. Each of the 13 users in the experiment was either a secretary or a computer scientist. All were expert users with the editors on which they were tested: Each had used the system for more than a year and had used the system within the week in which he was tested. About a quarter of the users had programmed or maintained one of the systems.

Design. Each editor was tested on three users. (Three is the smallest number that would give some notion of inter-user variability and the largest for which experts on the different editors were available.) Because few users were expert in more than one or two of these editors and to avoid the possibility of practice effects from repeated exposure to the tasks, each user was tested on a single editor. Only one user was tested on SOS because of its similarity to POET. Each of the four benchmarks was done with the POET, SOS, TECO, and RCG editors: only the

POET   (Russell, 1973). A version of QED (Deutsch and Lampson, 1967). "Line-oriented" (basic addressing unit is a line of text). Users select lines by giving text-strings contained on desired line or (more rarely) by giving line numbers, which change with each inserted or deleted line. Commands are single letters issued from the keyboard (example: D for the Delete command).

SOS   (Savitsky, 1969). A line-oriented editor with fixed line-numbers actually stored in the file with the text. The command language is similar to that of POET.

TECO   (BBN, 1973). A "character-oriented" editor (document is treated as one long string of characters, including RETURN characters). Pieces of text are referenced by search strings or character position numbers. TECO is distinguished by its very large repertoire of low-level commands, which can be combined into higher-level commands.

BRAVO   A display-oriented editor, designed by Charles Simonyi and Butler Lampson at Xerox PARC, which uses the mouse for pointing at text on the display. BRAVO contains a full repertoire of typefont and formatting capabilities. It right-justifies text on the display after each keystroke. The command invocation syntax in BRAVO is similar to that of POET. BRAVO was called DISPED in Card, Moran, and Newell (1976, 1980*a*, 1980*b*).

RCG   A display-oriented editor written by William Duvall; it is a descendent of the NLS editor (Englebart and English, 1968). This editor also uses a mouse for pointing, and a five-key chord device for input of commands.

Figure 3.3. Text-editors tested in Experiment 3A.

Letter Typing	The user is provided with a paper copy of a letter on which a few small changes are indicated in red ink. He is to type the corrected letter into the editing system and save it on a file.
Manuscript Modification	The user is provided with a paper copy of a letter stored on a file. There are 12 small modifications of one or two words each marked on the letter. He is to modify the file, using the editor, according to the markings on the letter.
Text Assembly	The user is to assemble a single file out of three files on the system, each of which contains a single paragraph of text, then type in a fourth paragraph copied from a supplied text.
Table Typing	The user is to type a table (photocopied from a book) into the system and store it on a file. The table contains a five-by-five array of three-digit numbers, plus labels for the rows and columns.

Figure 3.4. Benchmark tasks used for testing editors in Experiment 3A.

Manuscript Modification benchmark was done with BRAVO (which was run at a later date than the other editors). As a baseline against which to measure performance, one user was measured performing the tasks using an IBM Selectric II typewriter.

Procedure. Each user was tested individually. The user was seated in front of a 6 line/sec video display terminal as shown in Figure 3.1 and given a set of general instructions urging him to work as fast as possible without making errors. It was stressed that the editor, and not the user's abilities, was under examination. The user was given a warmup exercise on the editor of making some simple modifications, then each of the four benchmarks in the order: (1) Letter Typing, (2) Manuscript Modification, (3) Text Assembly, (4) Table Typing. The stimulus materials and instructions for each task were bound in a notebook, and the user was allowed to proceed through the benchmarks at his own pace. The experimental session was recorded on video tape with the time (to a sixtieth of a second) recorded on each video frame by means of a video clock.

RESULTS FOR EXPERIMENT 3A

How much of a time difference was there among editors? The answer was a factor of 1.4~2.3 between the fastest and slowest editors, depending on the benchmark. Figure 3.5 gives the total time required to perform each benchmark. The differences on the Letter Typing benchmark (after

	Task Type			
Text-editor	**Letter Typing** $M \pm SD$ (sec)	**Manuscript Modification** $M \pm SD$ (sec)	**Text Assembly** $M \pm SD$ (sec)	**Table Typing** $M \pm SD$ (sec)
POET	238±28	220±33	160±65	244±21
SOS	315	215	147	234
TECO	252±25	159±26	131±15	283±41
BRAVO	—	122±42	—	—
RCG	224±4	94±21	102±32	306±54
Typewriter	229	901	489	483
Ratio of slowest to fastest editor	1.4	2.3	1.6	1.3
Ratio of typewriter to fastest editor	1.0	9.6	4.8	2.1

Figure 3.5. Performance times for the benchmark tasks in Experiment 3A.
There were three users apiece for POET, TECO, BRAVO, and RCG, and only one user each for the typewriter and SOS. The SOS user was also measured on RCG; all other users were measured only once. The times for the Letter Typing benchmark were normalized to compensate for different users' typing rates by dividing the separate parts of the task (type inside address, etc.) by the ratio between a user's time to type the body of the letter and the all-user mean time to type the body.

normalizing for users' typing speeds) and of the Text Assembly benchmark generally reflected the setup costs of each system to do the task. The differences in the Table Typing benchmark mainly reflected the ingenuity of the users in capitalizing on features of the systems: methods varied from typing in the rows of the table directly (using fixed tabs provided by the system) to making many copies of the first line in the table and then substituting for each of the entries. The largest differences among the systems were in the Manuscript Modification benchmark, where the ratio of the slowest to fastest system was 2.3. Since there are small ways in which the RCG editor might be sped up and since some editors in common use are known to be even slower than POET, it is probably justified to say that, as a rough rule of thumb, the design of an editor can make a factor of 3 difference in the time to perform typical editing modifications.

Any of the editing systems was much faster to use than a typewriter. In Figure 3.5, the Manuscript Modification time was almost 10 times faster with the fastest editor. Of course, this ratio depends completely on the ratio of the length of text to be typed and number of modifications to be performed, so the number itself is not meaningful; but it does indicate the generally large advantage obtainable by using text-editors over typewriters.

How much of a difference was there among users? The answer here was a ratio of 1.3~1.9 between the lowest and fastest time/modification, depending on the editor. Figure 3.6 gives the mean time/modification for the Manuscript Modification benchmark. Since users made errors on 14% of the modifications (examples: substituting a misspelled word or invoking the wrong command) and the errors can severely distort comparisons (a single serious error can require a substantial amount of correction time), the mean time/modification for each user is also presented based only on the error-free tasks.

It is apparent from Figure 3.6 that no matter whether all modifications or only error-free modifications are considered, the times for users within an editor are more similar to each other than are the times among editors. The lower portion of the figure gives the average modification times over all the users on each editing system, along with the ratio of the slowest to fastest user on each system. The average slowest/fastest user ratio is about 1.5 when all modifications are considered and about 1.3 when only error-free modifications are considered. The editor BRAVO has the largest slowest/fastest user ratio—almost a

User (System)	All Modifications $M \pm SD\,(N)$ (sec)	Error-Free Modifications $M \pm SD\,(N)$ (sec)
S4 (POET)	16.7 ± 5.3 (10)	15.9 ± 4.9 (9)
S6 (POET)	21.6 ± 15.0 (10)	17.4 ± 7.1 (9)
S13 (POET)	16.9 ± 9.7 (12)	16.9 ± 9.7 (9)
S12 (SOS)	17.9 ± 10.8 (12)	10.4 ± 8.8 (6)
S18 (TECO)	13.9 ± 7.3 (12)	11.2 ± 3.9 (10)
S19 (TECO)	15.0 ± 10.1 (12)	11.5 ± 2.7 (10)
S20 (TECO)	10.8 ± 4.0 (15)	10.8 ± 4.0 (12)
S16 (BRAVO)	7.2 ± 2.8 (12)	7.2 ± 2.8 (12)
S30 (BRAVO)	9.2 ± 2.5 (11)	9.2 ± 2.5 (11)
S31 (BRAVO)	14.0 ± 10.9 (11)	13.9 ± 11.5 (10)
S12 (RCG)	7.4 ± 4.9 (12)	7.5 ± 5.4 (10)
S14 (RCG)	6.3 ± 2.4 (11)	6.0 ± 2.5 (9)
S15 (RCG)	9.7 ± 6.6 (12)	8.0 ± 2.7 (11)

		Ratio of slowest to fastest user		Ratio of slowest to fastest user
POET Users	18.5 ± 2.7 (3)	1.3	16.7 ± 0.8 (3)	1.1
SOS Users	17.9		10.4	
TECO Users	13.1 ± 2.1 (3)	1.4	11.2 ± 0.4 (3)	1.1
BRAVO Users	10.1 ± 3.5 (3)	1.9	10.1 ± 3.4 (3)	1.9
RCG Users	7.8 ± 1.7 (3)	1.5	7.2 ± 1.0 (3)	1.3

Ratio of slowest to fastest editor	2.4	2.3

Figure 3.6. Time per modification in the Manuscript Modification benchmark in Experiment 3A.

factor of 2—whereas all other editors have a factor of 1.5 or less. As a rule of thumb, it is probably fair to say that the difference between expert users is about a factor of 1.5—half the size of the difference between editors. The time differences among text-editors are thus substantial and about twice as large as the differences among expert users.

SOURCES OF THE TIME DIFFERENCES

What is the source of the observed differences in the time to use the different editors? A reasonable hypothesis is that the time for an expert to make modifications with a system is proportional to the amount of work required by the system as indexed by the number of keystrokes he types. This hypothesis appears to be partially, but only partially, correct. In Figure 3.7 the time per modification is plotted against the keystrokes per modification for the user who had the lowest error rates in each editor. Four editors—POET, SOS, TECO, and RCG—fall exactly on a line essentially through the origin:

$$T_{modification} = .26 + .57\,N_{keystrokes}\,\text{sec} \qquad (3.1)$$

($R^2 > .999$, $SE = .12$ sec). BRAVO, however, takes 4 sec longer per modification than predicted—about twice the time predicted by the above equation. More detailed comparison of the behavior of users using BRAVO suggests that the users spent more time than predicted at the beginning of each task and that the time required by the numerous pointing operations needs to be considered. A more definitive explanation requires additional experimentation. The real significance of Equation 3.1 is that a rational basis for the the time required by different editors appears within reach.

3.3. TIME DIFFERENCES AMONG NON-NOVICE USERS (EXPERIMENT 3B)

What about users who are not experts? How much will they vary in time to edit a manuscript? To find out, let us consider another experiment, this time using only the editor BRAVO from our previous set, but considering non-novice users with widely different levels of expertise. Rather than selecting different people and testing them, it is more

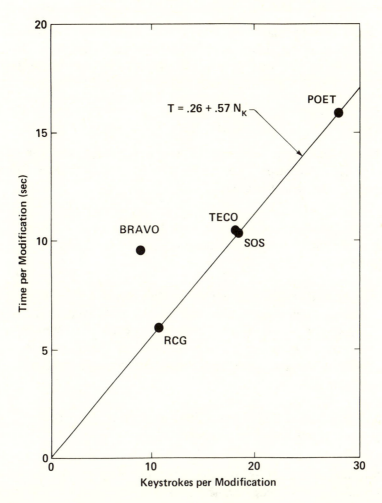

Figure 3.7. Mean time per modification for editors as a function of the number of keystrokes per modification in Experiment 3A.

efficient and insightful to hypothesize some of the characteristics thought to be relevant and to test people who have different combinations of those characteristics. Martin (1973) has suggested several user categories, of which the category Casual vs. Dedicated and the category Operator with Programming Skills look the most promising. (Several of his other categories, such as Active vs. Passive, are characteristics of systems rather than people and others, such as Rugged vs. Non-Rugged, are categories that could only be established with separate experiments.) Equation 3.1 suggests that typing ability may be an important variable. We were therefore led to test users on the following dimensions: Dedicated vs. Casual, Technical vs. Non-technical, and Fast-typist vs. Slow-typist.

METHOD FOR EXPERIMENT 3B

Users. Eight users were selected who were familiar with the BRAVO editor. They consisted of secretaries, computer scientists, and research managers. Users were classified as:

> *Dedicated* if they used the system at least once a day or *Casual* if they used the system only about every two weeks or less;

> *Technical* if they had written at least one major piece of code and had experience with several programming languages or *Non-technical* if they had had no programming experience (although the Non-technical users used computer systems for text generation, filing, and message sending); and

> *Fast-typists* if they typed at least 49 wpm or *Slow-typists* if they typed less than 40 wpm.

Each of the eight users tested represented a different one of the 2×2×2 = 8 combinations of these characteristics.

Task. The manuscript was a 22-page memo containing 66 modifications. The mixture of modifications on the manuscript was carefully balanced to include many different modification types (insertions, replacements, deletions, transpositions, and movements of text), many different sizes of text to be modified, and many different boundary conditions. The manuscript contained some very small modifications, such as inserting or deleting a word or replacing a few characters, as well as some very large tasks, such as switching two sentences on different manuscript pages or inserting a new paragraph of text. The modifications were grouped into four classifications:

> *Simple.* Modifications of 4 characters or less, requiring a single editor command.

Complex. Alterations or movement of phrases or sentences, usually requiring more than one editor command.

Long. Insertions of about 200 characters.

Other. Tasks that did not fall unambiguously into the above categories.

Procedure. The procedure was similar to that of Experiment 3A. First, the user was given a one-page memo containing ten modifications, as a warmup task, then he was given the main manuscript, containing 66 modifications (as described above). The user was instructed to make modifications as quickly as possible without making mistakes. Each session was videotaped, with each frame of the tape time-stamped.

RESULTS FOR EXPERIMENT 3B

According to Figure 3.8, there was about a factor of 3.5 between the slowest and the fastest user in the experiment. The column labeled "All" in the figure gives the mean time/modification for all modifications on which no user made an error (there were 41 of these out of the 66 tasks). The slowest user took an average of 67 sec/task, whereas the fastest took 19 sec/task.

Each dimension in the classification of the user seemed to have roughly the same size of effect: the slower category in each dimension increased user editing time by a factor of about 1.4 over the time required by the faster category. The lower part of Figure 3.8 gives the mean time/modification averaged over a single user category. The average time required by the Casual Users was 1.5 times greater than the time required by the Dedicated Users (53 sec/modification vs. 36 sec/modification). The average time of the Non-technical Users was 1.3 times greater than for the Technical users, and the Slow-typists were 1.4 times slower than the Fast-typists.

Surprisingly, the largest differences occur for the Simple modifications. The slowest user took 47 sec/modification for these, but the fastest required only 8 sec/modification, a factor of 5.9 difference. Dedicated Users got the largest advantage from the Simple modifications, suggesting that the differences on the Simple modifications derived from having the editing methods easily available in memory.

In view of the factor of 5.9 difference between the slowest and fastest users for Simple modifications, the factor of 2.5 difference on Long

User	Classification	Modification Type			
		All	**Sim-ple**	**Com-plex**	**Long**
		$N = 41$ (sec)	12 (sec)	14 (sec)	3 (sec)
S34	(73 wpm, Dedicated, Technical, Fast)	19	8	16	57
S32	(36 wpm, Dedicated, Technical, Slow)	30	10	24	104
S13	(88 wpm, Dedicated, Non-technical, Fast)	31	14	40	62
S37	(39 wpm, Dedicated, Non-technical, Slow)	65	34	74	134
S14	(49 wpm, Casual, Technical, Fast)	60	36	57	145
S1	(32 wpm, Casual, Technical, Slow)	48	19	51	129
S36	(59 wpm, Casual, Non-technical, Fast)	38	17	37	90
S35	(32 wpm, Casual, Non-technical, Slow)	67	47	74	140
Ratio of slowest to fastest		(3.5)	(5.9)	(4.6)	(2.5)
Ratio of 2nd slowest to 2nd fastest		(2.2)	(3.6)	(3.1)	(2.3)
All Users	(51 wpm)	45	23	47	108
Casual Users	(43 wpm)	53	30	55	126
Dedicated Users	(59 wpm)	36	16	39	89
Ratio		(1.5)	(1.9)	(1.4)	(1.4)
Non-Technical Users	(55 wpm)	50	28	56	106
Technical Users	(48 wpm)	39	18	37	109
Ratio		(1.3)	(1.6)	(1.5)	(1.0)
Slow-Typist Users	(35 wpm)	53	28	56	127
Fast-Typist Users	(66 wpm)	37	19	38	88
Ratio		(1.4)	(1.5)	(1.5)	(1.4)

Figure 3.8. Time per modification for each user and for each category of user in Experiment 3B.
Performance is for tasks without errors. N is the number of tasks in each category.

modifications, and the factor of 3.5 difference over all modifications, a reasonable rule-of-thumb would seem to be that non-novice users (experienced users, but including non-experts) differ by about a factor of 4. The dimensions we used for users each seemed to make about the same order of difference, in round numbers, a factor of 1.5. Thus differences among people are about the same size (factor of 4) as differences among different systems (factor of 3), contrary to Sackman's (1970) claim that "human differences are typically an order of magnitude larger than computer system differences." The discrepancy between this result and Sackman's is easily explained, however. The studies reviewed by Sackman involve problem-solving tasks of long duration (many hours), where it is possible for some users to spend considerable time exploring fruitless paths, resulting in large individual differences. The text-editing we have observed in this chapter, by contrast, is a skilled activity involving little problem solving and occurring over a short duration (measured in seconds). Also, none of our users were novices, further reducing inter-user differences.

3.4.   CONCLUSIONS

The exploratory experiments in this chapter have given us estimates for the effect of different text-editor designs and different users on performance time.

The design of an editor makes roughly a factor of 3 difference in the time to edit a manuscript, with display-oriented systems about twice as fast as teletypewriter-oriented systems. These differences among editors are at least partially traceable to the relative amounts of work required by alternative designs, such as the relative number of keystrokes required to accomplish a task.

The factor of 3 difference among editors compares to a factor of 1.5 among dedicated, expert users, or to a factor of 4 among non-novice users in general. The three dimensions of users tested each made a difference of about a factor of 1.5: (1) whether a user is a dedicated (frequent) user; (2) whether he is technically oriented; and (3) whether he is a fast typist.

The effects of text-editor design on speed, therefore, are comparable to, and not dominated by, the effects of individual differences.

Furthermore, the effects involved are substantial. There is an order of magnitude difference (an estimated factor of $3 \times 4 = 12$) in editing time between the fastest user on the best editor and the slowest user on the worst editor.

4. An Exercise in Task Analysis

4.1. SIMPLE MODELS OF TYPING AND EDITING
4.2. PREDICTION (EXPERIMENT 4A)
4.3. SENSITIVITY ANALYSIS
4.4. RESULTS
4.5. CONCLUSIONS

In the last chapter we reported exploratory experiments designed to give a rough estimate of the speed variability among users and text-editing systems. In this chapter we engage in exploratory modeling to discover how well we can predict editing time with a simple model based on the assumption that all editing tasks require a constant amount of time. This model should be of service to our later studies in two ways. First, it should reveal something of the characteristic difficulty of the problem of modeling user behavior in text-editing. Second, it should serve as a baseline against which to compare more complex models.

To make our modeling activity concrete, we address the following problem:

> **Problem.** The claim is made that it is faster to retype short texts on a typewriter than to modify them with a text-editor, and that the reverse is true for long texts. In order to find the crossover point between these two cases, an experiment is to be run measuring the times required to make modifications to five text manuscripts of varying lengths. The modifications are to be accomplished (1) by retyping them on an electric typewriter and (2) by using the WYLBUR text-editing system (Stanford, 1975), running on a time-shared computer. Given information about the marked-up texts to be modified, the problem is to predict the outcome of this experiment.

In order to ensure that the model is predicting, rather than rationalizing an already known result, the problem was arranged so that it corresponded to an actual experiment in progress by other researchers.[1] By agreement, the model's prediction and the experimental results were exchanged simultaneously after both had been completed.

4.1.   SIMPLE MODELS OF TYPING AND EDITING

The answer to the problem posed above can be derived from simple models of typing and editing. The time T_t to produce a new copy of a manuscript using a typewriter depends only on the length of the manuscript and the setup time of the typewriter:

$$T_t = T_{st} + LT_l,$$ (4.1)

where T_{st} is the time to set up the typewriter (in seconds), L is the length of the manuscript text (in lines), and T_l is the time to type a line (in sec/line).

The time T_e to edit a manuscript, on the other hand, is assumed to depend on the number of modifications to the manuscript. Suppose that every modification with an editing system takes a constant amount of time T_m to accomplish. Suppose furthermore that secondary effects, such as user fatigue and time spent turning pages, are negligible. Then the time to edit the manuscript would be given by

$$T_e = T_{se} + N_m T_m,$$ (4.2)

where T_{se} is the time to set up the editor (in sec), N_m the number of modifications to be made, and T_m the time per modification (in sec/mod). Expressing Equation 4.2 in terms of the modification density per unit line, $\rho = N_m/L$, makes it more comparable to Equation 4.1:

$$T_e = T_{se} + \rho LT_m.$$ (4.3)

We refer to this model of text-editing time as the Constant Time per Modification model.

---

[1] The problem was posed to us by I. Sutherland, then at the RAND Corporation. The experiment was run by F. Blackwell, also at RAND.

LENGTH CROSSOVER POINT L_c

If the typewriter is faster to set up ($T_{st} < T_{se}$), but the editor is faster in making modifications ($\rho T_m < T_l$), then there exists some document length L_c, called the *length crossover point*, such that

for $L > L_c$, the editor is faster, and
for $L < L_c$, the typewriter is faster.

To find L_c we use Equation 4.2 and Equation 4.3. The time for the editor and the typewriter will be the same when

$$T_{se} + \rho L_c T_m = T_{st} + L_c T_l \, ,$$

that is,

$$L_c = (T_{se} - T_{st}) / (T_l - \rho T_m) . \tag{4.4}$$

DENSITY CROSSOVER POINT ρ_c

Similarly there exists a certain density ρ_c, called the *density crossover point*, such that

for $\rho < \rho_c$, the editor is faster, and
for $\rho > \rho_c$, the typewriter is faster.

Solving for ρ in Equation 4.4 gives

$$\rho_c = T_l/T_m - (T_{se} - T_{st})/LT_m . \tag{4.5}$$

4.2. PREDICTION (EXPERIMENT 4A)

In order to calculate the outcome of the experiment, we need to have estimates for the parameters of the above equations.

From the videotapes of Experiment 3A, we determine that the average time to set up the typewriter in that experiment was

$$T_{st} \doteq 24 \sec .$$

		Manuscript					
		M1	M2	M3	M4	M5	All
L	(lines)	4	10	21	26	90	151
N_m	(mods)	2	6	8	14	58	88
ρ	(mods/line)	.50	.60	.38	.54	.64	.58

Figure 4.1. Modification density parameter values for the manuscripts used in the experiment.

The text-editor WYLBUR is similar to the POET and SOS editors in Experiment 3A. Again, from the videotapes we determine that the setup time of these editors averaged 12 sec. Add to that the 25 sec to log into the computer (measured time to telephone a local computer and log into the TENEX operating system), and we get as an estimate

$$T_{se} \doteq 37 \sec.$$

The modification density of the manuscripts can be obtained by counting lines and modifications of the text actually used in the experiment. As Figure 4.1 shows, the texts vary from $\rho = .38$ to $\rho = .64$, with an average of

$$\rho \doteq .58 \text{ mod/line}.$$

Again assuming that WYLBUR is similar to the POET and SOS editors, its modification time can be estimated from Experiment 3A (see Figure 3.6),

$$T_m \doteq 20 \sec.[2]$$

The average typing rate for the POET users in Experiment 3A was .22 sec/character. Since there were 63 characters per line in the test manuscripts,

---

[2] This number is slightly different from the numbers listed in Figure 3.6 for POET and SOS, since those numbers reflect a later re-analysis of the videotapes. In order to preserve the original predictions, the original estimate for T_m is used in this chapter.

$$T_l \doteq 14 \text{ sec/line}.[3]$$

Substituting these parameter estimates into Equation 4.4, the length crossover point is predicted to be:

$$
\begin{aligned}
L_c &= (T_{se} - T_{st}) / (T_l - \rho T_m) \\
&= (37 - 24) / (14 - .58 \times 20) \\
&= 5.4 \text{ lines}.
\end{aligned}
$$

From Equation 4.5 the density crossover point is predicted to be:

$$
\begin{aligned}
\rho_c &= (T_l/T_m) - (T_{se} - T_{st})/LT_m \\
&= .70 - .65/L.
\end{aligned}
$$

As $L \to \infty$, $\rho_c \to .7$ modifications/line. Another way of putting this result is to say: if there is more than one modification to be done every $1/.7 = 1.4$ lines, then it is better to retype the text.

Plotting the time to modify a text (from Equation 4.2 and Equation 4.3) as a function of the length of the text (Figure 4.2), it is apparent that the editor beats the typewriter immediately on any manuscript longer than about three lines. More importantly, Figure 4.2 reveals that as the length of the manuscript increases the editor does *not* continue to increase its superiority as much as might be expected.[4] Why not?

The answer is that the density chosen for the experiment, $\rho = .58$, is by chance near the critical crossover density $\rho_c = .70 - .65/L$. Had the experiment varied ρ, one manuscript at the critical value would not have been a problem. But, since each of the manuscripts had a density near this critical value, local fluctuations in T_m or ρ led to wavering of the length-crossover point. Another way to display the model's prediction is to plot the density crossover point ρ_c as a function of text length L using Equation 4.5 (see Figure 4.3). Note how close the manuscripts are to the

---

[3] It is interesting that in the time it takes to make one correction with WYLBUR, the user could have typed $T_m/T_l = (20 \text{ sec})/(14 \text{ sec/line}) = 1.4$ lines. Contrary to the usual assumption, it was more effective to type slowly but carefully on the editing system examined than it was to type at high speed and correct the errors later.

[4] The dip in the WYLBUR curve comes from the low modification density for manuscript M3.

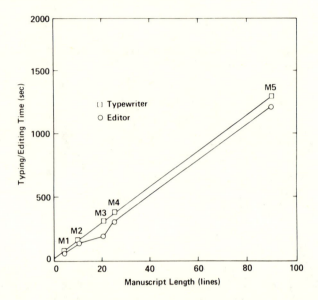

Figure 4.2. Predicted time to edit/type the experimental manuscripts as a function of the manuscript length.

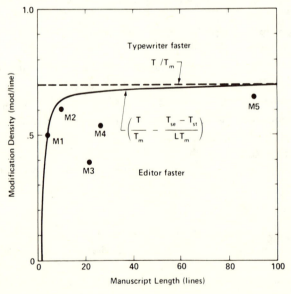

Figure 4.3. Density crossover point as a function of manuscript length.

The typewriter is faster for all manuscript length and modification density combinations above the solid line, slower for those below.

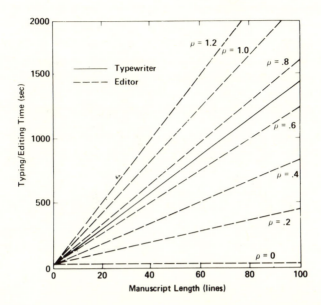

Figure 4.4. Predicted time to edit/type a manuscript as a function of the length of manuscript and of the modification density.

crossover density line. Because all the manuscripts sit relatively near the density crossover line, it can be predicted that the results of the experiment will be equivocal, that is, that the length crossover point will not be well-defined.

What about predictions at other values of ρ? The predicted task time as a function of the length of manuscript for different values of ρ is plotted in Figure 4.4. The typewriter either wins or loses immediately. This is true because the difference in setup times for the typewriter and for WYLBUR is (for manuscripts longer than five lines) only a small percentage of the time required to do the task.

4.3. SENSITIVITY ANALYSIS

There are several possible sources of error in our calculation. Only the manuscripts for the experiment were available; there was no information about the subjects, except that they were secretaries. Most of the parameters, including the typing rates of the users, were taken from pre-existing experiments by analogy. To what extent is the value of the predictions dependent on the accuracy of these parameter estimates? One way to determine the consequences of uncertainties in the parameter values is to see how sensitive the predictions of the equations are to small changes in the parameter values.

SENSITIVITY OF THE LENGTH CROSSOVER POINT L_c

Let the values of ρ, T_m, T_{se}, T_{st}, and T_l be as estimated previously. Let ρ', T_m', T_{se}', T_{st}', and T_l' have other, but nearby, values. Then we can use a Taylor expansion to approximate Equation 4.4 as

$$
L_c' \simeq L_c + \left[(\rho' - \rho)\frac{\partial}{\partial \rho} + (T_m' - T_m)\frac{\partial}{\partial T_m} + (T_{se}' - T_{se})\frac{\partial}{\partial T_{se}} \right.
$$
$$
\left. + (T_{st}' - T_{st})\frac{\partial}{\partial T_{st}} + (T_l' - T_l)\frac{\partial}{\partial T_l} \right] L_c
$$

$$
= L_c + \frac{\partial L_c}{\partial \rho}(\rho' - \rho) + \frac{\partial L_c}{\partial T_m}(T_m' - T_m) + \frac{\partial L_c}{\partial T_{se}}(T_{se}' - T_{se})
$$
$$
+ \frac{\partial L_c}{\partial T_{st}}(T_{st}' - T_{st}) + \frac{\partial L_c}{\partial T_l}(T_l' - T_l) .
$$

In order to normalize the magnitudes of the coefficients and the results, we express this equation in a ratio form:

$$
\frac{L_c' - L_c}{L_c} \simeq \frac{\rho}{L_c}\frac{\partial L_c}{\partial \rho}\left(\frac{\rho' - \rho}{\rho}\right) + \frac{T_m}{L_c}\frac{\partial L_c}{\partial T_m}\left(\frac{T_m' - T_m}{T_m}\right)
$$
$$
+ \frac{T_{se}}{L_c}\frac{\partial L_c}{\partial T_{se}}\left(\frac{T_{se}' - T_{se}}{T_{se}}\right) + \frac{T_{st}}{L_c}\frac{\partial L_c}{\partial T_{st}}\left(\frac{T_{st}' - T_{st}}{T_{st}}\right)
$$
$$
+ \frac{T_l}{L_c}\frac{\partial L_c}{\partial T_l}\left(\frac{T_l' - T_l}{T_l}\right) .
$$

Using δx for $(x' - x)/x$,

$$
\delta L_c \simeq \left(\frac{\rho}{L_c}\frac{\partial L_c}{\partial \rho}\right)\delta \rho + \left(\frac{T_m}{L_c}\frac{\partial L_c}{\partial T_m}\right)\delta T_m + \left(\frac{T_{se}}{L_c}\frac{\partial L_c}{\partial T_{se}}\right)\delta T_{se}
$$
$$
+ \left(\frac{T_{st}}{L_c}\frac{\partial L_c}{\partial T_{st}}\right)\delta T_{st} + \left(\frac{T_l}{L_c}\frac{\partial L_c}{\partial T_l}\right)\delta T_l .
$$

Evaluating the derivatives and substituting $(T_{se} - T_{st})/(T_l - \rho T_m)$ for L_c gives

$$\delta L_c = \left(\frac{1}{\dfrac{T_l}{\rho T_m} - 1} \right)(\delta \rho + \delta T_m) + \left(\frac{1}{1 - \dfrac{T_{st}}{T_{se}}} \right)\delta T_{se}$$

$$+ \left(\frac{1}{1 - \dfrac{T_{se}}{T_{st}}} \right)\delta T_{st} + \left(\frac{1}{\dfrac{\rho T_m}{T_l} - 1} \right)\delta T_l .$$

$$(4.6)$$

Equation 4.6 expresses relative changes in L_c as a linear combination of relative changes in the parameters of Equation 4.4. The percentage change in L_c is approximated as the sum of the percentage change due to each variable. The relative sensitivity of predicted L_c due to the different parameters may thus be assessed directly from the relative size of the coefficients. At $\rho = .6$, Equation 4.6 becomes

$$\delta L_c = 6.00 \, \delta\rho + 6.00 \, \delta T_m + 2.85 \, \delta T_{se}$$
$$- 1.85 \, \delta T_{st} - 7.00 \, \delta T_l .$$

That is, a 1% error in T_l will be amplified into a 7% error in L_c. The values of the coefficients for other values of ρ are plotted in Figure 4.5, as are those of the three following equations. The value of L_c is more sensitive to changes in T_m, ρ, and T_l than to changes in T_{se} and T_{st}, the ostensible parameters of interest. The sensitivity analysis makes it quite clear (1) that the prediction of $L_c = 5$ lines from the model is not robust over changes in the parameters and (2) that it will be difficult to maintain adequate control over the variables in the experiment at this level of ρ. Considerable variance in the measured value of L_c is predicted. Figure 4.5 shows that the coefficients for $\delta\rho$, δT_m, and δT_l are all very large in the region between $\rho = .06$ and $\rho = .08$. Conversely, had the experiment been designed with $\rho = .2$, then it would have been true that

$$\delta L_c = .40 \, \delta\rho + .40 \, \delta T_m + 2.85 \, \delta T_{se}$$
$$- 1.85 \, \delta T_{st} - 1.40 \, \delta T_l ,$$

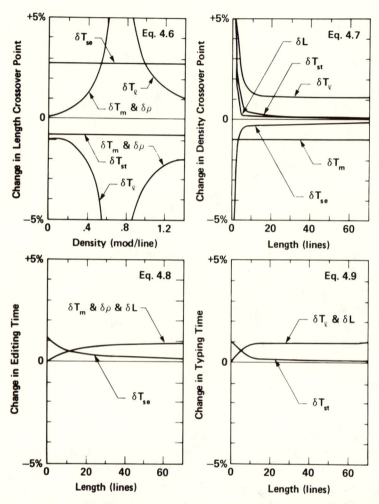

Figure 4.5. Coefficient values for the sensitivity equations.

in which case L_c would have been much less affected by parameters other than T_{se} and T_{st}.

SENSITIVITY OF THE DENSITY CROSSOVER POINT ρ_c

We examine the density crossover point ρ_c, by proceeding similarly for Equation 4.5:

$$\delta\rho_c \simeq \left(\frac{1}{1 - \dfrac{T_{se} - LT_l}{T_{st}}}\right)\delta T_{st} + \left(\frac{1}{\dfrac{T_{st} - LT_l}{T_{se}} - 1}\right)\delta T_{se}$$

$$+ \left(\frac{1}{\dfrac{LT_l}{T_{se} - T_{st}} - 1}\right)\delta L - \delta T_m + \left(\frac{1}{1 - \dfrac{T_{se} - T_{st}}{LT_l}}\right)\delta T_l \, . \tag{4.7}$$

At $L = 20$ lines, Equation 4.7 becomes

$$\delta\rho_c = .09\,\delta T_{st} - .13\,\delta T_{se} + .05\,\delta L$$
$$- 1.00\,\delta T_m + 1.05\,\delta T_l \, .$$

Hence, a 1% change in either T_m or T_l will produce about a 1% change in ρ_c, but a 1% change in the other parameters produces only a negligible change (.05%~.13%) in ρ_c. For manuscripts of reasonable length (longer than ten lines), ρ_c will depend mainly on T_m and T_l.

SENSITIVITY OF TOTAL TYPING TIME T_t

The total typing time T_e is examined by converting Equation 4.1:

$$\delta T_t \simeq \left(\frac{1}{1 + \dfrac{LT_l}{T_{st}}}\right)\delta T_{st} + \left(\frac{1}{1 + \dfrac{T_{st}}{Lt_l}}\right)(\delta L + \delta T_l) \, . \tag{4.8}$$

At $L = 20$ lines,

$$\delta T_t = .08\,\delta T_{st} + .92\,\delta L + .92\,\delta T_l \, .$$

The sensitivity of T_t to T_{st} fades quickly as L increases. A 1% change in the other parameters produces a little less than a 1% change in T_t.

SENSITIVITY OF TOTAL EDITING TIME T_e

The total editing time T_e is checked by converting Equation 4.2:

$$\delta T_e \simeq \left(\frac{1}{1 + \dfrac{\rho L T_m}{T_{se}}} \right) \delta T_{se} + \left(\frac{1}{1 + \dfrac{T_{se}}{\rho L T_m}} \right) (\delta \rho + \delta T_m + \delta L).$$

$$(4.9)$$

At $L = 20$ lines and $\rho = .6$ mod/line,

$$\delta T_e = .13 \, \delta T_{se} + .87 \, \delta \rho + .87 \, \delta T_m + .87 \, \delta L.$$

Again, the sensitivity of T_e to T_{se} fades quickly as L increases. And again, 1% change in the other parameters produces a little less than a 1% change in T_e.

The results of the confidence interval and sensitivity analyses tell us that, whereas it may be possible to predict the value of L_c functionally (that is, to produce an equation whose evaluation will give a reasonable value for L_c), it is not possible to predict the value of L_c numerically with any certainty on this group of manuscripts, because they are all set so near to ρ_c. Small errors in the parameter values will cause large errors in the predictions. The analyses tell us, furthermore, that the experiment is not likely to produce a well defined value of ρ_c against which to compare a prediction. On the other hand, the predictions of total time to process each text are likely to be reasonable and to depend very little on the setup times of the editor or the typewriter.

4.4.   RESULTS

Figure 4.6 shows the time to edit each manuscript, both for the typewriter and for WYLBUR, as a function of the length of manuscript plotted in the same manner as Figure 4.2. As predicted from the model, the crossover point was not well defined. Connecting the mean observed times produces three crossover points. The times for manuscripts M2, M3, and M5 were not reliably different from one another.

Accuracy of Parameters. Just how accurate were the simple models of typing and editing in Equation 4.2 and Equation 4.3? The comparison needs to be made in two ways. First, how accurate were the models at predicting the result in advance of any knowledge about the outcome? This *zero-parameter prediction* is usual in practice where reasonable values

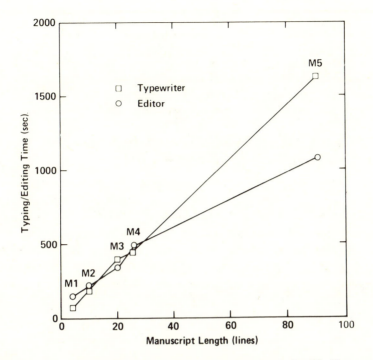

Figure 4.6.   Observed mean times to type/edit the manuscripts in the experiment.

The users tested were eight professional secretaries, each a proficient user of WYLBUR and of a typewriter. Each user was to edit all five texts twice, once with the typewriter, once with WYLBUR. Half the users used the typewriter first, half the editor. The order in which the texts were edited was varied systematically. Properties of test manuscripts are listed in Figure 4.1.

for the parameters are known. Second, how good were the models at predicting the result, given knowledge of the parameter values? This would be a *two-parameter prediction*, since two values must be estimated from the data. It allows an evaluation to be made of the accuracy of the functional form of the model, and it allows us to partition the prediction error into the error due to misestimating the parameters and error due to form of the model. In order to make two-parameter predictions, estimates of the parameters were made from regressions on the experimental data. A comparison between the parameters estimated in this way and the values assumed for making the predictions is given in Figure 4.7. The values we used for making our predictions were poor estimates (off

	Assumed	Observed	%Difference
T_{st} (sec)	24	5	– 85%
T_l (sec/line)	14	18	22%
T_{se} (sec)	37	179	649%
T_m (sec/mod)	20	16	– 20%

Figure 4.7. Comparison of the estimated parameter values with the values observed in the experiment.

by 649% and 86%) for the two setup times (T_{st} and T_{se}), but were within about 20% for the two rate parameters (T_l and T_m).

Accuracy of Typing Model. Figure 4.8 compares the predicted and observed times for the typing model (Equation 4.2). The zero-parameter prediction is indicated by a dotted line and the two-parameter prediction

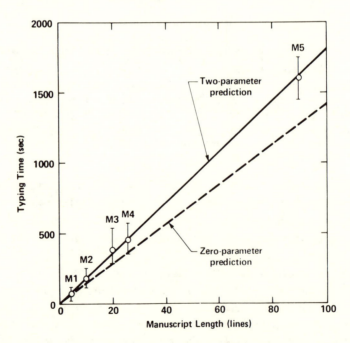

Figure 4.8. Comparison of the predicted typing time with the typing times observed in the experiment.
The vertical bars on the observed times extend one standard deviation up and down from the mean, based on the data from eight users.

by a solid line. When the actual typing rates of the subjects are used in the equation, the fit to the data is excellent. Using the sensitivity equation for this model (Equation 4.8), we can partition the sources of error in the zero-parameter prediction. The errors are tabulated for each manuscript in Figure 4.9. On the average, the prediction was about 11% too low. Almost all this error (10% of the 11%) resulted from the error in estimating the parameters; only 1% resulted from the form of the model. Although the estimate for T_{st} was much worse than the estimate for T_l, the latter was the source of twice as much error (19% for T_l, to 9% for T_{st}). Since the errors were in opposite directions, they partially offset each other.

	Sources of Error				
	Parameters			Model	Total
Manuscript	T_{st}	T_l	Subtotal		
M1	+ 27%	− 18%	9%	0%	9%
M2	+ 11%	− 19%	− 8%	− 2%	− 10%
M3	+ 4%	− 20%	− 16%	− 5%	− 20%
M4	+ 4%	− 20%	− 16%	+ 3%	− 13%
M5	0%	− 20%	− 20%	0%	− 20%
Mean	9%	− 19%	− 10%	− 1%	− 11%

Figure 4.9.   Partitioning the typing model's prediction error.

Accuracy of Editing Model. The editing model (Equation 4.3) is compared with the observed times in Figure 4.10. There is a good fit between the observed and the predicted editing times, even for the zero-parameter predictions. In Figure 4.11, the prediction error is partitioned using the sensitivity equation for the editing model (Equation 4.9). The model was about 24% too low; but, again, errors in estimating the input parameters were responsible for considerably more error (31%) than was the form of the model (7%). This time the major source of errors in estimating the parameters was underestimating the setup time of the editor. (It is instructive to note the frequency with which the various sources of errors partially cancel each other.)

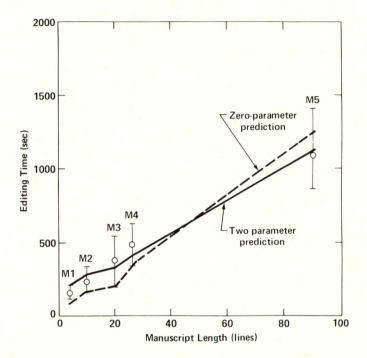

Figure 4.10. Comparison of the predicted editing time with the editing times observed in the experiment.
The vertical bars on the observed times extend one standard deviation up and down from the mean, based on the data from eight users.

4.5. CONCLUSIONS

The main point of this exercise was to explore how much insight could be gained from a simple model of text-editing, the Constant Time per Modification model, in which each editing modification is assumed to require the same amount of time. We investigated this model in a case study comparing the WYLBUR editor with a typewriter. There were two main results.

First, it was possible to produce several predictions leading to practical insight. A formula for the length crossover point showed its functional dependence on other associated variables. A related concept of modification density arose from the modeling effort, and the density

| Manuscript | Sources of Error | | | | |
| | Parameters | | | Model | Total |
	T_{se}	T_m	Subtotal		
M1	– 67%	+ 4%	– 63%	+ 46%	– 47%
M2	– 51%	+ 9%	– 42%	+ 17%	– 33%
M3	– 47%	+ 11%	– 36%	– 17%	– 46%
M4	– 35%	+ 14%	– 21%	– 13%	– 32%
M5	– 13%	+ 22%	+ 9%	+ 1%	+ 10%
Mean	– 43%	+ 12%	– 31%	+ 7%	– 24%

Figure 4.11. Partitioning the editing model's prediction error.

crossover point was expressed in functional form. It was then possible to predict some unfortunate consequences of an unlucky choice in modification density for an experiment. Without the insight of this derivation, the results of the experiment would have been difficult to interpret at all.

Second, the major errors in the predictions made by the simple editing and typewriting models did not result because they were too simple, but because of errors in estimating the values of the input parameters. For these predictions, a more sophisticated model would have been useful only to the extent that it reduced dependence of the prediction on the noisy parameters.

A sensitivity analysis identified those parts of the prediction from these models in which little confidence could be placed. It also allowed the prediction error to be quantitatively partitioned.

5. The GOMS Model of Manuscript Editing

In Chapter 4 we investigated a simple model of human text-editing performance. We now consider how our understanding might be improved by taking into account the cognitive information-processing activities of the user. Our starting point is the fundamental principle of task analysis, the Rationality Principle P8 from Chapter 2. According to the principle, users act rationally to attain their goals. To predict a user's behavior we must analyze the task to determine the user's goals and operators and the constraints of the task. From Chapter 2, we expect that underlying the detailed behavior of a particular user there is a small number of information-processing operators, that the user's behavior is describable as a sequence of these, and that the time the user requires to act is the sum of the times of the individual operators.

This, in outline, is the information-processing analysis of text-editing to be carried through in this chapter. We address several general issues: Is it possible to describe the behavior of a user engaged in text-editing as the repeated application of a small set of basic information-processing operators? Is it possible to predict the actual sequence of operators a person will use and the time required to do any specific task? In attempting to describe behavior in this way, the issue of the level of analysis is critical. How does the model's ability to describe and predict a person's behavior change as we vary the grain size of the analysis?

5.1. THE GOMS MODEL

In the models we describe, the user's cognitive structure consists of four components: (1) a set of Goals, (2) a set of Operators, (3) a set of Methods for achieving the goals, and (4) a set of Selection rules for choosing among competing methods for goals. We call a model specified by these components a GOMS model.

As an example of the basic concepts of a GOMS model and the notation used, let us consider a particular model (called Model F2) of manuscript editing with the line-oriented POET editor we studied in Chapter 3. According to the model, when the user begins editing he has the top level goal:

> **GOAL: EDIT-MANUSCRIPT .**

As we have seen, a user segments the larger task of editing the manuscript into a sequence of small, discrete modifications, such as to delete a word or to insert a character. Although it is often possible to predict the user's actual segmentation of the task into subtasks from the way the instructions are expressed on the manuscript, it is worth emphasizing that the definition of the subtasks is a decision of the user. We use the term *unit task* to denote these user-defined subtasks. Notationally, we write

> **GOAL: EDIT-MANUSCRIPT**
> .   **GOAL: EDIT-UNIT-TASK** *repeat until no more unit tasks*.

The indentation above indicates that **GOAL: EDIT-UNIT-TASK** is a subgoal of **GOAL: EDIT-MANUSCRIPT**, and the notation in italics says that the subgoal is to be invoked repeatedly until no more unit tasks remain to be done.

In order to edit a unit task, the user must first acquire instructions from the manuscript and then do what is necessary to accomplish them:

> GOAL: EDIT-UNIT-TASK
> . GOAL: ACQUIRE-UNIT-TASK
> . GOAL: EXECUTE-UNIT-TASK .

Each subgoal above will itself evoke appropriate methods. There is a simple method for acquiring a task:

> GOAL: ACQUIRE-UNIT-TASK
> . GET-NEXT-PAGE *if at end of manuscript page*
> . GET-NEXT-TASK .

The operator GET-NEXT-PAGE is invoked only if there are no more edit instructions on the current page of the manuscript. The bulk of the work towards the goal—looking at the manuscript, finding an editing instruction, and interpreting the instruction as an edit task—is done by the operator GET-NEXT-TASK.

In POET, like most line-oriented text-editors, to accomplish a unit task there is a two-step method:

> GOAL: EXECUTE-UNIT-TASK
> . GOAL: LOCATE-LINE
> . GOAL: MODIFY-TEXT .

In POET the editor must first be located at the line where the correction is to be made. Then the appropriate text on that line must be modified.

To locate POET at a line, there is a choice between two methods:

> GOAL: LOCATE-LINE
> . [select: USE-LF-METHOD
>   USE-QS-METHOD] .

To use the LF-METHOD, the LINEFEED key is pressed repeatedly, causing the editor to advance one line each time. To use the QS-METHOD (Quoted String), a string of characters is typed (between quotation marks) to identify the line. Usually the LF-METHOD is selected when the text for the new unit task is within a few lines of the text for the current unit task, and the QS-METHOD is selected when the new unit task is farther away.

Once the line has been located, there is a choice of how to modify the text:

GOAL: MODIFY-TEXT
.   [select: USE-S-COMMAND
            USE-M-COMMAND]
.   VERIFY-EDIT .

Either POET's Substitute command or its Modify command can be used to alter text on a line. A detailed example of the Substitute command has already been given (in Chapter 3). The Modify command allows the user to invoke a series of subcommands for moving forward and backward and for making modifications within a line. In either case, a VERIFY-EDIT operation is evoked to check that what actually happened matched the user's intentions.

Putting all the steps together into one structure, we have:

GOAL: EDIT-MANUSCRIPT
.   GOAL: EDIT-UNIT-TASK   *repeat until no more unit tasks*
. .   GOAL: ACQUIRE-UNIT-TASK
. . .   GET-NEXT-PAGE   *if at end of manuscript page*
. . .   GET-NEXT-TASK
. .   GOAL: EXECUTE-UNIT-TASK
. . .   GOAL: LOCATE-LINE
. . . .   [select: USE-QS-METHOD
                    USE-LF-METHOD]
. . .   GOAL: MODIFY-TEXT
. . . .   [select: USE-S-COMMAND
                    USE-M-COMMAND]
. . . .   VERIFY-EDIT .

The dots at the left of each line show the depth of the goal stack. To complete this model of manuscript editing, we must still add method selection rules for determining the actual submethods at the two occurrences of **select**.

The step-by-step behavior of the model in performing a unit task is traced in Figure 5.1. The user is assumed to have a goal stack with the current goal at its top. New subgoals are pushed onto the stack, and completed goals (whether satisfied or abandoned) are popped off the stack. The goals eventually cause operators to be executed. It is during

Step	Contents of Goal Stack	Operator Executed	External User Action
1	ED-MS		
2	ED-MS, ED-UT		
3	ED-MS, ED-UT, ACQ-UT		
4	ED-MS, ED-UT, ACQ-UT	GET-NEXT-TASK	*Looks at manuscript*
5	ED-MS, ED-UT		
6	ED-MS, ED-UT, EX-UT		
7	ED-MS, ED-UT, EX-UT, LOC-LINE		
8	ED-MS, ED-UT, EX-UT, LOC-LINE	USE-LF-METHOD	*Types* LINEFEED
9	ED-MS, ED-UT, EX-UT		
10	ED-MS, ED-UT, EX-UT, MOD-TEXT		
11	ED-MS, ED-UT, EX-UT, MOD-TEXT	USE-S-COMMAND	*Types* sldi RETURN idi RETURN RETURN
12	ED-MS, ED-UT, EX-UT, MOD-TEXT	VERIFY-EDIT	*Types* /
13	ED-MS, ED-UT, EX-UT		
14	ED-MS, ED-UT		
15	ED-MS		

Figure 5.1. Trace of Model F2 through one unit task.

The task being traced is the second one marked in Figure 3.2. In the goal stack column, the top of the goal stack (i.e., the current goal) is at the right. To save space, the symbol GOAL: is dropped from the beginning of goal expressions and the following abbreviations are used: ACQ = ACQUIRE, ED = EDIT, EX = EXECUTE, LOC = LOCATE, MS = MANUSCRIPT, MOD = MODIFY, and UT = UNIT-TASK.

execution of operators that interactions with the physical world take place. For example, the user executes the operator GET-NEXT-TASK by turning to the manuscript, scanning it until he finds the next task, reading the instructions, and turning back to the terminal.

Components of the GOMS Model

The above example provides specific instances to help us understand the information-processing components of GOMS models.

Goals. A goal is a symbolic structure that defines a state of affairs to be achieved and determines a set of possible methods by which it may be accomplished. In the example, the goals are GOAL: EDIT-MANUSCRIPT, GOAL: EDIT-UNIT-TASK, GOAL: ACQUIRE-UNIT-TASK, GOAL: EXECUTE-UNIT-TASK, GOAL: LOCATE-LINE, and GOAL: MODIFY-TEXT. The dynamic function of a goal is to provide a memory point to which the system can return on failure or error and from which information can be obtained about what is desired, what methods are available, and what has been already tried.

Operators. Operators are elementary perceptual, motor, or cognitive acts, whose execution is necessary to change any aspect of the user's mental state or to affect the task environment. In the example, the operators are: GET-NEXT-PAGE, GET-NEXT-TASK, USE-QS-METHOD, USE-LF-METHOD, USE-S-COMMAND, USE-M-COMMAND, and VERIFY-EDIT. The behavior of the user is ultimately recordable as a sequence of these operations. In the example traced in Figure 5.1, the sequence of operators in the user's behavior is:

> GET-NEXT-TASK
> USE-LF-METHOD
> USE-S-COMMAND
> VERIFY-EDIT .

A GOMS model does not deal with any fine structure of concurrent operations. Behavior is assumed to consist of the serial execution of operators.

An operator is defined by a specific effect (output) and by a specific duration. The operator may take inputs, and its outputs and duration may be a function of its inputs. An obvious example is the typing operator, whose input is the text to be typed, whose output is the keystroke sequence to the keyboard, and whose duration is (approximately) a linear function of the number of characters.

For a specific model, the operators define a grain of analysis. In general, they embody a mixture of basic psychological mechanisms and learned organized behavior, the mixture depending on the level at which the model is cast. The finer the grain of analysis, the more the operators reflect basic psychological mechanisms. The coarser the grain of analysis, the more the operators reflect the specifics of the task environment, such as the terminal, the physical arrangement, and the editor. The example model above is quite coarse, and its operators (e.g., USE-S-COMMAND) contain within themselves the specifics of POET's command language.

Methods. A method describes a procedure for accomplishing a goal. It is one of the ways in which a user stores his knowledge of a task. The description of a method is cast in a GOMS model as a conditional sequence of goals and operators, with conditional tests on the contents of the user's immediate memory and on the state of the task environment. In the example above, one of the methods was

> GOAL: ACQUIRE-UNIT-TASK
> .   GET-NEXT-PAGE   *if at end of manuscript page*
> .   GET-NEXT-TASK .

This method is associated with its GOAL: ACQUIRE-UNIT-TASK. It will give rise to either the operator sequence GET-NEXT-PAGE followed by GET-NEXT-TASK or the single operator GET-NEXT-TASK, depending on whether the test "*at end of manuscript page*" is true of the task environment at the time the test is performed.

In the manuscript-editing task, the methods are sure of success, up to the possibility of having been mis-selected, the occurrence of errors of implementation, and the reliability of the equipment. By contrast, in problem-solving tasks (such as a first attempt at solving the DONALD+ GERALD problem in Chapter 2), methods have a chance of success distinctly less than certain, because of the user's lack of knowledge or appreciation of the task environment. This uncertainty is a prime contributor to the problem-solving character of a task; its absence is a characteristic of a cognitive skill.

Methods are learned procedures that the user already has at performance time; they are not plans that are created during a task performance. They constitute one of the major ways in which familiarity (skill) expresses itself. The particular methods that the user builds up from prior experience, analysis, and instruction reflect the detailed structure of the task environment. In the manuscript-editing task, they reflect knowledge of the exact sequence of steps required by the editor to accomplish specific tasks.

Control Structure: Selection Rules. When a goal is attempted, there may be more than one method available to the user to accomplish the goal. The selection of which method is to be used need not be an extended decision process, for it may be that task environment features dictate that only one method is appropriate. On the other hand, a genuine decision may be required. The essence of skilled behavior is that these selections are not problematical, that they proceed smoothly and quickly, without the eruption of puzzlement and search that characterizes problem-solving behavior.

In a GOMS model, method selection is handled by a set of *selection rules.* Each selection rule is of the form "if such-and-such is true in the current task situation, then use method M." Selection rules for GOAL: LOCATE-LINE of the example model might read: *if the number of lines to the next modification is less than 3, then use the* LF-METHOD; *else use the* QS-METHOD. Such rules allow us to predict from knowledge of the task environment (in this case the number of lines to the target) which of several possible methods will be selected by the user in a particular instance.

Limitations of the GOMS Model

For error-free behavior, a GOMS model provides a complete dynamic description of behavior, measured at the level of goals, methods, and operators. Given a specific task (a specific instruction on a specific manuscript and a specific editor), this description can be instantiated into a sequence of operations (operator occurrences). By associating times with each operator, such a model will make total time predictions. If these times are given as distributions, it will make statistical predictions. But, without augmentation, the model is not appropriate if errors occur. Yet errors exist in routine cognitive skilled behavior. Indeed, error rates may not even be small, in the sense of having negligible frequency, taking negligible time, or having negligible consequences. What is true of skilled behavior is that the detection and correction of errors is mostly routine (we discuss this more later). It cannot be entirely routine, since the occurrence of rare types of errors for which the user is unprepared is always possible (the editor performing incorrectly, the terminal catching fire).   But, in the main, errors are quickly detected and result in additional time to correct the error.   The final effect of the behavior remains relatively error-free, and the behavior can be characterized solely by the time to completion.   Thus, errors can be converted to variance in operator times, so that the GOMS theory can be applied to actual behavior at the price of degraded accuracy.

For a general treatment of errors and interruptions of the user, the hierarchical control structure of a GOMS model is inadequate; a more general control structure is required.   The use of the stack-discipline GOMS model instead of a more general control structure, such as production systems (Newell and Simon, 1972), should be taken as an approximation especially appropriate for skilled cognitive behavior and preferred here because of its greater simplicity.

Design of the Experiments

The purpose of the experiments that follow is to describe the manuscript-editing task in information-processing terms.   The general technique is to observe a user in a close laboratory analogue of the task he commonly performs, to describe his behavior using a GOMS model, and to evaluate in various ways the adequacy of the description.   The experiments are directed specifically at three elements of this analysis: (1) description of how the user decides which method to use for a task, (2) description of the time course of events, and (3) an investigation of how the adequacy of the description varies as a consequence of the grain of analysis.

5.2. SELECTION RULES (EXPERIMENT 5A)

The purpose of this experiment was to discover how users choose which of several alternative methods to use and to determine if the method choices could accurately be described in terms of the selection rules of a GOMS model.

In the GOMS model for POET, we have seen two places where, for a given goal, the user has a choice of methods.   The first method selection came in deciding how to locate the line:

```
        GOAL: LOCATE-LINE
    .   [select: USE-LF-METHOD
                 USE-QS-METHOD] .
```

The second method selection came in choosing between commands for making the text modification:

```
        GOAL: MODIFY-TEXT
    .   [select: USE-S-COMMAND
                 USE-M-COMMAND]
    .   VERIFY-EDIT .
```

What we seek is a set of selection rules describing the conditions under which the user will choose one method over another.

METHOD FOR EXPERIMENT 5A

Users were given a manuscript, marked with corrections, and asked to use the POET text-editor to make the corrections. Although the experiment was performed in the laboratory, an effort was made to make the situation seem natural from the user's point of view: the physical surroundings, the task, the terminal, and the editor were all familiar as part of the user's daily activities. The manuscript and the modifications to be made on it were selected to be typical.

Users. Users were two professional secretaries and a Ph.D. computer scientist. All had at least one year of daily experience using POET.

Manuscript. The manuscript was an eleven-page memo. Each page was 8-1/2 by 11 inches, with 55 lines of text and 70 characters per line, printed unjustified in a 10-point fixed-pitch font. There were 73 different modifications marked with a red pen, giving an average density of one modification every 8.3 lines, or 6.6 modifications per page (from 3 to 11 on any one page). An effort was made to vary the number of lines between consecutive modifications and to place an equal number of modifications in each of the left, right, and middle portions of the page. The marked modifications were relatively short: four of them were deletions (of an average of 5.5 characters), 26 were insertions (of an average of 2.9 characters), and 40 were replacements (of an average of 4.1 characters to be replaced by an average of 4.4 characters). The manuscript fragment in Figure 3.2 was taken from the manuscript given to the users and illustrates the style in which modifications were indicated to the user.

Terminal.   Two terminals were used in the experiment: a Texas Instruments "Silent 700" teletypewriter (prints on paper at 30 characters/sec) and a video display, 8-1/2 inches wide by 10-3/4 inches high (42 lines, 72 characters per line, maximum display rate about 6 lines/sec). The display was programmed to operate according to a simple scrolling discipline (the same discipline used on the teletypewriter): each new line was displayed at the bottom of the screen with the other lines scrolling up to make room. The last 42 lines of an interaction were visible on the screen.

Procedure.   The user was seated before the terminal with the manuscript to his left. He first performed editing tasks on a one-page manuscript for warmup and for insurance that he understood what to do. Then he edited the manuscript described above. One user was run on the teletypewriter alone, one on the video display terminal alone, and one was run twice, first on the display and two weeks later on the teletypewriter. For two of the experimental sessions, users were instructed to proceed through the manuscript, inserting an asterisk at the beginning of each marked line (since these sessions were originally run only to investigate methods for locating the target line). In the other two experimental sessions, the users were instructed to edit the eleven-page manuscript. Editing the manuscript required approximately 20 minutes.

The users' keystrokes and the system's responses were recorded on a computer file. These data were used to infer the methods chosen for each task and the reasons for choosing them.

RESULTS OF EXPERIMENT 5A

Typescripts of the four experimental sessions were examined to identify the methods employed by the users. Figure 5.2 gives the methods observed and the frequencies with which the methods were selected. QS-METHOD and LF-METHOD are the methods previously described for GOAL: LOCATE-LINE. S-COMMAND and M-COMMAND are the methods previously described for GOAL: MODIFY-TEXT. The other methods were used less frequently and are described as follows:

+N-METHOD. The user estimates the number of lines n to the next unit task then types the command $+n/$, which causes POET to advance n lines and print the line. It is assumed that a correction may be

	User / Terminal Type			
	S1 (Comp. Sci.)	**S4** (Secy.)		**S22** (Secy.)
	(TTY)	(TTY)	(DISP)	(DISP)
Methods for **GOAL: LOCATE-LINE:**				
LF-METHOD	11 (16%)	14 (21%)	45 (68%)	25 (38%)
QS-METHOD	44 (65%)	1 (2%)	0	40 (62%)
+N-METHOD	2 (3%)	51 (77%)	20 (30%)	0
AN-METHOD	11 (16%)	0	1 (2%)	0
Methods for **GOAL: MODIFY-TEXT:**				
S-COMMAND	—	48 (73%)	—	57 (86%)
M-COMMAND	—	18 (27%)	—	9 (14%)

Figure 5.2.   Frequency of method selections for three subjects in Experiment 5A.
In two sessions no modifications were actually done, since only methods for GOAL: LOCATE-LINE were being studied at the time.

needed: the user may have to type a few LINEFEED commands (each of which moves him down a line), ↑ commands (each of which moves him up a line), or may even have to repeat the *+n/* command with a new *n*.

AN-METHOD. The user first selects an easily specified "anchor" line near the target line, such as a blank line (specified by the empty string **""**), the last line of a page (denoted by the special symbol $), or a line that has a short unique string, such as a paragraph number. Then the target line is reached by using LINEFEED's or ↑'s. For example, the command **""LINEFEED** locates POET at the first line of the next paragraph.

A striking feature of the method frequencies in Figure 5.2 is how each user clearly has a dominant method. By knowing only the dominant method of the user, his method selection can be predicted correctly about 66% of the time for **GOAL: LOCATE-LINE** and 80% of the time for

GOAL: MODIFY-TEXT.   Apparently, the user will use this dominant method unless it is obviously inefficient (such as LINEFEEDing a line at a time through ten pages of text to get to the next task).

That a user's selection of methods depends systematically on the features of the task environment is illustrated by the choice of method for GOAL: LOCATE-LINE. The most important characteristic of the task environment for this goal is the distance d (given in number of lines) between the Current Line and the line with the text to be next modified. As is clear from Figure 5.3, all users used the LF-METHOD if the next line was close enough. Where users differed was in the threshold for how far away the target had to be before they shifted to other methods. The time required to use the LF-METHOD was sensitive to the speed of the terminal, since the system prints out the new Current Line every time

User	Method	d										
		1	2	3	4	5	6	7	8	9	10-14	15+
S1	LF	8	3									
	QS		2	4	5	2		1	3	4	8	15
(TTY)	+N		1	1								
	AN			1		2			1	1	3	3
S4	LF	8	4	1		1						
	QS											1
(TTY)	+N		1	5	5	3		1	4	4	11	17
	AN											
S4	LF	6	7	6	5	3		1	3	2	2	10
	QS											
(DISP)	+N					1			1	2	9	7
	AN											1
S22	LF	6	5	6	5	1		1		1		
	QS		1		1	2			4	4	10	18
(DISP)	+N											
	AN											
Total Frequency		8	6	6	5	4	0	1	4	4	11	19

Figure 5.3.   Frequency of GOAL: LOCATE-LINE methods in Experiment 5A as a function of the distance d from the previous task.

The vertical bars indicate the thresholds where the LF-METHOD ceases being the preferred method in each session. The Total Frequency row gives the frequency of the different distances over the whole manuscript, taking the tasks in order. Since users often did some tasks in a different order, totals for different experiments in the same column are not necessarily equal.

LINEFEED is typed. It was not surprising, therefore, that the LF-METHOD was used less frequently by user S4 on the slower teletypewriter than on the faster display terminal (21% of the time on the teletypewriter vs. 68% of the time on the video display, according to Figure 5.2), or that the threshold for when to abandon the LF-METHOD was lower when S4 was using a slow terminal than when she was using a fast one ($d=3$ lines for the teletypewriter vs. $d=10$ lines for the display).

Figure 5.2 and Figure 5.3 make it clear that there are important individual differences in how users decide which method to use. Using the same terminal and doing the same task, S22 uses the QS-METHOD 62% of the time, but S4 never uses it. Averaging together the data for all users and attempting to write rules to describe the choices of the group would, therefore, produce inaccurate predictions, as well as be quite misleading. Yet, despite the existence of significant individual differences in methods for accomplishing this goal, each user's behavior taken individually was highly structured and amenable to a GOMS description.

The complete prediction of which method each user employed for GOAL: LOCATE-LINE is organized as a set of Selection Rules in Figure 5.4. Each row gives the results of the accumulation of Rule 1 to Rule n, adding rules one at a time. The "Hits" column shows the total number of cases correctly predicted, and the "Misses" column shows the number of cases in which the prediction was wrong (Hits + Misses = the total number of method selections). As each rule is added, the set of rules taken together predicts more cases correctly, but a few individual cases that were predicted correctly may now be missed. For example, adding Rule 2 for S1 (the second line of the figure) correctly predicts 11 method selections of the 24 that had been missed using Rule 1 alone, but at the cost of missing 2 of the 44 that were previously hits—a net gain of 9. As the figure shows, using from two to four simple rules, it is possible to predict a user's method selections an average of 90% of the time.

5.3.   TIME PREDICTIONS (EXPERIMENT 5B)

Experiment 5A showed that it is possible, using a GOMS model, to describe users' method selections. Experiment 5B was designed to examine chronometrically how users sequence operators to accomplish tasks. The technique was to observe users performing editing tasks,

User	Rule	This Rule		Cumulative		
		Gain	Loss	Hits	Misses	%Hits
S1 (TTY)	Rule 1: Use the QS-METHOD unless another rule applies.	44	0	44	24	65%
	Rule 2: If $d < 3$, use the LF-METHOD.	11	2	53	15	78%
	Rule 3: If the target line is the last line of the page, use the AN-METHOD (with $).	5	0	58	10	85%
	Rule 4: If the current method is to use paragraph numbers for search strings and the target line is near a paragraph number, use the AN-METHOD.	2	0	60	8	88%
S4 (TTY)	Rule 1: Use the +N-METHOD unless another rule applies.	51	0	51	15	77%
	Rule 2: If $d < 3$, use the LF-METHOD.	12	1	62	4	94%
S4 (DISP)	Rule 1: Use the LF-METHOD unless another rule applies	45	0	45	21	68%
	Rule 2: If $d > 9$, use the +N-METHOD.	16	12	49	17	74%
	Rule 3: If the target line is on the next page of the manuscript, use the LF-METHOD.	56	10	56	10	85%
S22 (DISP)	Rule 1: Use the QS-METHOD unless another rule applies.	40	0	40	25	62%
	Rule 2: If $d < 5$, use the LF-METHOD.	22	2	60	5	92%

Average Final %Hits   = 90%

Figure 5.4. Selection rules for GOAL: LOCATE-LINE in Experiment 5A.
Each row tallies the effect of adding its method selection rule to the rule set. With the addition of each rule, some more methods are predicted (Gain) and some previously predicted ones are now mispredicted (Loss), for a cumulative effect of so many predictions (Hits) and so many mispredictions (Misses).

recording (1) the sequence in which operators occurred and (2) the duration of each operator occurrence. These data allow testing of task time predictions calculated from the model.

METHOD FOR EXPERIMENT 5B

Users.   Users were two secretaries and two computer scientists familiar with POET. The terminal was similar to the video display of the previous experiment.

Measurement Apparatus.   The terminal was connected to a large computer running the POET editor under the TENEX time-sharing system. For this experiment, the terminal was modified to time-stamp and record all input events on a data file. Accuracy of time-stamping was to within 33 msec of the actual time of the event at the terminal.[1]   The average response time of the editor to commands during the experiment was .8 sec (SD = .6 sec).

Two television cameras were directed at the user, one camera giving an overall view of the user and terminal, the other a closeup of the user's face, from which it could be determined whether he was looking at the manuscript, the keyboard, or the display.   The user wore a lapel microphone, recording onto the soundtrack of the video tape.   A digital clock was electronically mixed with the video picture, time-stamping each frame.   The times measured from video frames were accurate to 33 msec (one video frame).

Procedure.   The procedure was similar to that for Experiment 5A. The user was first given a test to determine his typing rate and then several editing tasks as a warmup.   Finally, he edited the same manuscript that was used in Experiment 5A.

Data Sets.   The first three unit tasks were discarded before analysis to minimize any warmup effect.   The remaining 70 unit tasks were partitioned into two comparable data sets: a *Derivation* data set, consisting of the 36 unit tasks on the odd-numbered pages, and a *Crossvalidation* data set, consisting of the 34 unit tasks on the even-numbered pages.   This partition allowed basic operator statistics to be computed on the Derivation data, while preserving the Crossvalidation data for an attempt at prediction in a matched situation, no statistical advantage having been taken of chance.

The data were also partitioned into the set of *error-free* unit tasks and the set of *error* unit tasks, each of the latter containing at least one identifiable error.   The criterion for identifying an error was that the user

---

[1] The accuracy of the timing of events did not depend on the response of the time-sharing system.

	User				
	S34 (Comp. Sci.) (sec)	**S53** (Comp. Sci.) (sec)	**S50** (Secy.) (sec)	**S95** (Secy.) (sec)	*Mean* (sec)
Derivation data (36)	9.0 (25)	15.3 (27)	15.1 (28)	13.4 (21)	13.2 (25)
Crossvalidation data (34)	8.5 (23)	14.7 (25)	17.0 (27)	14.0 (24)	13.2 (25)

Figure 5.5.   Mean error-free unit task times for all users in Experiment 5B.
The values are the mean task times over all error-free tasks for each user. The numbers in parentheses are the number of error-free tasks. All differences between mean number of error-free tasks, mean error-free task time for Derivation data vs. Crossvalidation data, or computer scientists vs. secretaries, are non-significant by Mann-Whitney U-test, $p > .05$.

took some *overt corrective action*, defined as some action that undid the effect of a preceding action. All the analyses below use the error-free data.

Figure 5.5 gives statistics on the Derivation and Crossvalidation data sets and shows that both the Derivation and Crossvalidation data were comparable with respect to the number of tasks having errors and to the mean time per task for error-free tasks.

Protocols. The videotaped record of the user's behavior and the time-stamped file of keystrokes were coded into a protocol of operator sequences, using the operators of the GOMS Model F2 that was described in Section 5.1. Occurrences of the operators were identified according to the following operational definitions:

> GET-NEXT-PAGE.   Turning the manuscript page.   Starts when the user's eyes begin to turn towards the manuscript; ends when the turned page falls flat.
> GET-FROM-MANUSCRIPT.   Looking over to the manuscript to get the next task.   Starts when the user's eyes begin to turn towards the manuscript; ends when the user types a keystroke for the next operation or begins to look back to the display, whichever comes first.
> USE-LF-METHOD.   Using the LF-METHOD to locate the line of the task.   Starts when the user's eyes begin to turn

towards the screen or the user types the first
LINEFEED, whichever comes first; ends when the last
LINEFEED is typed.

USE-QS-METHOD. Using the **QS-METHOD** method to locate
the line of the task. Starts when the user's eyes
begin to turn toward the screen or the user types the
first keystroke, whichever comes first; ends when the
final character of the search command is typed.

USE-S-COMMAND. Using the Substitute command to
modify the text. Starts when the user types the first
keystroke of the command; ends when the final
character of the command is typed.

USE-M-COMMAND. Using the Modify command to modify
the text. Starts when the user types the first
keystroke of the command; ends when the final
character of the command is typed.

VERIFY-EDIT. Examining the output on the display to check
that the modification is correct. Starts when the
final character of the previous command is typed;
ends when the user's eyes turn to the manuscript for
the next task.

RESULTS OF OPERATOR SEQUENCE PREDICTIONS

Selection Rules. Selection rules were derived for each user by
examining their method selections in the Derivation data. The results of
using these rules to predict method selections replicated the results in
Experiment 5A. One or two selection rules (Figure 5.6) were sufficient
to predict 88% of the method choices in the Derivation data and 80% in
the Crossvalidation data. Accuracy of the rules was about the same for
the two different goals. Interestingly, the rules were better at predicting
the secretaries (90%) than at predicting the computer scientists (77%).

Accuracy of Sequence Predictions. In addition to wrong method
choices, there are other possible ways in which the model might make
errors in the prediction of operator sequences. Ultimately, these will be
registered as the intrusion into the observed data of unpredicted
operators or the non-occurrence of predicted operators.

Model F2 was used to calculate the predicted sequence of operators
for each task, and this sequence was matched against the sequence

User	Selection Rules	%Hits	
		Derivation Data	Crossvalidation Data
Rules For **GOAL: LOCATE-LINE**			
S34	Rule 1: Use the QS-METHOD as default. Rule 2: If $d < 3$, then use the LF-METHOD.	84%	74%
S50	Rule 1: Use the QS-METHOD as default. Rule 2: If $d < 3$, then use the LF-METHOD.	96%	93%
S53	Rule 1: Use the QS-METHOD as default. Rule 2: If $d < 3$, then use the LF-METHOD.	63%	72%
S95	Rule 1: Always use the LF-METHOD.	95%	71%
Rules For **GOAL: MODIFY-TEXT**			
S34	Rule 1: Use the M-COMMAND as default. Rule 2: If a word is to be replaced neither at the very beginning nor very end of the line, then use the S-COMMAND.	85%	83%
S50	Rule 1: Use the S-COMMAND as default. Rule 2: If the correction is at the very beginning or the very end of the line, then use the M-COMMAND.	84%	83%
S53	Rule 1: Use S-COMMAND as default. Rule 2: If the correction is at the very beginning or the very end of the line or is a double task or involves only punctuation, then use the M-COMMAND.	93%	60%
S95	Rule 1: Always use the S-COMMAND.	100%	100%
	Mean	88%	80%

Figure 5.6. Method selection rules for Experiment 5B.
Mean accuracy (%Hits) of the rules is significantly greater for Derivation data than for Crossvalidation data, Mann-Whitney $U(8,8) = 9$, $p = .014$; greater for secretaries (90%) than for computer scientists (77%), $U(8,8) = 12.5$, $p < .025$; but no different for GOAL: LOCATE-LINE (81%) than for GOAL: MODIFY-TEXT (86%), $U(8,8) = 25$, $p = .253$.

actually observed. There is no standard statistical technique for indexing how well one sequence matches another, so the following method was used. The sites of mismatches because of operator insertions, deletions, or replacements were determined using a simple dynamic programming algorithm (based on Hirschberg, 1975, and Sakoe and Chiba, 1978) to optimize the number of matches. Then the percentage of predicted

	User				
	S34 (Comp. Sci.)	**S53** (Comp. Sci.)	**S50** (Secy.)	**S95** (Secy.)	*Mean*
Derivation data	79%	81%	98%	94%	88%
Crossvalidation data	89%	83%	92%	93%	89%

Figure 5.7. Percentage of operator instances predicted in Experiment 5B.
The secretaries' operators were predicted significantly better (94%) than the computer scientists' (83%), Mann-Whitney $U(4,4) = 0$, $p = .014$; but the prediction of the model matched the Derivation data as well (88%) as the Crossvalidation data (89%), $U(4,4) = 8$, $p = .56$.

operator occurrences that matched observed operator occurrences was computed (see Appendix to this chapter for details). Sequences generated by the model were generally in good agreement with those observed (Figure 5.7). The percentage of matches varied from 79% to 98% with an average of 88%. There were no differences between the Derivation and the Crossvalidation data, but again, the model did better at calculating sequences for secretaries (94% of operators in sequences matched) than for computer scientists (83%). Except for the already noted method-selection errors (due to operator insertion, deletion, or replacement), the only error made by the model was to predict that users would always perform a VERIFY-EDIT operation, whereas users sometimes omitted it.

RESULTS OF TIME PREDICTIONS

The protocols contain times from which it is possible to compute chronometric statistics for each operator in each model. Estimates of the time to perform a specific unit task were computed in two ways: (1) *Given* the observed sequence of operators, sum the mean times for each operator in the sequence. This estimate, which we call a *Reproduction* of the data, corresponds with how well the models would do were there no sequence prediction errors. (2) Using the sequence of operators *predicted by the models*, sum the mean times for each operator in the sequence. This latter estimate, which we call a *Prediction*, should correspond more with what we might expect to find applying the models in practice. Error can enter into the estimates either because an operator actually takes longer in some contexts than others or, in the prediction case,

Operator	User														
	S34 (Comp. Sci.) (TR = .16)			S53 (Comp. Sci.) (TR = .30)			S50 (Secy.) (TR = .16)			S95 (Secy.) (TR = .12)			All Users	
	M (sec)	CV	N	M (sec)	CV	N	M (sec)	CV	N	M (sec)	CV	N	M (sec)	CV
GET-NEXT-PAGE	2.50	.23	5	1.18	.45	4	1.81	.41	5	3.31	—	1	2.20	.42
GET-NEXT-TASK	1.29	.41	25	2.11	.41	27	2.07	.46	28	1.25	.44	21	1.68	.28
USE-QS-METHOD	2.07	.24	18	3.32	.37	12	4.48	.36	22	—	—	—	3.29	.37
USE-LF-METHOD	2.10	.76	4	1.85	.53	4	3.47	.49	5	5.40	.53	17	3.21	.51
USE-+N-METHOD	2.10	.40	3	4.07	.48	8	—	—	—	—	—	—	3.09	.45
USE-AN-METHOD	—	—	—	8.18	.33	2	—	—	—	10.06	.21	3	9.12	.15
USE-S-METHOD	2.94	.29	5	6.60	.34	12	6.78	.40	21	4.66	.35	21	5.25	.35
USE-M-METHOD	4.38	.29	20	8.12	.44	15	8.52	.45	7	—	—	—	7.01	.33
VERIFY-EDIT	.64	.30	11	.96	.31	21	.76	.37	18	.85	.68	18	.80	.17

Average between-user CV = .36

Mean CV   .37   .41   .33   .44

Average within-user CV = .40

Figure 5.8. Operator duration statistics for all users in Experiment 5B.

TR is the typing rate in sec/keystroke. The *Mean* and *CV* for all users (rightmost column) is based on user means.

because the model predicts the wrong sequence of operators, and this sequence takes a different amount of time than does the correct sequence.

Operator Times. The durations of all occurrences of each operator type in the Derivation data were used to estimate the operator times, shown in Figure 5.8. Since the data come from a quasi-natural situation and since a rare method may appear only once in the data, there is a fair chance that some extreme times may show up in the distributions of operator times. Though these must be accepted in any prediction test, it is appropriate to avoid them in estimating the characteristics of the operators. Consequently, in Figure 5.8 we have dropped outliers that fall beyond two standard deviations from the raw mean and then recomputed the mean and coefficient of variation *CV* for each operator.[2]

---

[2] Here and elsewhere we report the coefficient of variation $CV = SD/Mean$ as a way of partially normalizing the *SD* to make it more comparable for operators of different durations.

	User				
	S34 (Comp. Sci.)	S53 (Comp. Sci.)	S50 (Secy.)	S95 (Secy.)	*Mean*
Derivation data:					
Reproduction	32%	31%	29%	29%	30%
Prediction	31%	32%	29%	34%	32%
Crossvalidation data:					
Reproduction	35%	35%	36%	35%	35%
Prediction	33%	36%	37%	39%	36%

Figure 5.9. Prediction error for task times in Experiment 5B.
The prediction error measure is the *RMS* (root mean square) error as a percentage of the observed mean task time. The prediction error is less for the Derivation data (31%) than for the Crossvalidation data (36%), Mann-Whitney $U(8,8) = 0$, $p = .01$; but there is no difference between Reproduction (33%) and Prediction (34%), $U(8,8) = 24$, $p > .25$; or between computer scientists (33%) and secretaries (34%), $U(8,8) = 26$, $p > .40$.

Whereas there are moderate differences between users in their operator times, the variation in times between users is comparable to the variation of times within a user. The average *CV* between users is .36, whereas the average *CV* within a user is .40.

Accuracy of Time Predictions. Comparing the time per task calculated from the model with the observed times gives an *RMS* (root mean square) error of 33% of the mean observed time. As shown in Figure 5.9, there were no differences in prediction accuracy between computer scientists and secretaries or between Reproduction and Prediction, but the Derivation data was slightly more accurately predicted (*RMS* error of 31%) than the Crossvalidation data (36%).

If the *RMS* error measure is interpreted as the average model error, 33% error may seem high. But predicting editing times unit task by unit task for a single user is a very stringent test. If the unit of prediction were the whole manuscript rather than the unit task, then the prediction error would drop considerably, since the high and low predictions of the various unit tasks would tend to cancel each other. The *RMS* error approximately obeys a square root of *n* law, where *n* is the number of unit tasks.[3] So the *RMS* error for predicting the time to edit the whole

---

[3] $RMS(e) = \sqrt{[\Sigma e_i^2/n]}$, where e_i is the prediction error on the ith unit task. The *RMS* error is the standard deviation *SD* of *e* about zero, instead of the actual mean of *e*, which is *M(e)*, and thus $RMS(e) \geq SD(e)$. If $M(e) = 0$, then $RMS(e) = SD(e)$, and

manuscript (70 tasks) would be 33% $\times$ $\sqrt{70}$ = 4% (neglecting, of course, the effects of users' mistakes, which are not addressed by this model). The error for these models of variable-sequence cognitive activity would thus seem to be in the same range (about 5%) as that sometimes cited for predetermined time system predictions of invariable-sequence physical activity by industrial engineers (Eady, 1977; Maynard, 1971).

5.4.   GRAIN OF ANALYSIS (EXPERIMENT 5C)

The model discussed above is not the only possible GOMS model for the manuscript-editing task. Because models could be constructed with either more or less detail, there is an important issue of the appropriate *grain* of the analysis.

A priori, it is not possible to know which grain size is appropriate. As the grain of the analysis becomes finer, the model successively accumulates opportunities for conditional behavior (either optional application of some method or differentiation into cases). Thus, from one point of view, models at a finer grain should be more accurate. But opposing forces are also at work. At a finer grain, operators will be likely to appear in a larger number of contexts. In combining low-level operators to form functional units that a coarser grain would reflect directly, one may miss setup or other operations that are properties of the unit as a whole. The duration of operators may depend on other operators in the sequence (Abruzzi, 1956). And finally, there is typically greater error in the measurement of finer grain operators than of coarser grain operators.

A direct test of how the grain of analysis affects the accuracy of a GOMS model is to recast the analysis at several levels of detail. There appear to be two essentially independent dimensions along which the grain of analysis can be made finer or coarser. The primary dimension involves duration of the operators. The second dimension involves variations among operators of approximately the same duration.

We explore variations of GOMS models along both of these dimensions. Figure 5.10 describes briefly the family of nine manuscript-

---

the *RMS* error is equivalent to the standard error. The square root law argument should actually be made with respect to *SD(e)* about *M(e)*, but the use of the *RMS* error is approximately correct if *M(e)* is close to zero.

UNIT-TASK LEVEL:

Model UT Constant time per unit task. Only one operator: EDIT-UNIT-TASK. (This model is like the Constant Time per Modification model of Chapter 4, except for the substitution of unit tasks for modifications.)

FUNCTIONAL LEVEL:

Model F1 Single operator for each functional step in unit task sequence: GET-NEXT-TASK, LOCATE-LINE, MODIFY-TEXT, VERIFY-EDIT.

Model F2 Like Model F1, but with operators LOCATE-LINE and MODIFY-TEXT broken into separate cases based on the methods used to accomplish them.

ARGUMENT LEVEL:

Model A1 Like Model F2, but with operators at the level of typing a system command (SPECIFY-COMMAND) or typing an argument to a command (SPECIFY-ARG).

Model A2 Like Model A1, but with SPECIFY-COMMAND and SPECIFY-ARG broken into separate cases according to whether they involve an implicit need to get information from manuscript (suffix = /G) or not (suffix = /NG).

Model A3 Like Model A1, but with SPECIFY-COMMAND and SPECIFY-ARG broken into separate cases according to four method contexts: quoted string method (suffix = /Q), first argument to Substitute command (suffix = /S1), second argument to Substitute command (suffix = /S2), or Modify command (suffix = /M).

Model A4 Like Model A1, but with all the distinctions in both Model A2 and Model A3 combined multiplicatively.

KEYSTROKE LEVEL:

Model K1 Like Model A2, but with operators at the level of basic perceptual, cognitive, and motor actions: LOOK-AT, HOME, TURN-PAGE, TYPE, and MOVE-HAND. All mental actions not overlapped with motor operations are represented as the MENTAL operator.

Model K2 Like Model K1, but with MENTAL broken down into SEARCH-FOR, COMPARE, CHOOSE-COMMAND, and CHOOSE-ARG.

───

Figure 5.10. Description of the family of GOMS models investigated for POET.

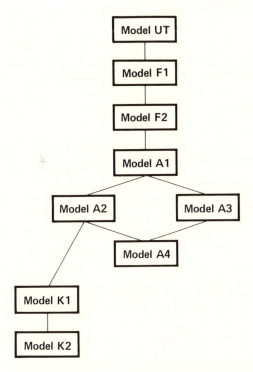

Figure 5.11.   Graph of the family tree of GOMS models investigated for POET.
The links in the tree show how the models outlined in Figure 5.10 can be derived from each other by making further distinctions. Distinctions are made by either case analysis (as in Model F1 into Model F2) or by splitting operators (as in Model F1 into Model A1).

editing models that we consider in this experiment. Figure 5.11 shows the family tree, where the links in the tree show which models are further elaborations of each other (either by splitting or by differentiating operators). Finally, Figure 5.12 lays out the full models themselves.

Each model is given a name of the form "Model *level number.*" We distinguish models in Figure 5.11 at four levels: the Unit-Task Level, the Functional Level, the Argument Level, and the Keystroke Level. We begin at the Unit-Task Level with Model UT (see Figure 5.12), which consists of a single operator, EDIT-UNIT-TASK. The goal of manuscript editing is accomplished by repeating this operator for each unit task. With only a single operator, Model UT always predicts that it takes the same amount of time to do a unit task. Functional Level models come from decomposing the unit task into its functional cycle: (1) get the next

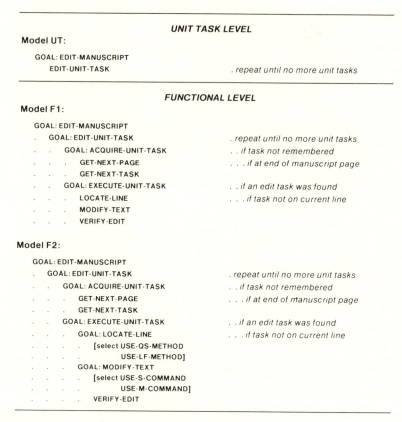

Figure 5.12. GOMS models for POET.

edit task, (2) locate the editor at the line on which the modification is to be made, (3) make the modification, and (4) verify that the edit was done correctly. The model used to analyze Experiment 5B, Model F2, is a Functional Level model. Argument Level models arise by decomposing the methods used at the Functional Level into the individual steps of specifying commands and arguments. Both Functional Level models and Argument Level models are driven by the structure of the POET commands. These are themselves reflections of the demands of the task as it is defined in the manuscript.

At the Keystroke Level, an entirely different set of operators comes into view, defined not by their functional role in the command language, but by reference to the basic physical and mental actions of the user: typing, looking, moving a hand, plus various mental operations. These operators are more task-independent than the operators at other levels.

ARGUMENT LEVEL

Model A1:

```
GOAL: EDIT-MANUSCRIPT
.    GOAL: EDIT-UNIT-TASK                              . repeat until no more unit tasks
.    .    GOAL: ACQUIRE-UNIT-TASK                      . . if task not remembered
.    .    .    GET-NEXT-PAGE                           . . . if at end of manuscript page
.    .    .    GET-NEXT-TASK
.    .    GOAL: EXECUTE-UNIT-TASK                      . . if an edit task was found
.    .    .    GOAL: LOCATE-LINE                       . . . if task not on current line
.    .    .    .    [select  GOAL: USE-QS-METHOD
.    .    .    .                  .    SPECIFY-COMMAND
.    .    .    .                  .    SPECIFY-ARG
.    .    .    .                  GOAL: USE-LF-METHOD
.    .    .    .                  .    SPECIFY-COMMAND]     . . . . . repeat until at line
.    .    .    VERIFY-LOC
.    .    .    GOAL: MODIFY-TEXT
.    .    .    .    [select  GOAL: USE-S-COMMAND
.    .    .    .                  .    SPECIFY-COMMAND
.    .    .    .                  .    SPECIFY-ARG
.    .    .    .                  .    SPECIFY-ARG
.    .    .    .                  GOAL: USE M-COMMAND
.    .    .    .                  .    SPECIFY-COMMAND
.    .    .    .                  .    SPECIFY-COMMAND     . . . . . repeat until at text
.    .    .    .                  .    SPECIFY-ARG
.    .    .    .                  .    SPECIFY-COMMAND]
.    .    .    .    VERIFY-EDIT
```

Model A2: *as in Model A1 but substitute*

SPECIFY-COMMAND/G *or* SPECIFY-COMMAND/NG *for* SPECIFY-COMMAND

SPECIFY-ARG/G *or* SPECIFY-ARG/NG *for* SPECIFY-ARG

Model A3: *as in Model A1 but substitute*

SPECIFY-ARG/Q *or* SPECIFY-ARG/M *or*
SPECIFY-ARG/S1 *or* SPECIFY-ARG/S2 *for* SPECIFY-ARG

Model A4: *as in Model A1 but substitute*

SPECIFY-COMMAND/G *or* SPECIFY-COMMAND/NG *for* SPECIFY-COMMAND

SPECIFY-ARG/Q/G *or* SPECIFY-ARG/Q/NG *or*
SPECIFY-ARG/M/G *or* SPECIFY-ARG/M/NG *or*
SPECIFY-ARG/S1/G *or* SPECIFY-ARG/S1/NG *or*
SPECIFY-ARG/S2/G *or* SPECIFY-ARG/S2/NG *for* SPECIFY-ARG

The cost of obtaining the estimates of all the different operators and selection rules increases as the size of the operators decrease, because more data are required for a given level of robustness and because the observation and measurement problems increase at the lower levels. A possible compensation for the greater cost of using the Keystroke Level operators is that, unlike the larger operators, it may not be necessary to determine lower-level operators for each new application.

Model K2:

```
GOAL: EDIT-MANUSCRIPT
.    GOAL: EDIT-UNIT-TASK                              . repeat until no more unit tasks
.    .    GOAL: ACQUIRE-UNIT-TASK                      . . if task not remembered
.    .    .    GOAL: TURN-PAGE* (see below)           . . . if at end of manuscript page
.    .    .    GOAL: GET-FROM-MANUSCRIPT*
.    .    GOAL: EXECUTE-UNIT-TASK                      . . if an edit task was found
.    .    .    GOAL: LOCATE-LINE                       . . . if task not on current line
.    .    .    .    CHOOSE-COMMAND
.    .    .    .    [select  GOAL: USE-QS-METHOD
.    .    .    .                  .    GOAL: SPECIFY-COMMAND*
.    .    .    .                  .    GOAL: SPECIFY-ARG*
.    .    .    .             GOAL: USE-LF-METHOD
.    .    .    .                  .    GOAL: SPECIFY-COMMAND*]    . . . . . repeat until at line
.    .    .    .    GOAL: VERIFY-LOC*
.    .    .    GOAL: MODIFY-TEXT
.    .    .    .    CHOOSE-COMMAND
.    .    .    .    [select  GOAL: USE-S-COMMAND
.    .    .    .                  .    GOAL: SPECIFY-COMMAND*
.    .    .    .                  .    GOAL: SPECIFY-ARG*
.    .    .    .                  .    GOAL: SPECIFY-ARG*
.    .    .    .             GOAL: USE-M-COMMAND
.    .    .    .                  .    GOAL: SPECIFY-COMMAND*
.    .    .    .                  .    GOAL: SPECIFY-COMMAND*    . . . . . repeat until at text
.    .    .    .                  .    GOAL: SPECIFY-ARG*
.    .    .    .                  .    GOAL: SPECIFY-COMMAND*]
.    .    .    GOAL: VERIFY-EDIT*
```

* Expansion of goals appearing several times:

```
GOAL: TURN-PAGE
. LOOK-AT-MANUSCRIPT                 . repeat twice
. ACTION
. MOVE-HAND                          . repeat twice
. TURN-PAGE
GOAL: GET-FROM-MANUSCRIPT
. LOOK-AT-MANUSCRIPT
. SEARCH-FOR
. LOOK-AT-DISPLAY                    . optional
GOAL: SPECIFY-COMMAND
. GOAL: GET-FROM-MANUSCRIPT*         . if not already selected
. CHOOSE-COMMAND                     . if not already selected
. GOAL: TYPE-STRING*
GOAL: SPECIFY-ARG
. GOAL: GET-FROM-MANUSCRIPT*         . optional
. CHOOSE-ARG
. GOAL: TYPE-STRING*
GOAL: VERIFY
. LOOK-AT-DISPLAY
. GOAL: GET-FROM-MANUSCRIPT*         . optional
. COMPARE
GOAL: TYPE-STRING
. HOME                               . optional
. LOOK-AT-KEYBOARD                   . optional
. LOOK-AT-DISPLAY                    . optional
. TYPE-STRING
```

(Figure 5.12. Conclusion.)

METHOD FOR EXPERIMENT 5C

User. A single user, S13, was employed for this experiment, because of the amount of data analysis required at the fine-grained levels. The user was a highly skilled secretary (typing rate, 103 words per minute) with two years experience on the POET editor, much of it on the type of terminal used in this experiment

Procedure. The procedure was the same as in Experiment 5B.

Protocol. A protocol of the user's behavior was coded directly from the videotape record and the time-stamped keystroke file, using a set of descriptive operators not related a priori to any model. The overwhelming bulk of behavior was coded by the operators TYPE, LOOK-AT, and MENTAL, which are defined as follows:[4]

> TYPE (Key1, Key2, ...). A burst of typewriting starting with the beginning of the finger trajectory toward the first key and ending when the last key makes contact. A "burst" is defined as a sequence of keystrokes with no more than .30 sec between successive key contacts and is based on studies (Kinkead, 1975) showing that keystrokes for skilled typists doing copy typing usually do not take more than this time.
>
> LOOK-AT (Place). The act of looking from one place to another, where Place is either the video display, the keyboard, or the manuscript. LOOK-AT includes the physical head movement and the gross eye movement, but does not include any perceptual scanning within a place (such as searching a manuscript page for a new task).
>
> MENTAL. The generic operator for any mental activity that does not overlap with physical operations. MENTAL operations are identified as pauses between physical operations.

Figure 5.13 shows a fragment of the protocol, which describes S13's behavior on the last unit task in Figure 3.2 in terms of these descriptive operators.

---

[4] Other operators, used infrequently, were HOME (Hand, Place) for moving a hand to the keyboard preparatory to typing, MOVE-HAND (Hand, Place) for other hand movements, TURN-PAGE, ACTION (Description), and EXPRESSION (Description). The last two were miscellaneous categories for recording other behavior.

| Start | Stop | ΔT | Operator |
(min:sec)	(min:sec)	(sec)	
18:56.33	18:56.73	.40	LOOK-AT-MANUSCRIPT
18:56.73	18:58.89	2.16	MENTAL
18:58.89	18:59.41	.52	HOME (LEFT-HAND)
18:59.41	18:59.66	.25	MENTAL
18:59.66	18:59.94	.23*	LOOK-AT-KEYBOARD
18:59.89	19:00.14	.25	TYPE (")
19:00.14	19:00.24	.10	MENTAL
19:00.24	19:00.48	.24	LOOK-AT-DISPLAY
19:00.48	19:01.11	.63	MENTAL
19:01.11	19:01.43	.32	LOOK-AT-KEYBOARD
19:01.43	19:01.70	.27	MENTAL
19:01.70	19:01.82	.12	TYPE (e)
19:01.82	19:01.92	.10	MENTAL
19:01.92	19:02.66	.07*	TYPE (x i s RETURN /)
19:01.99	19:02.34	.35	LOOK-AT-DISPLAY
19:02.34	19:04.16	1.82	MENTAL
19:04.16	19:04.53	.37	LOOK-AT-MANUSCRIPT
19:04.53	19:05.48	.95	MENTAL
19:05.48	19:05.83	.15*	LOOK-AT-DISPLAY
19:05.63	19:05.91	.28	TYPE (. s)
19:06.06	19:06.40	.13*	LOOK-AT-KEYBOARD
19:06.19	19:06.50	.31	TYPE (c o)
19:06.50	19:06.74	.24	MENTAL
19:06.74	19:06.86	.07*	TYPE (-)
19:06.81	19:07.18	.32*	LOOK-AT-MANUSCRIPT
19:07.13	19:07.25	.12	TYPE (e)
19:07.25	19:07.51	.26	MENTAL
19:07.51	19:07.63	.12	TYPE (x)
19:07.63	19:09.46	1.83	MENTAL
19:09.46	19:09.65	.19	TYPE (RETURN)
19:09.65	19:09.92	.27	MENTAL
19:09.92	19:10.04	.12	TYPE (e)
19:10.04	19:10.11	.07	MENTAL
19:10.11	19:10.46	.00*	LOOK-AT-DISPLAY
19:10.11	19:10.72	.61	TYPE (x RETURN RETURN /)
19:10.72	19:11.76	1.04	MENTAL

Figure 5.13. Segment of the protocol record for one unit task in Experiment 5C.

This part of the protocol describes S13's performance of the last unit task shown in Figure 3.2. On those cases marked with an asterisk, the time ΔT charged to an operator is less than the difference between the Start and Stop clock times because the operator overlaps with the next operator.

Data Sets. As in Experiment 5B, the first three tasks were discarded and the remaining 70 tasks were partitioned into a Derivation data set and a Crossvalidation data set. The two data sets were found to be comparable with respect to time per unit task (Mann-Whitney $U(19,26)$ = 180.5, $p > .05$).

Fitting the Models to the Data. The protocol record for the error-free Derivation unit tasks was re-coded into a sequence of operators for each model. For example, the protocol fragment in Figure 5.13 is encoded into Model F2 as follows:

18:56.33 – 18:59.94	3.61 sec	GET-UNIT-TASK
18:59.94 – 19:04.16	4.22 sec	USE-QS-METHOD
19:04.16 – 19:10.72	6.56 sec	USE-S-COMMAND
19:10.72 – 19:11.23	.51 sec	VERIFY-EDIT .

To encode each operator requires a recognizer that determines whether the operator occurs in the data and, if so, what its boundary times are. Such recognizers are insensitive to many of the details of what happens. An odd MENTAL operator within a SPECIFY-COMMAND (at the Argument Level), a USE-QS-METHOD (at the Functional Level), or an EDIT-UNIT-TASK (at the Unit-Task Level) is quite consistent and is accepted by the recognizers for these operators. Thus, the higher-level models account for all the descriptive operators in the protocol. But these odd descriptive operators (e.g., the odd MENTAL) are not without consequence; they may show up as sequence errors in the lower-level models and, in chronometric analysis, as variance in the higher-level operator times.

The Keystroke Level models, on the other hand, must map one-to-one onto the protocol, since the Keystroke Level operators are at the same level of aggregation as the protocol operators. Many of the protocol operators (such as TYPE) are identical to the Keystroke Level operators and are identified directly, whereas other protocol operators (such as MENTAL) must be relabeled (e.g., SEARCH-FOR or CHOOSE-COMMAND in Model K2) to fit the models. The possibility then exists that there will be descriptive operators in the protocol that are not accounted for by the models. More often, a descriptive operator, though a possible operator type in the model, may not correspond to any possible operator produced by the model at that point. This happens for 78 of the 581 operator instances in the protocol. The most significant kind of unaccounted-for operators are instances of MENTAL that cannot be interpreted as one of the Model K2 operators and are labeled **UNKNOWN**. These mostly arise from our stringent rule of coding the occurrence of a **MENTAL** operator whenever there is a pause in the protocol. The mean time of the **UNKNOWN** operators is only .28 sec. Of the unaccounted-for operators, 71 are **UNKNOWN**s, 6 are **MOVE-HANDS**, and one is an **ACTION**.

It sometimes happens that two mental operators (such as **VERIFY-LOC** and **SPECIFY-COMMAND** in Model A1) are predicted by the model to occur in succession. In these cases there is a problem determining the boundary between them, for there is no overt indication in the data. Each operator type involved in such cases (e.g., **VERIFY-LOC**) was compared to instances of the operator where the boundaries were observable (instances where it was surrounded by non-mental operators). This comparison showed clearly that the operator times of these adjacent mental operators are not additive—that the time for **VERIFY-LOC** plus the time for **SPECIFY-COMMAND** when each is surrounded by non-mental operators is not the same as the combined time for the pair when they occur together in sequence. These cases are listed later in Figure 5.15 as if they were separate operator types (and are given combined names like **VL + SC**). In all, there are four different combined operator types, two at the Argument Level (**GFM + SC** and **VL + SC**) and two at the Keystroke Level (**SF + CM** and **C + CC**). For purposes of predicting task times, the values of the non-combined versions of these operators were used, thus counting their non-additivity against the models.

RESULTS OF OPERATOR SEQUENCE PREDICTIONS

Selection Rules. Analysis of the Derivation data yielded selection rules for the user very similar in form and in accuracy to those for users in the previous experiments. The rules for S13 are:

> *Selection rules for* **GOAL: LOCATE-LINE:**
> Rule 1. Use the **QS-METHOD** as default.
> Rule 2. Use the **LF-METHOD** if $d < 5$ lines.
> *Selection rules for* **GOAL: MODIFY-TEXT:**
> Rule 1. Use the **S-COMMAND** as default.

The selection rules for **GOAL: LOCATE-LINE** were correct 88% of the time. The rule for **GOAL: MODIFY-TEXT** was correct 92% of the time.

Accuracy of Sequence Predictions. For some of the models it was necessary to fix the conditions under which the "optional" operators would be invoked. These operators mainly center around the question of when to invoke extra GET-FROM-MANUSCRIPT operators, either implicitly (the /G versions of the SPECIFY operators in Model A2 and Model A4, see Figure 5.12) or explicitly (the GOAL: GET-FROM-MANUSCRIPT goal in Model K2). Since the conditions that cause extra GET-FROM-MANU-SCRIPT operators were not clear from the data, each option was decided such that exactly one extra GET-FROM-MANUSCRIPT was predicted for each unit task.

The match between predicted and observed sequences was comparable to that obtained in Experiment 5B for the comparable Model F2 (96% in the present experiment vs. 88% in Experiment 5B). As expected, the match declined as the grain of analysis became finer (see Figure 5.14). The decline in accuracy for Model A2 and Model A4 resulted mainly from their inability to predict the exact sites in the protocol at which the user would glance back at the manuscript for more information and how often the user would consult the manuscript. Models at the Keystroke Level encountered two other difficulties as well. First, it happened that this particular user would always move her hand to her mouth and lick her fingers before turning the page. In fact, she would usually also lick her fingers one task too early (a true case of "fractional anticipatory goal response" *in vivo*). Because this action was not in the model, it caused mismatched operators. The second difficulty at the Keystroke Level was that the UNKNOWN operators counted as mismatches.

RESULTS OF TIME PREDICTIONS

Operator Times. Durations of the operators for all models, as empirically determined from the Derivation data, appear in Figure 5.15 along with the percentage of the time spent in each operator. Since manuscript editing has the appearance of a motor-intensive task, it is interesting that 60% of the time for the manuscript-editing task was mental time; only 22% of the time was actually spent in typing.

All operators, except the TYPE operator, are assumed to take constant time. Although it is obvious that TYPE should be parameterized by the number of characters to be typed, we must be able to predict the search strings and the substitution strings the user will employ in order to

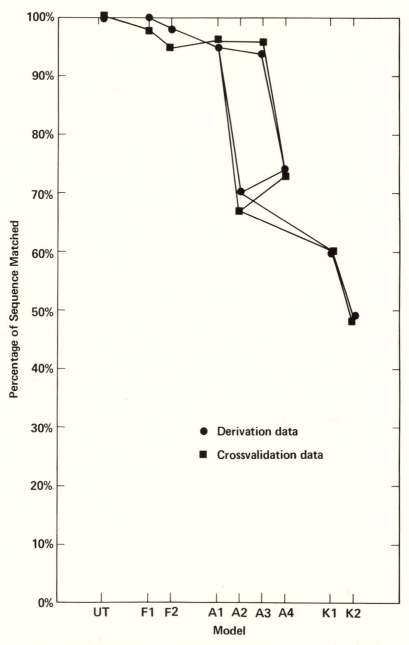

Figure 5.14. Percentage of operators correctly predicted by each model in Experiment 5C.

capitalize on the parameterization. The time for TYPE was parameterized by the number of shift characters N_{shift}, carriage returns N_{cr}, and other characters N_{other}, according to the equation

$$T = .05 + .17N_{shift} + .19N_{cr} + .11N_{other} \text{ sec}.$$

The equation is based on the regression fit of 157 short typing bursts from the Experiment 5A (1 to 18 characters in a burst, mean 3.8 characters).[5] The equation explains 92% of the variance. The operator times of this user for Model F2 were comparable to the times for the same operators observed in Experiment 5B.

Accuracy of Time Predictions. The main result is that time calculations based on all the GOMS models were about equally accurate (except for Model UT, which was somewhat less accurate). Accuracy of the Functional Level Model was comparable to that obtained in Experiment 5B. There the *RMS* error was 33%; here it was 29% for the comparable Model F2. Various combinations of models, data sets, and calculation methods varied in the range of 20% to 40% *RMS* error, as can be seen in Figure 5.16. Finer grain models did better on Reproduction, but not on Prediction, of the Derivation data. The finer grain models were no better at either Reproduction or Prediction of the Crossvalidation data.[6]

A study of the prediction errors on unit tasks with different task environment features revealed that the only task environment feature that allowed gain in prediction was the one in which the unit task shared the same line on the manuscript with another unit task (i.e., $d = 0$). There were two tasks with this feature in the Derivation data, and they were the

---

[5] Fitting this same data with only one parameter, the number of characters in a burst, yields the equation $T = .06 + .12N_{char}$ sec, which explains 89% of the variance. The .12 sec per character rate is equivalent to 91 words per minute, which is quite close to S13's typing test speed of 103 wpm. Thus, the user types at almost her highest typing rate even on these short bursts.

[6] In fact, prediction of the Crossvalidation data is worse at the Keystroke Level than at the Argument Level. This occurs because in one of the tasks the user compares information on the display with information on the manuscript much more often than the model predicts, resulting in a large underprediction. Recomputation of the points in Figure 5.16 using the mean absolute error (an index not as sensitive to single outliers as the *RMS* error) gave a graph similar to Figure 5.16, but with the prediction of the Derivation data indistinguishable from the curves for the Crossvalidation data, confirming the general stability of the results.

Operator	Duration			Percentage of Total Time in Operator								
	M (sec)	CV	N	Model UT	Model F1	Model F2	Model A1	Model A2	Model A3	Model A4	Model K1	Model K2
EDIT-UNIT-TASK	11.38	.30	26	100%	—	—	—	—	—	—	—	—
GET-NEXT-PAGE	2.14	.64	5	—	3%	3%	3%	3%	3%	3%	—	—
GET-NEXT-TASK	1.92	.33	24	—	16%	16%	—	—	—	—	—	—
LOCATE-LINE	3.98	.29	24	—	32%	—	—	—	—	—	—	—
MODIFY-TEXT	3.85	.40	26	—	35%	—	—	—	—	—	—	—
VERIFY-EDIT	1.49	.57	26	—	14%	14%	14%	14%	14%	14%	—	—
USE-QS-METHOD	3.94	.30	21	—	—	28%	—	—	—	—	—	—
USE-LF-METHOD	4.27	.25	3	—	—	4%	—	—	—	—	—	—
USE-S-COMMAND	3.63	.37	24	—	—	29%	—	—	—	—	—	—
USE-M-COMMAND	9.72	.63	2	—	—	6%	—	—	—	—	—	—
GET-FROM-MANUSCRIPT	2.06	.44	5	—	—	—	4%	4%	4%	4%	—	—
GFM+SC	1.80	.22	18	—	—	—	12%	12%	12%	12%	—	—
VERIFY-LOC	1.94	.45	17	—	—	—	12%	12%	12%	12%	—	—
VL+SC	2.00	.44	7	—	—	—	4%	4%	4%	4%	—	—
SPECIFY-COMMAND	1.47	.77	28	—	—	—	13%	—	13%	—	—	—
SPECIFY-ARG	1.46	.57	76	—	—	—	38%	—	—	—	—	—
SPECIFY-COMMAND/NG	.40	.88	11	—	—	—	—	2%	—	2%	—	—
SPECIFY-COMMAND/G	2.03	.49	17	—	—	—	—	11%	—	11%	—	—
SPECIFY-ARG/NG	1.29	.54	63	—	—	—	—	29%	—	—	—	—
SPECIFY-ARG/G	2.28	.45	13	—	—	—	—	10%	—	—	—	—
SPECIFY-ARG/Q	2.07	.28	21	—	—	—	—	—	14%	—	—	—
SPECIFY-ARG/S1	1.34	.70	24	—	—	—	—	—	12%	—	—	—
SPECIFY-ARG/S2	.94	.31	24	—	—	—	—	—	8%	—	—	—
SPECIFY-ARG/M	2.04	.67	7	—	—	—	—	—	5%	—	—	—
SPECIFY-ARG/Q/NG	1.94	.22	14	—	—	—	—	—	—	9%	—	—

Operator	Mean	CV	N	%	%
SPECIFY-ARG/Q/G	2.29	.33	7		5%
SPECIFY-ARG/S1/NG	1.12	.65	21		9%
SPECIFY-ARG/S1/G	2.79	.34	3		3%
SPECIFY-ARG/S2/NG	.93	.32	23		8%
SPECIFY-ARG/S2/G	1.20	—	1		0%
SPECIFY-ARG/M/NG	2.05	.59	5		3%
SPECIFY-ARG/M/G	2.02	1.13	2		1%
MENTAL	.62	.88	260	60%	—
TYPE	.39	.31	173	22%	22%
LOOK-AT	.31	.32	139	13%	13%
HOME	.52	.22	9	2%	2%
TURN-PAGE	.67	.32	5	1%	1%
MOVE-HAND	.19	.91	17	1%	1%
ACTION	.13	1.56	6	0%	0%
EXPRESSION	.23	—	1	0%	0%
SEARCH-FOR	.72	.71	28	—	7%
SF + CC	1.07	.52	20	—	7%
CHOOSE-COMMAND	.74	.57	8	—	2%
CHOOSE-ARG	.41	.81	56	—	9%
COMPARE	1.01	.82	59	—	22%
C + CC	1.14	.60	18	—	7%
UNKNOWN	.28	.92	71	—	8%

Figure 5.15. Operator duration statistics for all models in Experiment 5C.

In computing the *Mean* and *CV* of each operator, all instances greater than 2 *SD*'s from the mean were discarded, and the *Mean* and *CV* were recomputed with the remaining instances. In the combined-operator abbreviation, GFM = GET-FROM-MANUSCRIPT, SC = SPECIFY-COMMAND, SF = SEARCH-FOR, CC = CHOOSE-COMMAND, and C = COMPARE.

175

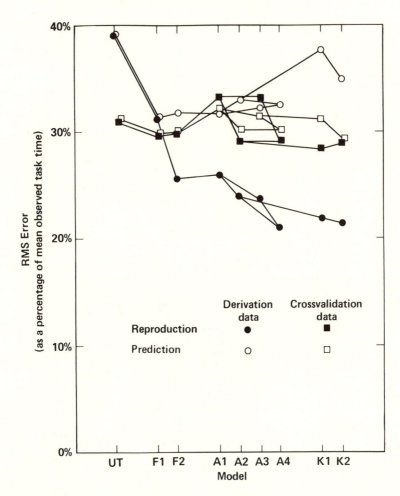

Figure 5.16. Task time predictions by all models in Experiment 5C.

reason why Model UT predicted the Derivation data less well than the Crossvalidation data.

ERROR BEHAVIOR

So far we have concentrated on error-free behavior. But errors have a significant effect on the efficiency with which text-editing is done. Overall, about 26% of the total time spent in all the experimental tasks is due to error. As Figure 5.17 shows, errors were frequent, occurring on

	N	M (sec)	CV
All unit tasks:			
Derivation data	36	13.37	.66
Crossvalidation data	34	19.46	1.18
All data	70	16.33	1.06
Error unit tasks:			
Derivation data	10	16.96	.89
Crossvalidation data	15	29.46	1.09
All data	25	24.46	1.10
Error unit tasks with error time removed:			
Derivation data	10	10.69	.25
Crossvalidation data	15	13.72	.47
All data	25	12.51	.43
Error-free unit tasks:			
Derivation data	26	11.99	.38
Crossvalidation data	19	11.57	.31
All data	45	11.81	.35

Figure 5.17. Unit task time statistics for error and error-free unit tasks in Experiment 5C.

36% of the tasks (25 out of 70), and errors doubled the time to perform the tasks in which they occurred (from 12.5 sec to 24.4 sec).

The longer time required for tasks in which errors occur is accounted for by the extra operations that must be performed by users on these occasions. When an error occurs, the user progresses through a sequence of distinct stages:

1. *Error.* The user makes a mistake.
2. *Detection.* He becomes aware of the error.
3. *Reset.* He resets the editor to allow correction.
4. *Correction.* He undoes the effects of the error.
5. *Resumption.* He resumes error-free activity.

Error Type	N	M (sec)	CV	%N	%T
Typing errors	7	1.53	.51	27%	4%
Method-abortion errors	8	4.17	.51	31%	11%
Method-failure errors	7	4.41	.41	27%	10%
"Big" errors	3	71.89	.60	12%	72%
Unclassifiable error	1	8.18	—	4%	3%
All errors	26	11.49	25.45	100%	100%

Figure 5.18. Error times in Experiment 5C partitioned into different error types.
Column %N gives the percentage of occurrences of each error type, and Column %T gives the percentage of the total error time in each error type.

The occurrence of an error requires additional time for Error, Detection, Reset, and Correction stages over the time otherwise required for the task. The time spent in these four stages is called *error time*. When the error time is subtracted from S13's protocol, the adjusted times are similar to the times for error-free tasks (11.81 sec vs. 12.51 sec).

The errors for S13 can be classified into four categories: typing errors, method abortion errors, method failure errors, or "big" errors (Figure 5.18). Simple typing errors required about 1.5~3.0 sec for recovery. There was minor variation in the choice of method, leading to a small variation in the correction time. The user detected mistyped characters immediately, canceling the bad character by typing CONTROL-A.[7] But this action printed a BACKSLASH followed by the canceled letter, messing up the displayed line of typing. This, in turn, caused S13 to sometimes redisplay a clean version of the line with another command.

Method-abortion errors, in which the user abandoned a command part-way through by pressing the DELETE key, required about 2~7 sec for recovery. There were many reasons for aborting a method: the user decided it was the wrong method, that there was a better method, or that

---

[7] The notation CONTROL-A indicates the typing of the key A while holding down the CONTROL key, as is done with a SHIFT key.

the argument strings (to the Substitute command) would not work. Once, the method was aborted as the result of mistyping a command character. Method abortion was even used for its effect in cleaning up the display after it had been made messy by too many CONTROL-A's. All the abortions except one were done to the Substitute command; the exception was a QS-METHOD being aborted in favor of an LF-METHOD.

Method-failure errors, in which a correctly executed method produced an unintended result, required 2~8 sec for recovery. All these method failures were with the Substitute command—either no substitutions or too many substitutions were made—and in all cases the user was able to correct the error by issuing one additional Substitute command.

The above three categories of errors occurred with about equal frequency. Together they accounted for 22 of the 25 classifiable errors, but only for 25% of the error time. In contrast, the remaining 3 big errors accounted for the remaining 72% of the total error time. Although these big errors were method failures, they were classed separately because their times (43, 52, and 121 sec) were an order of magnitude larger than simple (4 sec) method-failure errors. The important characeristic of these errors is that their correction involved real problem-solving activity, mostly having to do with the user finding her place in a large text file.[8]   These results suggest two radically different sorts of errors that system designers should consider: The first are small, frequent, routine errors that can be corrected quickly in a skilled manner. The second are big, infrequent, but enormously time-consuming errors that require problem solving to correct.

5.5. DISCUSSION

Assessment of the Models

Description of Behavior. From the three experiments, it is apparent that descriptions of a user's error-free behavior in the manuscript-editing task can be constructed from a reasonably small number of components.

---

[8] It is not hard to describe where the time goes in the big errors. The 43-sec error was straightforward: S13 modified the wrong line and had to undo the modification and then find and modify the right line. In the 52-sec error, she issued two bad quoted string commands and had to wait 41 sec for the system to search the entire file and respond that these strings did not exist anywhere in the file. The 121-sec error was the only occasion on which S13 was genuinely confused. She modified the wrong line, which was on a different page of the manuscript than the target line, and then could not find the correct line. She spent most of the 121 sec moving back and forth in the file.

Depending on the grain of analysis, the behavior of each of the seven users observed in these experiments has been described by 1~20 goals, 1~13 operators, 4~6 methods, and 1~4 selection rules. Moreover, this description is a reasonably accurate account of each user's error-free behavior in the task. The selection rules were able to predict the user's choice of methods about 90% of the time using the data on which they were derived and 80% of the time on new data. The various versions of the GOMS model were able to predict 80~100% of the operators in sequence for the manuscript-editing task at the Functional Level. But models at the Argument Level or at the Keystroke Level that attempted to predict the exact site and number of looks at the manuscript or that attempted to account for pauses on the order of a quarter of a second were considerably less accurate. Other work on visual feedback for skilled keying (Long, 1976) indicates that users routinely look to the manuscript for information concerning errors and to the keyboard to locate unfamiliar keys. Undoubtedly, users also look to the manuscript because they forget what they are supposed to do. Successful modeling of this behavior would either require (1) models contingent on the contents of the text, such as the familiarity of the user with certain words or the clarity of particular editing instructions, or (2) stochastic models.

Prediction of Task Times. The GOMS models likewise provide a reasonable prediction for the amount of time taken by error-free tasks. In Experiment 5B, the model was able to predict, on new data, the time for a single task to within 36% (Figure 5.9, Crossvalidation). This prediction included the times for all the operators as well as the operator sequence. In Experiment 5C, the equivalent prediction on Crossvalidation data was within about 30% (Figure 5.16).

Even when the model fails to predict the sequence of operators exactly, the resulting time prediction may sometimes not be far off. The reason is that there is a certain amount of continuity in the space of methods. If the model predicts the user will look to the manuscript and he does not actually look until after the next operation, the time prediction will not suffer, since the frequencies of the operators remain unchanged. If the user inserts one extra operator into a sequence of 15 operators, the time prediction will be degraded only slightly. Even if the user chooses the wrong method, there is a reasonable chance that the substituted method will not be wildly different in time, because the model is also likely to err in choosing among methods whose times are comparable.

Grain of Analysis. How do the abilities of the GOMS models to predict the behavior of the user vary as a function of the grain of analysis? In the current experiment, the rather surprising answer was that accuracy at the Functional Level and finer levels was essentially independent of the grain.

Two factors seem to be at work. First, the gain in chronometric predictive power arising from new opportunities for conditional behavior in the finer grain models seems to have been canceled by the difficulties in predicting the sequence of operations (Figure 5.14). Second, there seems to have been insufficient task variability for the finer grain models to display their advantage. With respect to the latter, if the models could predict operator sequences perfectly, then the prediction curve in Figure 5.16 for the Derivation data would drop to the reproduction curve. That the prediction curve is essentially horizontal implies that refining the grain of analysis did not tap the sources of time variability. In the models, variability is expressed in the method selection rules and optional operator choices, both of which are triggered by features of the task environment. Thus, either the models did not capitalize on all the available features in the task environment or there were no task environment features that gave clues to the variability. In the case of the Crossvalidation data, the gains made by the finer grain models were not sufficient to overcome the error in predicting operator duration arising from the determination of operator times from independent data. Both the reproduction and prediction curves are essentially flat (and in a few instances, the prediction is actually slightly better than the reproduction).

It is important to note that variability in the set of error-free unit tasks in Experiment 5C is quite small (Figure 5.17), both with respect to the user's performance times and with respect to the possible range of editing tasks—all are small edits of about the same complexity. This low variance occurred because the experiment tried not to manipulate the task environment, but to assess the natural variability in the data and the ability of various models to deal with them. It appears that, whereas the models as a whole were not bad at predicting the average time per unit task, there was insufficient variation within the editing tasks to trigger increased responsiveness from the finer grain models.

Status of Goals and Operators

What psychological reality is to be ascribed to the various components and features of the GOMS model?

The occurrence of goals in a GOMS model is one of its primary cognitive features. Goals are required in generating the model and in supporting its rational character as behavior directed towards the end of editing the manuscript. As it stands, however, the goals do not make any distinguishable contribution to the time calculations of the various models. Technically, this arises from a confounding of goals and operators: any time assigned to creating a goal or to cleaning up and disposing of a goal is not distinguishable from additional time in the associated operators. Goal-manipulation operations should not take longer than about .5 sec, so that goal operators should not show up at any level above the Keystroke Level, in any event.

The confounding of goal-manipulation times results in part from GOMS being a model of error-free skilled behavior, so that the overt record contains evidence only of the sequence of effective actions. For our users, there are essentially no verbal expressions that indicate goal activity. However, protocols from inexperienced users are sprinkled with goal statements that correspond to the goals in a GOMS model. In one such experiment, when the model predicted the processing of the GOAL: USE-QS-METHOD, the user would almost invariably make comments like: "Okay, I want to get down to a line that starts with 'Food store'." When the model predicted the GOAL: USE-S-COMMAND, the user would say: "Now I want to substitute '30' for '39'." But no verbalizations were recorded in connection with low-level operators like TYPE.

Operator Variability

The order of precision of our operators, as measured by the CV, ranges from about .9 at the Keystroke Level to .3 at the Unit-Task Level. In general, CVs should be expected to decrease with increasing mean when operators are composed of suboperators, a relationship that might be called "Abruzzi's Law" (Abruzzi, 1952, 1956). It is easy to see why such a relationship is reasonable. Suppose a composite operator of mean duration M were simply composed of strings of n identical elementary operators of mean duration m. Then, $M = nm$ and

$$\text{var}(M) = n\,\text{var}(m)\,.$$

Recasting this equation in terms of the CV gives

$$CV_M = \beta M^{-.5}, \text{ where } \beta = m^{.5}CV_m\,.$$

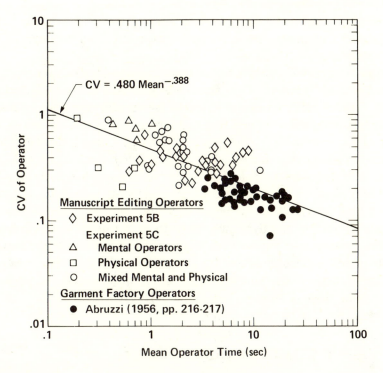

Figure 5.19. Coefficient of variation for a variety of operators as a function of mean operator time.

That is, the *CV* is inversely proportional to the square root of the mean operator time. The actual decrease is illustrated in Figure 5.19, which plots *CV* against operator mean *M*. Each point on the graph is based on multiple observations of a single person. The open symbols are manuscript-editing operators from Figure 5.8 and Figure 5.15 that occurred more than five times (excluding combination operators, TYPE, ACTION, and UNKNOWN). The solid circles are operators from Abruzzi (1956, pp. 216-217), such as cutting and stitching clothing patterns in a ladies' garment factory. In log-log coordinates, the relationship between mean and *CV* is essentially linear. A regression fit of the points in Figure 5.19 gives

$$\ln CV = -.735 - .388 \ln M$$

$(R^2 = .55, SE = .38,$ coefficient $\neq 0$ at $p < 10^8$), or

$$CV = .480 M^{-.388}.$$

Figure 5.19 suggests that, in absolute terms, the *CV*s observed in our experiment are roughly what would be expected from the size of the operations alone.

As *CV* increases, the number of observations needed to estimate the mean operator duration to a fixed precision also increases (Abruzzi, 1956). This is reflected in the figure as greater dispersion for operators having small durations and in the fact that many of the points on the outlying edge are those with the lowest *N*s. As the time for the operators becomes shorter, approaching the grain of characteristic physiological events, the operators tend to become more purely physical or mental. Since the physical operators are easier to identify and measure, these should have lower *CV*s. In Figure 5.19, the outlying points below the regression line are mostly simple physical acts (indicated by squares), such as LOOK-AT and TURN-PAGE. The outlying points above the line are mostly mental actions (indicated by triangles), such as CHOOSE-COMMAND or VERIFY-EDIT. If the purely physical and purely mental operators are ignored, the slope of the line becomes $-.433$, even closer to the $-.5$ for the ideal case of simple composition.

Extending GOMS to Cover Errors

It is important to ask how a GOMS model might be extended to cover errors and associated behavior. As we have seen, skilled behavior does not preclude the existence of a substantial number of errors, with an appreciable fraction of the total time spent correcting the errors. In Experiment 5C, about half the time was spent in error unit tasks, and about half that time (about a quarter of the total time) was error time, time the user actually spent committing and correcting errors. Compared with other experienced secretarial users we have run in our laboratory, S13 produces a higher percentage of error unit tasks, but is faster in overall performance (and also in performing error-free unit tasks). S13 thus gives up accuracy in favor of speed, since she is able to recover rapidly from errors.

We can model the states of an error unit task by a slight extension of the GOMS theory. First, we must allow operators to fail as well as succeed. Then we must specify how the user corrects the failure. The extension to the GOMS theory is to add the provision that when an operator fails, it produces an error condition, which can be represented as the *correction goal*. This goal is accomplished by selecting a *correction method*.

We have said that a skilled user committing an error proceeds through the stages: (1) Error, (2) Detection, (3) Reset, (4) Correction, and (5) Resumption of error-free activity. The Detection stage occurs when the correction goal becomes active immediately after a failed operator, thus causing an interruption in the error-free behavior sequence.

The action in the Reset stage can be modeled by a new operator, **ABORT-COMMAND**. This operator denotes more than just the physical striking of the DELETE key to reset the editor to accept commands again; the user's mental goal stack also is cleared back to the last USE goal—a "mental reset." Although such an operator can be provided within the general spirit of a GOMS model, it should be noted that this operator is the first departure from the simple stack discipline for goal control.

The new unit task(s) in the Correction stage can be modeled simply as error-free unit tasks with one exception—a new operator, **GENERATE-UNIT-TASK**, is needed in place of the **ACQUIRE-UNIT-TASK** operator.

Let us now consider how an extended GOMS model would handle the three types of errors noted in Experiment 5C: typing errors, method-abortion errors, and method-failure errors. The method for handling typing errors is the simplest. When a typing error occurs, the user becomes aware that the last character typed may be wrong. In terms of the model, the TYPE operator produces a goal to correct the bad character. The method for accomplishing this goal is as follows:

```
GOAL:  CORRECT (BadCharacter)
.  LOOK-AT-DISPLAY
.  COMPARE
.  TYPE (CONTROL-A)
.  TYPE (CorrectCharacter) .
```

That is, the user is to look at the display (if not already looking at it), compare the last typed character with the intended one, delete the bad one (if they are different), and type the correct one (if they are different). The user may then resume typing the string in which the error occurred. The predicted time for this method, using the operator times in Figure 5.15 and the typing formula in Equation 5.1, is from 1.36 sec to 1.80 sec (depending on the specifics of the situation, such as whether the

CorrectCharacter is in the touch-typing zone). The predicted time compares favorably with the observed mean typing error time of 1.53 sec (Figure 5.18).

The COMPARE operator in the above method may, of course, determine that the bad character is correct, in which case it is not changed. There is no real error in this case, only a goal to check for one, but the goal still causes an error-like interruption. Such behavior may account for some of the UNKNOWN operators in Experiment 5C.

Method-abortion and method-failure errors also evoke routine correction methods. A method-abortion error is triggered by the failure of some operator subordinate to the goal of using some POET command. For example, when specifying the second argument of the Substitute command, the user may notice that the first argument will not work and must be respecified:

> GOAL: RESPECIFY-ARG
> . ABORT-COMMAND
> . GOAL: USE-S-COMMAND .

Method-failure errors are even simpler in structure. This kind of error is produced by a failure signal from the VERIFY-EDIT operator. For the Substitute command, the failure is caused by either no substitutions or too many substitutions. For the former, the corrective method is to establish the goal of redoing the original modification using the Substitute command.

> GOAL: MODIFY-TEXT
> . GOAL: USE-S-COMMAND
> . VERIFY-EDIT .

A more general method would not specify which command to use, in which case a command other than the Substitute command could be selected for the second try. The remedy for extra, uncorrect substitutions is to generate a new unit task to remove them:

> GOAL: REDEFINE-UNIT-TASK
> . GENERATE-UNIT-TASK
> . GOAL: EXECUTE-UNIT-TASK .

These two methods cover all observed errors in Experiment 5C that were classified as method-failure errors.

As a final note, we observe that the control structure of a GOMS model begins to break down during error behavior. Some method failures, such as the big errors in Experiment 5C, seem to require genuine problem-solving behavior for recovery. For example, the failure of a **GOAL: LOCATE-LINE** method in POET can leave the user in a state of confusion as to just where in the file POET is currently located. And if the user does not detect the error until after making the modification (on the wrong line), then the user must undo the modification before searching for the correct line.[9] Such errors are rather rare events, and when they occur the user embarks on a correction course without employing a routine method and without planning an optimal method, leaving the user in a problem-solving mode of behavior. We assume that the user would acquire a routine and nearly optimal method for correcting this kind of error if it were to happen often enough.

Manuscript Editing as a Cognitive Skill

Our analysis of user behavior in the manuscript-editing task leaves little doubt about its characterization as a *cognitive skill*, however that phrase is ultimately defined. The *cognitive* apparatus is much in evidence, epitomized by the GOMS models, which dictates that there be selection of the course of action in accordance with the demands of the task, mediated by hierarchical goal structures. The GOMS models give a reasonable account of error-free user behavior and may be extended to routine error-correction behavior.

It is likewise obvious that the users we observed were *skilled*. Applied to physical motions, skill connotes smoothness, control, and economy of effort (Bartlett, 1958; Welford, 1968). Although 60% of the time was spent in non-physical activities, these descriptions certainly are appropriate for the users we observed. One indicator of skill is the time taken to perform the same task by those who are obviously unskilled. In Chapter 3, we saw that low typing skills, lack of technical background, and limited experience combined to make a factor of three difference in text-editing time.

---

[9] In the biggest observed error in Experiment 5C, however, the user went searching for the correct line before undoing the bad modification and then had to later return to undo it again.

The notion of skill is intimately related to the routine character of a task, for people generally become skilled in whatever becomes routine for them. Observation of our users demonstrates, if any additional demonstration is needed, that, just as in sensory-motor tasks, skill is highly evident in cognitively-dominated routine tasks.

Learning. The absence of significant learning during performance can often be taken as a characteristic of skilled routine performance. In Experiment 5C, S13 seems to be engaged in a steady-state performance. Within the experimental session, there is no evidence of learning; if anything, rather than the increasingly faster times characteristic of learning, there is a slight slowdown over the course of the 20-minute experimental session. Nor is there evidence of S13 learning over extended time. Five months earlier, S13 used the same terminal and system to edit a different manuscript at the rate of 11.0 sec per unit task (compared with 11.8 sec in this experiment). Our assertion that absence of learning characterizes routine skilled behavior must be qualified. Though there is no appearance of skill learning over a single session, it is only through repeated sessions that a user becomes skilled, and much of this happens after the user's time is far enough out on the Power Law of Learning curve to give the appearance of being very skilled. Furthermore, substantial learning does take place within a single session about the specific manuscript being edited (which is, of course, entirely new to the user).

Unit-Task Structure. Perhaps the most important feature to emerge from our analysis of manuscript editing is its unit task structure. Manuscript editing is broken into a sequence of almost-independent unit tasks. Within each unit task the user's behavior is highly organized and under the control of well-learned methods, which are quickly triggered into action by the dynamic features of the task situation. Unit tasks take only about 12 sec each with the POET editor, and even less with faster editors. This provides an extremely short time horizon for the integration of behavior.

5.6. CONCLUSIONS

It is possible to describe the behavior of the user of a computer text-editing system by a cognitive theory composed of a small number of goals, operators, methods, and selection rules. In this chapter we have exhibited models composed of these elements that give a reasonable quantitative account of the behavior.

A GOMS model for the manuscript-editing task predicted the sequences of user actions in the task reasonably well. It predicted a user's choices of methods about 80~90% of the time; and it predicted the actual operators in sequence 80~100% of the time in models at the Argument Level; but the accuracy for predicting operator occurrences in sequence was reduced to 50% at the most detailed level, the Keystroke Level.

The model also made reasonably good predictions for the amount of time necessary to make individual modifications to the text. It was able to predict time to within about 35% on new (Crossvalidation) data. This is comparable to achieving 4% error on the whole 20-min task of editing the manuscript (neglecting user errors).

It is important to consider at what level of behavior a GOMS model will operate—the issue of the grain size of the analysis. In this chapter we answered this question directly, repeating our analyses with nine different GOMS models. In general, there appears to be a gain in accuracy when refining the model at the Unit-Task Level (which is similar to the model of Chapter 4), but further gains in accuracy with finer grains of analysis were hard to achieve. Accuracy in predicting the sequences of user actions fell off as the model grain became finer, whereas accuracy in predicting time remained constant.

We have argued that manuscript editing can be characterized as a cognitive skill, at least for expert users. Even the user's behavior immediately after the occurrence of routine errors has the character of cognitive skill. One of the characteristic features of this skill is its unit task structure.

Appendix to Chapter 5:
MATCHING OPERATOR SEQUENCES

The problem is to put two sequences of operators, which may be of different lengths, into correspondence and then to assign a value to how well they match. For example, if GFM, SC, SA, SE, VE, and VL are acronyms for operators, the algorithm to be described takes as input both a Predicted sequence and an Observed sequence of operators:

> Predicted: GFM SC SC VL SC SA SA VE
> Observed: GFM SC SA SC SA SA VE .

It inserts dummy x operators to bring them into correspondence:

> Predicted: GFM SC SC VL SC SA SA VE
> Observed: GFM SC SA X SC SA SA VE .

There are now 6 matches out of a possible 8, or a 75% match. The algorithm inserts dummy operators in such a way as to maximize the number of matches.

The following procedure is a translation of the Interlisp function that was used for computations in Experiments 5B and 5C into an informal Algol-like notation. The algorithm is based on Hirschberg (1975) and Sakoe and Chiba (1978). It takes as input predicted and observed sequences and returns the percentage of matches and new versions of the input sequences resulting from the addition of dummy operators.

```
procedure matchSeqs(PredSeq, ObsSeq):
```

Step 1. Initialize.
```
    PredLength ← length(PredSeq);
    ObsLength ← length(ObsSeq);
    array   PredSeq[1:PredLength],      ...Predicted sequence of operators
            ObsSeq[1:ObsLength],        ...Observed sequence of operators
            Score[0:PredLength, 0:ObsLength]←0,   ...Working space
            PredSeqResult[1:PredLength + ObsLength],
            ObsSeqResult[1:PredLength + ObsLength];
```

Step 2. Compute scores for a matrix with one row for every operator in the
predicted sequence and one column for every operator in the observed
sequence.

```
for i from 1 to PredLength do
    for j from 1 to ObsLength do
        if (PredSeq[i] = ObsSeq[j])
        then Score[i,j] ← Score[i - 1,j - 1] + 1;
        else Score[i,j] ← max(Score[i - 1,j], Score[i,j - 1]);
```

Step 3. Traverse the matrix backward along the path of highest scores.

```
    i ← PredLength;  j ← ObsLength; k ← 1;
    until (i = 0 and j = 0) do
        if (i≠0 and (j = 0 or (Score[i - 1,j] > Score[i - 1,j - 1]))
        then PredSeqResult[k] ← PredSeq[i];
            ObsSeqResult[k]  ← "X";
            k ← k + 1; i ← i - 1;
        elseif (j≠0 and (i = 0 or (Score[i,j - 1] > Score[i - 1,j - 1]))
        then PredSeqResult[k] ← "X";
            ObsSeqResult[k]  ← ObsSeq[j];
            k ← k + 1; j ← j - 1;
        else PredSeqResult[k] ← PredSeq[i];
            ObsSeqResult[k]  ← ObsSeq[j];
            k ← k + 1; i ← i - 1; j ← j - 1;
    %Match ← Score[PredLength, ObsLength] / (k - 1);
    return(%Match, PredSeqResult, ObsSeqResult); end;
```

6. Extensions of the GOMS Analysis

There are several directions in which the GOMS models might be extended. In this chapter we consider the issues involved in three of these. The first extension is to another editor. In particular, we would like assurance that a GOMS description can be given for a display-oriented editor (the editor in Chapter 5 was line-oriented). A display-oriented editor may cause new issues to arise concerning the interaction of the user with the display.

The second extension concerns the accuracy of a GOMS model for predicting a user's action. We saw in Chapter 5 that, as the detail of the GOMS models increased, it became more difficult to predict the precise operator sequence the user would employ on a specific occasion; it was especially difficult to predict when the user would consult the manuscript for information. Actually, for our purposes, it would be sufficient to predict the distribution of operator sequences over a set of similar occasions. But the GOMS notation would have to be extended to incorporate stochastic elements of two types: (1) operator times expressed as probability distributions rather than as single numbers and (2) probabilistic selection rules and conditionalities for predicting which method the user will employ and for expressing probabilistic conditionality within those methods.

Third, we continue our information-processing task analysis of text-editing by defining a symbolic representation for the instructions on the manuscript and by further explicating how these instructions lead to the behavior we observe.

EDITING COMMANDS

To delete text: Select old text with mouse
                                 Type **D**

To insert text: Select insertion point with mouse
                                 Type **I**
                                 Type new text
                                 Type **ESC**

To replace text: Select old text with mouse
                                 Type **R**
                                 Type new text
                                 Type **ESC**

SELECTIONS WITH MOUSE

To select a character: Point to character with mouse
                                 Push **MOUSE-BUTTON-1**

To select a word: Point to word with mouse
                                 Push **MOUSE-BUTTON-2**

To select a string of characters: Point to first character with mouse
                                 Push **MOUSE-BUTTON-1**
                                 Point to last character with mouse
                                 Push **MOUSE-BUTTON-3**

To select a string of words: Point to first word with mouse
                                 Push **MOUSE-BUTTON-2**
                                 Point to last word with mouse
                                 Push **MOUSE-BUTTON-3**

Figure 6.1. Subset of BRAVO editor commands.

As a vehicle for discussing these extensions, we sketch a GOMS model simulation of a user for the display-oriented editor BRAVO, one of the editors tested in Chapter 3. This editor is similar to POET in command structure (see Figure 6.1), but uses a mouse for selection of text on a full-page video display (30~50 lines of text are displayed on the page, depending on the typefont used and the spacing between lines). Stochastic predictions are derived from the model by assuming probability distributions for operator times and method choices, then running Monte Carlo simulations. The expanded task analysis is accomplished by having the model operate on a symbolic representation of the manuscript instructions.

6.1. TASK ANALYSIS

The purpose of a task analysis is to map out the constraints imposed on behavior by the nature and features of the task environment. Here we add two pieces of task analysis to that already developed in Chapter 5: (1) a description of the elements of knowledge a user can have about the editing tasks he is to do and (2) a partial description of the physical environment.

Editing Tasks

What information does the user know about the editing task he performs and when does he know it? Take, for example, Task A2 of Figure 6.2. The instructions marked on the manuscript indicate that the character "a" is to be inserted as a word in front of the word "necessary." At some point during the execution of this task, the user must know that the task is an insertion, where the insertion point is, what new text is to be inserted, and, perhaps, information about where the task is, relative to other tasks on the page (this knowledge may take the form of: the task is the second task on the page; it is on line 12; it is before Task A3; it is after Task A1). To keep track of these bits and pieces of information, we can represent the user's knowledge as a network, with chunks as nodes and relations between the chunks as links. This is done for Task A2 in Figure 6.3. Of course, at any moment the user might actually possess only part of the knowledge indicated in the diagram, some of the links or nodes being missing. The diagram shows the maximum knowledge we

Chapter I: INTRODUCTION

While the official, chartered purpose of this Subcommittee on Data Base Management systems is to investigate the potential for standardization in the area of data base management systems, a necessary first step of the work of the Subcommittee has been the development of a set of requirements for effective data base management systems. These requirements have emerged as the work of the Subcommittee ~~manifested~~ proceeded and have themselves in the form of a generalized model for the description of data base management systems. As no existing or proposed implementation of a data base management system completely satisfies these requirements nor comprises all of the concepts involved, a preliminary to any necessary discussion of standards is an explanation of this model. The bulk of this Report provides such an explanation.

As a preliminary, it is appropriate to discuss briefly the order of events that has led to this report ~~within reason to explain what events led to the preparation of this document.~~ Among the responsibilities of the Specifications Planning and Requirements Task Force of the Ad Hoc Marketing Committee for Computers and Information Processing is the generation of recommendations for action by the parent Task Force on appropriate areas for the initiation of specifications development efforts. For some time, starting in about 1969, the task force has been aware that data base management systems are becoming central elements of information processing systems, and that there is less than full agreement in the community on appropriate design. In addition to the existence of a number of implementations of such systems, a list that continues to grow, there are several documents generated out of the collective wisdom of some segment of the information processing community which are either proposals for specific systems (SMITH 1971) or more general statements of requirements (JAYME 1970), (HO 1971). As is well known, there is a debate in the community on whether existing and proposed implementations meet the indicated requirements, or whether the requirements as drawn are all really necessary ~~and entirely useful.~~ Further, there have been serious questions about the economics of systems meeting all the stated requirements.

Figure 6.2. Sample page of a marked-up manuscript.

The labels A1 through A5 identify the unit tasks for this study. The user did not see these task labels.

presume it is possible for the user to have about the task, regardless of whether he actually has it at a given instant.

Figure 6.3 is a diagram for a specific task. We can also describe some of the general editing concepts possessed by the user by making a distinction between (1) general notions, such as the general notion of an **# INSERTION**, and (2) particular cases that exemplify the general notions, such as the Task **A2**, which is a particular instance of an **# INSERTION** task. The general notions are called *concepts* (notationally we begin concept names with a **#** to distinguish them), and the particular cases that exemplify the concepts are called *exemplars.* Concepts are defined by *schemata,* which give the *attributes* and *values* the exemplars of a concept may have and the higher-level *superconcepts* to which a concept is related.

Figure 6.4 shows the schemata for the general concepts used in Figure 6.3 and their relationships to the parts of Task **A2**: The exemplar **A2** has as its concept **# INSERTION**. The concept **# INSERTION** has as its superconcept the concept **# BASIC-TASK**. The concept **# BASIC-TASK** has an attribute **LINE-NUMBER:**, whose value is some (unknown) exemplar of the concept **# INTEGER**. Therefore **A2** (which is also an exemplar of the concept **# BASIC TASK**) also has an attribute **LINE-NUMBER:**, with a value **12** (which, in turn, is an exemplar of the concept **# INTEGER**).

The diagram in Figure 6.4 also shows the relationship between other parts of the exemplar **A2** and the schemata of the concepts it references. The relationships quickly become complex. Whereas such diagrams are illuminating for small networks of knowledge, they rapidly become unreadable (and undrawable) as the number of elements increases. It is therefore necessary to use a text-language notation for a description of any complexity. The text-language equivalent for **A2** in Figure 6.3 is:

$$
\begin{array}{llll}
\textbf{A2} = \textbf{\# INSERTION} & (& \textbf{FUNCTION:} & \textbf{INSERT} \\
& & \textbf{INSERTION-POINT:} & \lambda_4 \\
& & \textbf{NEW-TEXT:} & \chi_2 \\
& & \textbf{REL-TASK-NO:} & \textbf{2} \\
& & \textbf{LINE-NO:} & \textbf{12} \\
& & \textbf{PREVIOUS:} & \textbf{A1} \\
& & \textbf{NEXT:} & \textbf{A3}\ )\ ,
\end{array}
$$

where

$$
\begin{array}{llll}
\chi_2 = \textbf{\# CHARACTER} & (& \textbf{TEXT-TYPE:} & \textbf{CHARACTER} \\
& & \textbf{BOUNDARY:} & \textbf{WORD} \\
& & \textbf{LENGTH:} & \textbf{1}\ )\ .
\end{array}
$$

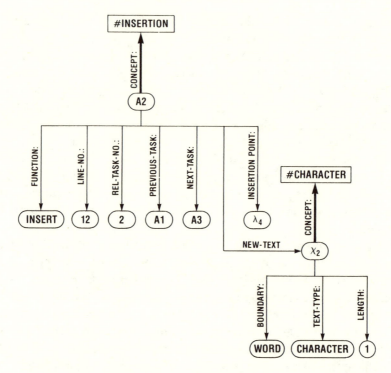

Figure 6.3. Symbolic representation of Task A2.

This can be read as "**A2** is an exemplar of the concept **#INSERTION**, with **FUNCTION: INSERT** and **INSERTION-POINT:** λ_4 and ...", or more succinctly, "**A2** is an **#INSERTION** with ...". The text-language definitions for the schemata that define the user's concepts describe the space of editing tasks addressed by the model (see Figure 6.5). For purposes of our simulation, we are concerned with only the common sort of text manipulations, such as those in Figure 6.2, excluding formatting tasks, such as specifying typefonts or leading between text lines. The tasks we consider are: (1) insertion of new text, (2) deletion of old text, (3) replacement of old text by new text, (4) movement of text to a new location, and (5) transposition of two adjacent pieces of text.

Given the schemata in Figure 6.5, the tasks on the manuscript page in Figure 6.2 can now be described (see Figure 6.6). Figure 6.6 is an example of the symbolic description of editing tasks we use as input to our simulation.

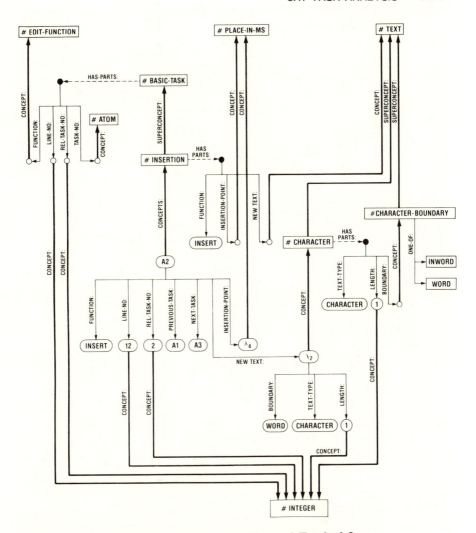

Figure 6.4. Symbolic representation of Task A2.

This representation shows the relationship between the exemplars and the concepts of which they are instances.

Physical Environment

We have described the logical elements of the editing task, but there are also physical elements of the user interface that affect the user's behavior. The video display might not be legible, for instance, or the

```
concept # BASIC-TASK
        SUPERCONCEPT: NIL
        HAS-PARTS:      (TASK-NO:           {a  # ATOM}
                        REL-TASK-NO:        {a  # INTEGER}
                        LINE-NO:            {a  # INTEGER}
                        FUNCTION:           {a  # EDIT-FUNCTION})

concept # TASK
        IS-ONE-OF:      ({a  # DELETION}
                        {a  # INSERTION}
                        {a  # REPLACEMENT}
                        {a  # TRANSPOSITION})

concept # DELETION
        SUPERCONCEPT: # BASIC-TASK
        HAS-PARTS:      (FUNCTION:           DELETE
                        OLD-TEXT:           {a  # TEXT-IN-MS})

concept # INSERTION
        SUPERCONCEPT: # BASIC-TASK
        HAS-PARTS:      (FUNCTION:           INSERT
                        INSERTION-POINT:    {a  # PLACE-IN-MS}
                        NEW-TEXT:           {a  # TEXT})

concept # REPLACEMENT
        SUPERCONCEPT: # BASIC-TASK
        HAS-PARTS:      (FUNCTION:           REPLACE
                        NEW-TEXT:           {a  # TEXT}
                        OLD-TEXT:           {a  # TEXT-IN-MS})

concept # MOVE
        SUPERCONCEPT: # BASIC-TASK
        HAS-PARTS:      (FUNCTION:           MOVE
                        OLD-TEXT:           {a  # TEXT-IN-MS}
                        INSERTION-POINT:    {a  # PLACE-IN-MS})

concept # TRANSPOSITION
        SUPERCONCEPT: # BASIC-TASK
        HAS-PARTS:      (FUNCTION:           TRANSPOSE
                        LEFT-TEXT:          {a  # TEXT-IN-MS}
                        RIGHT-TEXT:         {a  # TEXT-IN-MS})

concept # BOUNDS
        HAS-PARTS:      (START:             {a  # PLACE-IN-MS}
                        END:                {a  # PLACE-IN-MS})

concept # CHARACTER
        SUPERCONCEPT: # TEXT
        HAS-PARTS:      (TEXT-TYPE:          CHARACTER
                        BOUNDARY:           {a  # CHARACTER-BOUNDARY}
                        LENGTH:             1)
```

```
concept # CHARACTER-BOUNDARY
        IS-ONE-OF:      (IN-WORD
                        WORD)

concept # CHARACTER-IN-MS
        SUPERCONCEPT: # CHARACTER
        HAS-PARTS:      (LOCATION:       {a  # PLACE-IN-MS})

concept # TEXT
        IS-ONE-OF:      ({a  # WORD}
                        {a  # CHARACTER}
                        {a  # TEXT-SEG})

concept # TEXT-IN-MS
        IS-ONE-OF:      ({a  # WORD-IN-MS}
                        {a  # CHARACTER-IN-MS}
                        {a  # TEXT-SEG-IN-MS})

concept # TEXT-SEG
        SUPERCONCEPT: # TEXT
        HAS-PARTS:      (TEXT-TYPE:       TEXT-SEG
                        LENGTH:          {a  # INTEGER}
                        BOUNDARY:        {a  # TEXT-SEG-BOUNDARY})

concept # TEXT-SEG-BOUNDARY
        IS-ONE-OF:      (LINE
                        SPLIT-LINES
                        SPLIT-PAGES)

concept # TEXT-SEG-IN-MS
        SUPERCONCEPT: # TEXT-SEG
        HAS-PARTS:      (START-LOC:       {a  # PLACE-IN-MS}
                        END-LOC:         {a  # PLACE-IN-MS})

concept # WORD
        SUPERCONCEPT: # TEXT
        HAS-PARTS:      (TEXT-TYPE:       WORD
                        BOUNDARY:        WORD
                        LENGTH:          {a  # INTEGER})

concept # WORD-IN-MS
        SUPERCONCEPT: # WORD
        HAS-PARTS:      (LOCATION:        {a  # PLACE-IN-MS})
```

Figure 6.5. Concepts that define the space of editing tasks addressed by the simulation model.

```
PAGE-1 = # PAGE        ( PAGE-NO:            1
                         TASKS:              (LIST: A1 A2 A3 A4 A5)
                         NEXT:               PAGE-2 )

A1 = # MOVE            ( FUNCTION:           MOVE
                         OLD-TEXT:           $\chi_1$
                         INSERTION-POINT:    $\lambda_1$
                         REL-TASK-NO:        1
                         LINE-NO:            8
                         NEXT:               A2 )

A2 = # INSERTION       ( FUNCTION:           INSERT
                         INSERTION-POINT:    $\lambda_4$
                         NEW-TEXT:           $\chi_2$
                         REL-TASK-NO:        2
                         LINE-NO:            12
                         PREVIOUS:           A1
                         NEXT:               A3 )

A3 = # INSERTION       ( FUNCTION:           INSERT
                         INSERTION-POINT:    $\lambda_5$
                         NEW-TEXT:           $\chi_3$
                         REL-TASK-NO:        3
                         LINE-NO:            12
                         PREVIOUS:           A2
                         NEXT:               A4 )

A4 = # REPLACEMENT     ( FUNCTION:           REPLACE
                         NEW-TEXT:           $\chi_4$
                         OLD-TEXT:           $\chi_5$
                         REL-TASK-NO:        4
                         LINE-NO:            16
                         PREVIOUS:           A3
                         NEXT:               A5 )

A5 = # DELETION        ( FUNCTION:           DELETE
                         OLD-TEXT:           $\chi_6$
                         REL-TASK-NO:        5
                         LINE-NO:            33
                         PREVIOUS:           A4
                         NEXT:               NIL )
```

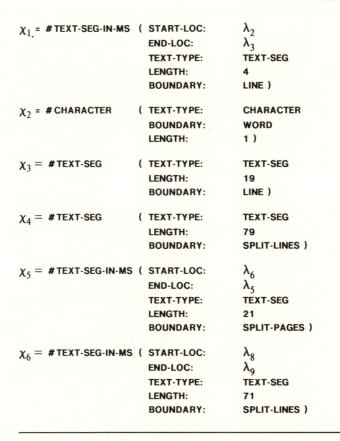

Figure 6.6. Symbolic representation of the manuscript page shown in Figure 6.2.

user might not be able to see certain information not on the current page of the manuscript. Four main entities of the physical environment are of interest to us: the *user*, the *editor* (including its input devices, the keyboard and mouse), the editor's *video display*, and the marked-up *manuscript*. We take the point of view that a description of this environment should permit any of these elements to be altered (the page of the manuscript to be changed, for example) without altering the description of the other three. A technique (based on the simulation language Smalltalk: Kay, 1977, and Ingalls, 1978) that will permit such a separation is to describe the physical environment in terms of a set of *transactions* between these entities, each transaction consisting of a message and its reply, if it has one (see Figure 6.7). The user's act of consulting the manuscript to get the next task after Task A1, for example, is implemented in the simulation by having the model of the user send the message *READ-NEXT-LOCATION: to the manuscript and having the

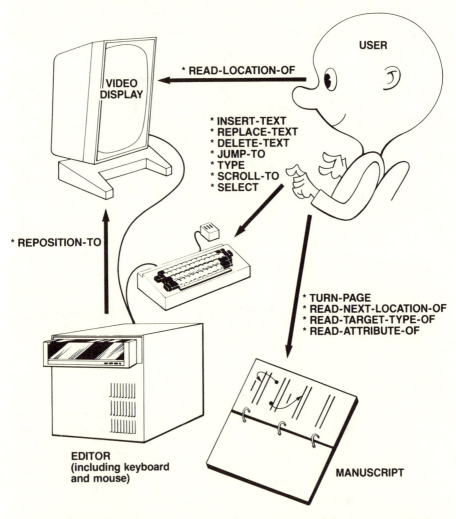

Figure 6.7. Analysis of the physical environment as entities and transactions.

manuscript reply with the message λ_4, where λ_4 is a symbol denoting the physical location of Task A2.

There are four types of transactions:

(1) The user consults the manuscript to find a new task or to discover more details about the current task (User $\Rightarrow$ Manuscript transactions, where the symbol $\Rightarrow$ shows that the user initiates a transaction to the editor).

(2) The user issues commands to the editor (User $\Rightarrow$ Editor transactions).

(3) The editor changes the display (Editor $\Rightarrow$ Display transactions).

(4) The user consults the display to locate a piece of text (User $\Rightarrow$ Display transactions).

Two entities, the user and the editor, are active, able to initiate transactions. The two other entities, the manuscript and the display, are passive, only replying to messages sent them. In the simulation model, each of these four entities is represented as a separate process, interacting via the transactions.

We can describe the physical environment (according to this model) by listing the model transactions available between the entities in the environment. Return messages, when they exist, are listed following the symbol $\rightarrow$.

User $\Rightarrow$ Editor Transactions:

> *INSERT-TEXT
> *REPLACE-TEXT
> *DELETE-TEXT
> *JUMP-TO ({a #TASK})
> *TYPE ({a #TEXT})
> *SCROLL-TO ({a #TASK})
> *SELECT .

These transactions reflect the commands available in the editor BRAVO. *INSERT-TEXT is the insertion command and denotes the command portion of the interaction (typing the key I to begin the command and ESC to terminate it). The actual typing of the text to be inserted is denoted by the TYPE transaction. The full series of keys that the user would actually type to insert the letter "a" in Task A2 is

I A SPACE ESC .

In the simulation, this action would be modeled as sending two messages from User ⇒ Editor:

> *INSERT-TEXT
> *TYPE (A SPACE).

*REPLACE-TEXT is the command to replace some text with other text, and *DELETE-TEXT is the command to remove text. *SCROLL-TO is the command to reposition ("scroll") the text on the display so that the cursor, controlled by the mouse, is at the top of the display. *JUMP-TO is similar to the scroll command, except that the text to be positioned at the top of the display is specified by a search string. *SELECT is the command (invoked by pressing a button on the mouse) to make the text indicated by the cursor the current selection.

The expression {a #TASK} should be read: "an exemplar of the concept #TASK."

Editor ⇒ Display Transaction:
> *REPOSITION-TO ({a #TASK}).

Only one editor ⇒ display transaction is modeled: repositioning the display in response to a *SCROLL-TO or *JUMP-TO transaction from the user.

User ⇒ Display Transactions:
> *READ-LOCATION-OF ({a #TASK})
> → {a #MAIN-PART-OF-SCREEN} or
> {a #BOTTOM-PART-OF-SCREEN} or
> {a #OFF-SCREEN}.

When the user acts to find the text of a task on the video display, he looks at the display and searches for the text. We describe this activity in terms of the model as the user sending a message *READ-LOCATION to the display and the display making a reply giving the location of the task. The exact location of the text on the screen is of little use in predicting the user's performance and is below the grain of the model. What is important is whether the text is in the main (middle to top) part of the screen, in the bottom part, or not on the screen at all. The user has his own internal representation of where the task is, which may or may not correspond with the display's state.

User ⇒ *Manuscript Transactions:*

*TURN-PAGE ({a #DIRECTION})
  → OK or
    NO-MORE-PAGES
*READ-NEXT-LOCATION-OF ({a #TASK})
  → {a #TASK} or
    NO-MORE-TASKS-THIS-PAGE
*READ-TARGET-TYPE-OF ({a #TASK})
  → INSERTION-POINT or
    {a #CHARACTER} or
    {a #WORD} or
    {a #TEXT-SEG}
*READ-ATTRIBUTE-OF ({a #TASK}, {a #TASK-ATTRIBUTE})
  → {an #ATTRIBUTE}

According to these transactions, the user can turn the pages of a manuscript either forward or backward. If he tries to turn past the last page, he discovers immediately that there are no more pages. The user can look for the next task on the manuscript, he can note what sort of target he must select, or he can read the new text that is to be inserted or other attributes of a task.

All the interactions between the display, the editor, the manuscript, and the user are described in terms of the listed transactions. The strict partitioning of the physical environment into independent processes that communicate through messages reflects the structure of the physical environment itself. If the manuscript is changed, for example, the other entities should work as before.

6.2. MODEL OF THE USER

We have thus far described the task environment, the editing task and its physical surroundings. Using the results of our task analysis, we can now set out on a general GOMS analysis of a BRAVO user. But before the GOMS model can be completed, it is necessary to augment our task analysis with the details of how a user consults the manuscript for information and how he scrolls the display. This latter must be determined by observation of users. Consequently, we first sketch an outline for the model, delaying the full presentation until observational data have been presented.

General GOMS Analysis

To specify a GOMS model for a user we must, as discussed in Chapter 5, specify goals, operators, methods, and selection rules. Figure 6.8 lists these elements of a GOMS model for a BRAVO user. The goals are organized into a hierarchy as pictured in Figure 6.9. Eventually each goal (a rectangular box in the figure) terminates on a set of operators (rounded boxes in the figure). Associated with each goal is a set of alternative methods (not shown in the Figure) by which the goal can be achieved and a set of selection rules (also not shown) for selecting among the methods.

GOALS

As in Chapter 5, the user is assumed to have a top-level goal to edit the manuscript one unit task at a time:

> GOAL: EDIT-MANUSCRIPT
> . GOAL: EDIT-UNIT-TASK
> . . GOAL: ACQUIRE-UNIT-TASK .

The accomplishment of GOAL: EDIT-UNIT-TASK is again broken into an acquisition part and an execution part, just as in the POET model; in the simulation, though, EXECUTE is treated as an operator that causes a subgoal to be set up for the task, based on the instructions acquired from the manuscript during GOAL: ACQUIRE-UNIT-TASK. The subgoal to be set up is one of the following:

> GOAL: REPLACE or
> GOAL: DELETE or
> GOAL: INSERT (InsertionPoint, NewText) or
> GOAL: MOVE (InsertionPoint, OldText) .

Unlike the simpler POET model, some of the goals are parameterized. For example, the goal

$$\text{GOAL: INSERT ({a \# APPROXIMATE-TARGET}, } X_2) \tag{6.1}$$

represents the user's goal to insert, in an approximately-known location, the text described by X_2, where X_2 is (as given earlier) a single character of text on a word boundary:

$$X_2 = \text{\# CHARACTER} \quad (\quad \text{TEXT-TYPE:} \quad \text{CHARACTER}$$
$$\text{BOUNDARY:} \quad \text{WORD}$$
$$\text{LENGTH:} \quad 1 \quad) \, .$$

The parameters for goals and operators are listed in Figure 6.8. These parameters are the memory chunks that must be maintained in Working Memory at the time the goal is executed. Each chunk is given a symbolic slot name (for accounting purposes) in the model. In Expression 6.1, InsertionPoint is the slot name associated in the model with an exemplar of **# APPROXIMATE-LOCATION**, and NewText is the slot name associated with X_2. Expression 6.1 is short for:

GOAL: INSERT (InsertionPoint = {a **# APPROXIMATE-TARGET**},
NewText = X_2) .

The precise form in which the user has these pieces of knowledge represented in memory, however, is not specified.

Also, unlike the POET model, the BRAVO model must contain a set of goals related to the use of the mouse for selecting text:

GOAL: SELECT-TARGET (MSPosition, PositionType,
VisualSearchTarget),
. GOAL: POINT-TO-TARGET (MSPosition, VisualSearchTarget,
Select?)
. . GOAL: POINT-THERE (ScreenPosition, TextType, Select?) .

OPERATORS

The operators are cast roughly at the Argument Level. For example, the user can gather information from the manuscript and the display:

GET-FROM-MANUSCRIPT (DesiredInformation, Attribute) and
GET-FROM-DISPLAY (DesiredInformation, Attribute, MSPosition) .

He can point at a certain TextType in a certain ScreenPosition, then (optionally) select it (notationally, the Select? parameter takes on the value SELECT) by pressing a button on the mouse:

POINT (ScreenPosition, TextType, Select?) .

Goals:

GOAL: EDIT-MANUSCRIPT
GOAL: EDIT-UNIT-TASK
GOAL: ACQUIRE-UNIT-TASK
GOAL: INSERT (InsertionPointKey, NewText)
GOAL: DELETE (OldTextKey)
GOAL: REPLACE (OldTextKey, NewText)
GOAL: MOVE (InsertionPointKey, OldTextKey)
GOAL: SELECT-TARGET (MSPosition, PositionType, VisualSearchTarget)
GOAL: POINT-TO-TARGET (MSPosition, VisualSearchTarget, Select?)
GOAL: POINT-THERE (ScreenPosition, TextType, Select?)

Operators:

GET-FROM-MANUSCRIPT (DesiredInformation, Attribute)
GET-FROM-DISPLAY (DesiredInformation, Attribute, MSPosition)
SCROLL-TO (LineInMS)
JUMP-TO (LineInMS)
POINT (ScreenPosition, TextType, Select?)
INSERT-TEXT
DELETE-TEXT
REPLACE-TEXT
TYPE (NewText)
EXECUTE (Task)
VERIFY-EDIT

Methods:

ONE-AT-A-TIME-METHOD
ACQUIRE-EXECUTE-VERIFY-METHOD
READ-TASK-IN-MS-METHOD
INSERT-COMMAND-METHOD
DELETE-COMMAND-METHOD
REPLACE-COMMAND-METHOD
DELETE-INSERT-METHOD
ZERO-IN-METHOD
ROUGH-POINT-METHOD
CHAR-POINT-METHOD
WORD-POINT-METHOD
TEXT-SEG-POINT-METHOD
INSERTION-POINT-METHOD
POINT-WITHOUT-SCROLLING-METHOD
SCROLL-AND-POINT-METHOD
JUMP-METHOD

Selection Rules:

ROUGH-LOC-RULE
TEXT-SEG-RULE
CHAR-POINT-RULE
WORD-POINT-RULE
INSERTION-POINT-RULE
TOP-2/3-RULE
BOTTOM-1/3-RULE
OFF-SCREEN-RULE

Figure 6.8.   Outline of the GOMS model for a BRAVO user.

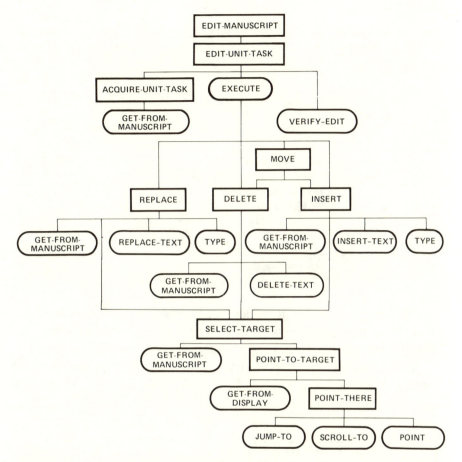

Figure 6.9. Hierarchy of goals and operators in the GOMS model for BRAVO.

Goals are shown as square boxes and operators as round boxes.

And he can issue commands to the editor:[1]

SCROLL-TO (LineInMs)
JUMP-TO (LineInMs)
INSERT-TEXT
DELETE-TEXT
REPLACE-TEXT
TYPE .

[1] These user operators should not be confused with the user ⟹ editor transactions with similar names. For example, **INSERT-TEXT** is a user *operator* that eventually causes the *transaction* ***INSERT-TEXT** to occur.

METHODS

The simulation contains the methods for particular goals, expressed in a formal notation. Some of the methods are essentially the same as for the editor POET in Chapter 5:

> **Method for GOAL: EDIT-MANUSCRIPT**
> **ONE-AT-A-TIME-METHOD =**
> **until NoMorePages = TRUE do GOAL: EDIT-UNIT-TASK .**

> **Method for GOAL: EDIT-UNIT-TASK**
> **ACQUIRE-EXECUTE-VERIFY-METHOD =**
> **GOAL: ACQUIRE-TASK**
> **EXECUTE (Task)**
> **with-probability .4 do VERIFY-EDIT .**

> **Method for GOAL: ACQUIRE-UNIT-TASK**
> **READ-TASK-IN-MS-METHOD =**
> **GET-FROM-MANUSCRIPT ({slot Task}) .**[2]

The first method breaks the manuscript into unit tasks, the second breaks a unit task into the Acquire-Execute-Verify cycle, and the third acquires the instructions for a task by reading them from the manuscript.

Other methods, such as the method for performing an insertion, are more detailed than in Chapter 5, taking into account the display-oriented pointing operations:

> **Method for GOAL: INSERT**
> **INSERT-COMMAND-METHOD =**
> **if no InsertionPoint**
> **then GET-FROM-MANUSCRIPT ({slot InsertionPointKey})**
> **GOAL: SELECT-TARGET(MSPosition, {slot InsertionPoint},**
> **InsertionPointKey)**
> **INSERT-TEXT**
> **if no NewText then GET-FROM-MANUSCRIPT ({slot NewText})**
> **if NewText ≠ DEFAULT then TYPE (NewText) .**

According to this method, if the user does not know where to make the insertion, he looks over to the manuscript to find out. Then he selects the location he found and issues the Insert command to the editor. If he

---

[2] If we think of the upper/lower case symbols, such as **Task**, as variables representing a pointer to particular types of information held in Working Memory, the expression **{slot Task}** means the pointer itself rather than its contents.

cannot remember the text to be inserted, he consults the manuscript. Finally, the user types in the new text (except in the special "default" case, where the text to be inserted is the argument to a previous command, such as a Delete command).

SELECTION RULES

Just as in the simulation model, the selection rules for choosing among methods available for a particular goal are expressed in a formal notation. A simple example is **GOAL: POINT-TO-TARGET**. There are at least three methods in BRAVO to accomplish this goal, depending on the kind of target: (1) to select a character, the user moves the mouse (to position the cursor at the character) and presses the first mouse button; (2) to select a word, he moves the mouse (to position the cursor at any part of the word) and presses the second button; (3) to select a text segment, the user first does either (1) or (2) to point to the beginning of a the text segment and then moves the mouse to point to the end of the segment and presses the third button to select all text between the two points. The corresponding selection rules are written:

> **Selection rules for GOAL: POINT-TO-TARGET**
> **CHAR-POINT-RULE =**
>   if VisualSearchTarget isa #CHARACTER
>   then CHOOSE (CHAR-POINT-METHOD)
> **WORD-POINT-RULE =**
>   if VisualSearchTarget isa #WORD
>   then CHOOSE (WORD-POINT-METHOD)
> **TEXT-SEG-RULE =**
>   if VisualSearchTarget isa #TEXT-SEG
>   then CHOOSE (TEXT-SEG-POINT-METHOD) .

The expression VisualSearchTarget isa #CHARACTER is true if the VisualSearchTarget is an exemplar of the concept #CHARACTER (or the exemplar of any concept that has #CHARACTER as its superconcept).

Observational Studies

The modeling of two local sequences of user behavior requires additional empirical observation. First, there is the question of how frequently a user will consult the manuscript during a task? This is not a new question. Difficulty in predicting the answer lowered the accuracy of the POET models in Chapter 5. Second, there is the question of when will the user scroll the display? This is a new question that derives from the user's interactions with the video display in BRAVO. Fortunately, answers to both questions can be derived from a re-examination of the videotaped protocols of the dedicated BRAVO users in Experiment 3B.

CONSULTING THE MANUSCRIPT

The user usually consults the manuscript several times during the course of a task. How frequently does he look? What does he look for? How can methods be written to describe this process?

To answer these questions, 40 instances were observed of the GET-FROM-MANUSCRIPT operator (as performed by one user, S13, during the course of 16 insertion tasks). Three different kinds of information that the user sought from the manuscript could be identified: Getting the instructions for the Next Task (GNT), Getting the Location of the task on the manuscript (GL), and Getting the New text to be inserted (GN). Of course, from a single glance at the manuscript the user often acquires more than a single piece of information. Figure 6.10 shows the inferred distribution of reasons for S13's consultation of the manuscript during insertion tasks, grouped by number of characters in the inserted string. Each row in the table describes a separate task. The three middle columns tally how many times S13 looked at the manuscript for each task for each reason.

On Task A1, for example, the user consulted the manuscript once at the beginning of the task. Since she proceeded to point at the target and then to insert the new text without further consultations, she must also have obtained the information for each of these operators on that first consultation.

On Task A18, she consulted the manuscript once at the beginning of the task, then twice more before finally pointing to the target, and a fourth time before beginning to type the new text. While typing the new text, she looked at the manuscript, but glanced at the keyboard twice more. From the first consultation, she probably learned the approximate location of the task and the operation to be performed. On the second look, she probably obtained a better, but still approximate, location for the target insertion point. On the third look, she must have learned the exact target position. And on the fourth look, she probably got the

Task ID	$N_{chars.}$	Reason for Looking at Manuscript			Total	While-typing
		GNT	GL	GN		
A1	1	1			1	
A21	1	1		1	2	
B8	1	1	3	1	5	
B26	1	1			1	
A6	4.7	1	1		2	
A32	4.7	1	2		3	
B2	4.7	1			1	
B23	4.7	1			1	
A3	18.2	1		1	2	
A14	18.2	1	1	1	3	1
B1	18.2	1	1	1	3	
B16	18.2	1	1	2	4	1
A18	75	1	2	1	4	2
B6	75	1	1	1	3	2
A30	522	1	1	1	3	4
B10	522	1			1	7

Figure 6.10. Frequency of manuscript consultations.
Each row of the table describes a different unit task. All tasks are insertion tasks, and they are grouped by the number of characters $N_{chars.}$ being inserted. The number of consultations of the manuscript are tallied by reason for consultation: GNT = Get Next Task, GL = Get Location, GN = Get New text. "Total" is the column sum of GNT + GL + GN columns. "While-typing" is the number of times the user consulted the manuscript while typing.

beginning of the text to be inserted. At this point she proceeded to type while watching the manuscript, taking small glances back to the display or keyboard to check for suspected errors or to locate different keys (cf. Long, 1976). Consultations of the manuscript while typing text passages are tallied in the "While-typing" column in Figure 6.10. These GET-FROM-MANUSCRIPT operations overlap with the TYPE operation and can be ignored for the present analysis.

S13's procedure for locating a target on the display (reflected in the GL column in Figure 6.10) is especially interesting. First, she extracts a few words from the manuscript to use as a visual search target. The words may be either the exact target or an approximate target in the form of some other words or characters. In either case, she points to the visual search target she has extracted from the manuscript. If the visual search target is only an approximate target, she does not select it, but looks over to the manuscript again and repeats the procedure. Otherwise, she selects the visual search target and moves on to the next step of the task. This method of locating the target is called the ZERO-IN-METHOD and is described as follows:

> **Method for GOAL: SELECT-TARGET**
> **ZERO-IN-METHOD =**
> **while VisualSearchTarget isa # APPROXIMATE-TARGET**
> **do POINT-TO-TARGET (MSPosition, VisualSearchTarget,**
> **DON'T-SELECT)**
> **GET-FROM-MANUSCRIPT ({slot VisualSearchTarget},**
> **PositionType)**
> **finally POINT-TO-TARGET (MSPosition, VisualSearchTarget,**
> **SELECT) .**

VisualSearchTarget is the identifying visual search target extracted from the manuscript by the user. MSPosition represents the user's memory for which task she is doing. PositionType identifies which of several possible targets she is considering (for example, a move task has an InsertionPoint and an OldText).

Although it is not known for any task how many times GET-FROM-MANUSCRIPT will be invoked in succession (and in an engineering analysis, a prediction would usually need to be done in the absence of a particular manuscript), the numbers in the GL column of Figure 6.10 are well approximated by a Poisson distribution of mean .81 (see Figure 6.11). This fact tells us that the GET-FROM-MANUSCRIPT operator in the simulation should be constructed so as to pick up approximate targets (as opposed to exact targets) in such proportion that the number of iterations will be Poisson distributed with the above mean.

SCROLLING THE DISPLAY

Before a user can make a modification with BRAVO, he must get the task onto the screen using either the SCROLL-TO or JUMP-TO commands. Even if the task is already on the screen, the user sometimes prefers to move it closer to the top (nearer to eye level), which he does by scrolling.

Number of GL Operations in a Unit Task, N	Frequency	
	Observed	Predicted $16(.81^N e^{-.81}/N!)$
0	7	7.1
1	6	5.8
2	2	2.3
3	1	.6
4	0	.1

Figure 6.11.   Comparison of the observed number of GL operations in a unit task with the number predicted by the Poisson distribution.

Data are from GL column of Figure 6.10. GL stands for Get Location, or more strictly, GET-FROM-MANUSCRIPT({slot Task}), which obtains the location of the task.

Thus, on a given task, the user may or may not scroll the text on the screen. How can a set of selection rules be written that will predict the user's choice?

For a detailed examination of scrolling, S13's performance on all the tasks in the first half of the manuscript were examined. For each task, the following were recorded: the number of lines from the top of the screen to the target, whether her move repositioned the text on the display, and what method she used.

Selection Based on Manuscript Positions.   Figure 6.12 shows the number of times the user adopted each of these methods as a function of the distance of the target from the top of the screen. The selection rules used by S13 may be simply expressed: If the target is in the top two-thirds of the screen, do not reposition the screen; if the target is in the bottom third of the screen, scroll; and if the target is not on the screen, use the jump command (defining the **#MAIN-PART-OF-SCREEN** as the first 19 lines, and the **#BOTTOM-PART-OF-SCREEN** as lines 20 to 24). In our notation, this can be written:

Selection rules for GOAL: POINT-THERE
>   **TOP-2/3-RULE =**
>>   **if ScreenPosition isa #MAIN-PART-OF-SCREEN**
>>   **then CHOOSE (POINT-WITHOUT-SCROLLING-METHOD)**

	Number of Lines from Top of Screen	Method		
		PWSM	SAPM	JM
On Screen	1-4	6		
	5-8	9		
	9-12	3		
	13-16	2	1	
	17-20	1	2	
	21-24		2	1
Off Screen	25-28		1	1
	29-32			0
	33-36			1
	37-40			3

Figure 6.12.   Frequency of alternative methods for the GOAL: POINT-THERE as a function of distance of the target from the top of the screen.
The methods are abbreviated as follows: PWSM = POINT-WITHOUT-SCROLLING-METHOD, SAPM = SCROLL-AND-POINT-METHOD, JM = JUMP-METHOD.

```
BOTTOM-1/3-RULE =
      if ScreenPosition isa # BOTTOM-PART-OF-SCREEN
         then CHOOSE (SCROLL-AND-POINT-METHOD)
OFF-SCREEN-RULE =
      if ScreenPosition isa # OFF-SCREEN
         then CHOOSE (JUMP-METHOD) .
```
(6.2)

These rules predict the user's method choices 85% of the time.

Why did the user go to the expense of scrolling the display, simply because the target was in the bottom third? The answer is apparently that the bottom third of the screen was outside her comfortable vision zone. Normal comfortable vision is about 15° below the horizon, which, with the comfortable head inclination of about 20°, gives a total of 35° (Van Cott and Kinkade, 1972, p. 393; Cakir, Hart, and Stewart, 1980, p. 171). The bottom third of the screen was probably (on the basis of later measurements) outside this comfort region. The mismatch of screen height with user, a common phenomenon, apparently can reduce effective screen size and, consequently, time-efficiency by causing more scrolling operations.

The set of selection rules above (6.2) has the advantage that it makes clear the mechanism whereby S13 makes her choices.   It has the disadvantage that it demands knowledge of the state of the screen at any arbitrary point in the editing process.   This disadvantage could be overcome if the selection rules did not demand such detailed knowledge of the dynamics of the situation.

Selection Rules Based on Manuscript Positions. The number of lines d between tasks on the manuscript is easily determined by inspection of the manuscript alone.   Figure 6.13 shows the method selections of four users as a function of d, the distance (in lines) from the site of the previous unit tasks).   A set of selection rules based on d is as follows:

> **Selection rules for GOAL: POINT-THERE**
> **LITTLE-d-RULE =**
>   if $d \leq 16$
>   then CHOOSE (POINT-WITHOUT-SCROLLING-METHOD)
> **MEDIUM-d-RULE =**
>   if $16 < d \leq 25$
>   then CHOOSE (SCROLL-AND-POINT-METHOD)
> **BIG-d-RULE =**
>   if $d > 25$
>   then CHOOSE (JUMP-METHOD) .   (6.3)

These rules correctly predict 85% of S13's method selections, the same percentage as the rules based on screen positions (Rule Set 6.2).

Selection Rules for Other Users. The results for selection rules based on manuscript positions (Rule Set 6.3) encourage us to use the manuscript distance between tasks as the measure by which to examine the behavior of other users, to see how stable these rules are across users. Figure 6.13 shows the frequency with which three other users in Experiment 3B used the different pointing methods. The main difference between these users and S13 is that they do not have the JUMP-METHOD in their repertoires.   All three switch from the POINT-WITHOUT-SCROLLING-METHOD to the SCROLL-AND-POINT-METHOD as the distance between tasks increases.   The crossover point, at which users switched from one to the other of these methods, varied from a distance of 4 lines between targets to a distance of 11 lines. The following rules characterize their selections:

Distance from Previous Task d	S13			S32		S34		S37	
	PWSM	SAPM	JM	PWSM	SAPM	PWSM	SAPM	PWSM	SAPM
0	12			11	2	13		12	1
1	16			15	1	15	1	15	1
4	11	2		10	3	8	5	7	6
16	3	6	7	4	11	2	13	2	13
32		7	7	1	6		7	3	4

Figure 6.13. Frequency of alternative methods for the GOAL: POINT-THERE as a function of the distance between tasks on the manuscript.

The methods are abbreviated as follows: PWSM = POINT-WITHOUT-SCROLLING-METHOD, SAPM = SCROLL-AND-POINT-METHOD, JM = JUMP-METHOD.

Selection rules for GOAL: POINT-THERE
LITTLE-d-RULE2 =
    if d $\leq$ 8
       then CHOOSE (POINT-WITHOUT-SCROLLING-METHOD)
BIG-d-RULE2 =
    if d > 8
       then CHOOSE (SCROLL-AND-POINT-METHOD) .   (6.4)

These two rules explain 85~94% of the selections for the three users.

If scrolling is the only means employed to move the text on the display, then the amount of scrolling will be determined by the manuscript length almost independent of the distribution of tasks on the manuscript. From examination of the data from S32, S34, and S47, users move the text approximately 16 lines each time they scroll the display. Thus, the number of scrolls a user (who does not use the JUMP-TO or FIND command) can be expected to perform is given by:

$$Total\ number\ of\ scrolls = (Lines\ in\ manuscript) / 16 .$$

Figure 6.14 shows the number of lines per scroll computed for individual users. Reasonable scrolling behavior may be approximated by having the model scroll 16 lines at a time.

Estimation of Parameters

In order to make time predictions with the model, it is necessary to make numerical estimates of several of its parameters. Estimates for the parameters are summarized in Figure 6.15. The time for the operator

User	Number of Lines per Scroll (All tasks)
S13	20.0
S32	17.7
S34	11.6
S37	16.1
Average	16.4

Figure 6.14. Observed average number of lines per scroll.

Parameter	Estimated Time		Source
	M (sec)	CV	
User parameters			
GET-FROM-MANUSCRIPT	2.1	.44	Figure 5.15
SCROLL	2.6	.54	Measurement of 10 instances
POINT	1.7	.76	Measurements of S13
TYPE	.127	.50	Average of two typing tests,
			$SD = .5 M$ (Kinkead, 1975)
VERIFY-EDIT	1.1	.91	Measurement of 12 instances
System parameters			
*INSERT-TEXT	1.1	.36	Measured response, 25 instances
*DELETE-TEXT			
*REPLACE-TEXT			
*JUMP-TO	1.0	1.0	Measured response, 10 instances
*SCROLL-TO	1.7	.71	Measured response, 10 instances

Figure 6.15. Parameter estimates for the simulation model.

GET-FROM-MANUSCRIPT is taken from Chapter 5. The time for POINT is from measurements of S13 in Experiment 3B. The time for VERIFY-EDIT is based on the time previously measured for BRAVO in Experiment 3A. TYPE time is based on an average of two typing tests embedded in an editing exercise given to the user before the start of Experiment 3B as a warmup. The standard deviation for the TYPE time is estimated by multiplying the mean time per keystroke by a typical coefficient of variation for typing of .5 (Kinkead, 1975).

In order to estimate response times of the system, 25 each of the command invocations *INSERT-TEXT, *REPLACE-TEXT, and *DELETE-TEXT were measured. Since there were no obvious differences between the times taken by these commands, their measured times were pooled to give a common estimated time. Ten invocations of the *JUMP-TO command and ten of *SCROLL-TO were also measured.

The assumption is made that operator times are gamma-distributed. The assumption is reasonable for at least three reasons: (1) The sum of a sequence of gamma-distributed operators is also gamma-distributed. Thus the distribution for smaller, more elementary operators has the same shape as for larger, more composite operators. (2) The gamma distribution is commonly found appropriate for operators in the industrial engineering literature (Nanda, 1968; Johnson, 1965). (3) The basic shape (skewed to the right) of the gamma distribution is correct, so that even if there were to be second-order difficulties in the fit of the distribution, the gamma distribution would still be a reasonable approximation of the real distribution shape.

Simulation of User Behavior

The full model can now be stated. The GOMS elements of the simulation model are listed in full in Figure 6.16, where the methods and selection rules are grouped together with the goals they address. We can illustrate the workings of the model by tracing out its behavior on a task. Figure 6.17 shows a trace of the model for Task A2 (see Figure 6.1 and Figure 6.3). Writing the sequence of operators from Figure 6.17, we get:

> GET-FROM-MANUSCRIPT (Task, NIL)
> GET-FROM-DISPLAY (ScreenPosition, {a # APPROXIMATE-TARGET})
> POINT ({a # MAIN-PART-OF-SCREEN}, WORD, DON'T-SELECT)
> GET-FROM-MANUSCRIPT (VisualSearchTarget, INSERTION-POINT)
> GET-FROM-DISPLAY (ScreenPosition, λ_4, A2)
> POINT ({a # MAIN-PART-OF-SCREEN}, CHARACTER, SELECT)
> INSERT-TEXT
> TYPE (NewText)
> VERIFY-EDIT .

The sequence in Figure 6.17 is only one of the possible sequences the model predicts for this task. If the simulation were run again it would make different method selections, and it would eliminate conditional operators (that appear in **with-probability** statements). For example, it might predict the sequence

> GET-FROM-MANUSCRIPT (Task, NIL)
> POINT ({a # MAIN-PART-OF-SCREEN}, CHARACTER, SELECT)
> GET-FROM-MANUSCRIPT (NewText, INSERTION-POINT)
> INSERT-TEXT
> TYPE (NewText) .

GOAL: EDIT-MANUSCRIPT
    METHOD:
        ONE-AT-A-TIME-METHOD =
            until NoMorePages = TRUE do GOAL: EDIT-UNIT-TASK

GOAL: EDIT-UNIT-TASK
    METHOD:
        ACQUIRE-EXECUTE-VERIFY-METHOD =
            GOAL: ACQUIRE-UNIT-TASK
            EXECUTE (Task)
            with-probability .4 do VERIFY-EDIT

GOAL: ACQUIRE-UNIT-TASK
    METHOD:
        READ-TASK-IN-MS-METHOD =
            GET-FROM-MANUSCRIPT ({slot Task})

GOAL: INSERT (InsertionPointKey, NextText)
    METHOD:
        INSERT-COMMAND-METHOD =
            if no InsertionPointKey then GET-FROM-MANUSCRIPT ({slot InsertionPointKey})
            SELECT-TARGET (MSPosition, {slot InsertionPoint}, InsertionPointKey)
            INSERT-TEXT
            if no NewText then GET-FROM-MANUSCRIPT ({slot NewText})
            if NewText ≠ DEFAULT then TYPE (NewText)

GOAL: DELETE (OldTextKey)
    METHOD:
        DELETE-COMMAND-METHOD =
            if no OldTextKey then GET-FROM-MANUSCRIPT (OldTextKey)
            SELECT-TARGET (MSPosition, {slot OldText}, OldTextKey)
            DELETE-TEXT

GOAL: REPLACE (OldTextKey, NewText)
    METHOD:
        REPLACE-COMMAND-METHOD =
            if no OldTextKey then GET-FROM-MANUSCRIPT ({slot OldTextKey})
            SELECT-TARGET (MSPosition, {slot OldText}, OldTextKey)
            REPLACE-TEXT
            if no NewText then GET-FROM-MANUSCRIPT ({slot NewText})
            if NewText ≠ DEFAULT then TYPE (NewText)

GOAL: MOVE (InsertionPointKey, OldTextKey)
    METHOD:
        DELETE-INSERT-METHOD =
            DELETE (OldTextKey)
            INSERT (InsertionPointKey, DEFAULT)

GOAL: SELECT-TARGET (MSPosition, PositionType, VisualSearchTarget)
    METHOD:
        ZERO-IN-METHOD =
            while   VisualSearchTarget isa # APPROXIMATE-TARGET
            do   GOAL: POINT-TO-TARGET (MSPosition, VisualSearchTarget, DON'T-SELECT)
                     GET-FROM-MANUSCRIPT ({slot VisualSearchTarget}, PositionType)
            finally GOAL: POINT-TO-TARGET (MSPosition, VisualSearchTarget, SELECT))

224

```
GOAL: POINT-TO-TARGET (MSPosition, VisualSearchTarget, SELECT)
    SELECTION-RULES:
        ROUGH-LOC-RULE =
            if VisualSearchTarget isa # APPROXIMATE-TARGET then CHOOSE (ROUGH-POINT-METHOD)
        TEXT-SEG-RULE =
            if VisualSearchTarget isa # TEXT-SEG then CHOOSE (TEXT-SEG-POINT-METHOD)
        CHAR-POINT-RULE =
            if VisualSearchTarget isa # CHARACTER then CHOOSE (CHAR-POINT-METHOD)
        WORD-POINT-RULE =
            if VisualSearchTarget isa # WORD then CHOOSE (WORD-POINT-METHOD)
        INSERTION-POINT-RULE =
            if VisualSearchTarget isa # PLACE-IN-MS then CHOOSE (INSERTION-POINT-METHOD)
    METHODS:
        ROUGH-POINT-METHOD =
            GET-FROM-DISPLAY ({slot ScreenPosition}, VisualSearchTarget, MSPosition)
            GOAL: POINT-THERE (ScreenPosition, WORD, SELECT)
        CHAR-POINT-METHOD =
            GET-FROM-DISPLAY ({slot ScreenPosition}, LOCATION: (VisualSearchTarget), MSPosition)
            GOAL: POINT-THERE (ScreenPosition, CHARACTER, SELECT)
        WORD-POINT-METHOD =
            GET-FROM-DISPLAY ({slot ScreenPosition}, LOCATION: (VisualSearchTarget), MSPosition)
            GOAL: POINT-THERE (ScreenPosition, WORD, SELECT)
        TEXT-SEG-POINT-METHOD =
            GET-FROM-DISPLAY ({slot ScreenPosition}, START-LOC: (VisualSearchTarget), MSPosition)
            GOAL: POINT-THERE (ScreenPosition, CHARACTER, SELECT)
            GET-FROM-DISPLAY ({slot ScreenPosition}, CHARACTER, SELECT)
            GOAL: POINT-THERE (ScreenPosition, CHARACTER, SELECT)
        INSERTION-POINT-METHOD =
            GET-FROM-DISPLAY ({slot ScreenPosition}, VisualSearchTarget, MSPosition)
            GOAL: POINT-THERE (ScreenPosition, CHARACTER, SELECT)

GOAL: POINT-THERE (ScreenPosition, TextType, SELECT)
    SELECTION-RULES:
        TOP-2/3-RULE =
            if ScreenPosition isa # MAIN-PART-OF-SCREEN
                then CHOOSE (POINT-WITHOUT-SCROLLING-METHOD)
        BOTTOM-1/3-RULE =
            if ScreenPosition isa # BOTTOM-PART-OF-SCREEN
                then CHOOSE (SCROLL-AND-POINT-METHOD)
        OFF-SCREEN-RULE =
            if ScreenPosition isa # OFF-SCREEN
                then CHOOSE (JUMP-METHOD)
    METHODS:
        POINT-WITHOUT-SCROLLING-METHOD =
            POINT (ScreenPosition, TextType, SELECT)
        SCROLL-AND-POINT-METHOD =
            SCROLL-TO (MSPosition)
            POINT (ScreenPosition, TextType, SELECT)
        JUMP-METHOD =
            JUMP-TO (MSPosition)
            POINT (ScreenPosition, TextType, SELECT)
```

Figure 6.16.   Methods and selection rules for BRAVO.

GOAL: EDIT-MANUSCRIPT
The only method is ONE-AT-A-TIME-METHOD
Use ONE-AT-A-TIME-METHOD
. GOAL: EDIT-UNIT-TASK
. The only method is ACQUIRE-EXECUTE-VERIFY-METHOD
. Use ACQUIRE-EXECUTE-VERIFY-METHOD
. . GOAL: GET-TASK
. . The only method is READ-TASK-IN-MS-METHOD
. . Use READ-TASK-IN-MS-METHOD
    GET-FROM-MANUSCRIPT (Task NIL)
     □ User $\Rightarrow$ Manuscript message *READ-NEXT-LOCATION-OF (A1)
     □ Manuscript $\Rightarrow$ User; Reply: $\rightarrow$ A2
     □ User $\Rightarrow$ Manuscript message *READ-ATTRIBUTE-OF (A2, FUNCTION:)
     □ Manuscript $\Rightarrow$ User; Reply: $\rightarrow$ INSERT
     □ User $\Rightarrow$ Manuscript message *READ-ATTRIBUTE-OF (A2, NEW-TEXT:)
     □ Manuscript $\Rightarrow$ User; Reply: $\rightarrow \chi_2$
. . GOAL: INSERT ({a # APPROXIMATE-TARGET}, χ_2)
. . The only method is INSERT-COMMAND-METHOD
. . Use INSERT-COMMAND-METHOD
. . . GOAL: SELECT-TARGET (A2, InsertionPoint, {a # APPROXIMATE-TARGET})
. . . The only method is ZERO-IN-METHOD
. . . Use ZERO-IN-METHOD
. . . . GOAL: POINT-TO-TARGET (A2, {a # APPROXIMATE-TARGET}, DON'T-SELECT)
. . . . ROUGH-LOC-RULE recommends ROUGH-POINT-METHOD
. . . . Use ROUGH-POINT-METHOD
    GET-FROM-DISPLAY (ScreenPosition, {a # APPROXIMATE-TARGET}, A2)
     □ User $\Rightarrow$ Display message *READ-LOCATION-OF (A2)
     □ Display $\Rightarrow$ User; Reply: $\rightarrow$ {a # MAIN-PART-OF-SCREEN}
. . . . . GOAL· POINT-THERE ({a # MAIN-PART-OF-SCREEN}, WORD, DON'T-SELECT)
. . . . . TOP-2/3-RULE recommends POINT-WITHOUT-SCROLLING-METHOD
. . . . . Use POINT-WITHOUT-SCROLLING-METHOD
    POINT ({a # MAIN-PART-OF-SCREEN}, WORD, DON'T-SELECT)
    GET-FROM-MS (VisualSearchTarget, InsertionPoint)
     □ User $\Rightarrow$ Manuscript message *READ-ATTRIBUTE-OF (A2, INSERTION-POINT:)
     □ Manuscript $\Rightarrow$ User; Reply: $\rightarrow \lambda_4$
. . . . GOAL: POINT-TO-TARGET (A2, λ_4, BUG)
. . . . INSERTION-POINT-RULE recommends INSERTION-POINT-METHOD
. . . . Use INSERTION-POINT-METHOD
    GET-FROM-DISPLAY (ScreenPosition, λ_4, A2)
     □ User $\Rightarrow$ Display message *READ-LOCATION-OF (A2) $\rightarrow$ {a # MAIN-PART-OF-SCREEN}
     □ Display $\Rightarrow$ User; Reply: $\rightarrow$ {a # MAIN-PART-OF-SCREEN}
. . . . . GOAL: POINT-THERE ({a # MAIN-PART-OF-SCREEN}, CHARACTER, SELECT)
. . . . . TOP-2/3-RULE recommends POINT-WITHOUT-SCROLLING-METHOD
. . . . . Use POINT-WITHOUT-SCROLLING-METHOD
    POINT ({a # MAIN-PART-OF-SCREEN}, CHARACTER, SELECT)
    □ User $\Rightarrow$ Editor message *SELECT
    INSERT-TEXT
    □ User $\Rightarrow$ Editor message *INSERT-TEXT
    TYPE (χ_2)
    □ User $\Rightarrow$ Editor message *TYPE (χ_2)
    VERIFY-EDIT

Figure 6.17. Trace of the simulation model for Task A2.

The sequence of user operators produced in this trace correspond to sequence number 8 in Figure 6.18. Traces of transactions are marked with a □.

Even if the model were to make all the same choices, the times for the different operators would be different according to their distributions. By running the simulation model several times on the same task, the model can be used to make Monte Carlo predictions of (1) the set of possible operator sequences the user will employ to do an editing task, (2) the relative frequency with which the different operator sequences will be employed, (3) the distribution of time for each sequence, and (4) the distribution of times for all the sequences.

As an illustration, the simulation model was run 100 times on task A2 (Figure 6.18). The runs generated 14 different operator sequences, containing from 4 to 11 operators, and having tasks times from about 3 to about 20 sec. The model predicted a mean task time of 9.5 sec. The distribution of run times was characterized by a 5th percentile of 4.3 sec, a 95th percentile of 17.4 sec, and a CV of .41 sec.

Seq. No.	User Operator Sequence	Freq.	M (sec)	CV (sec)	5% (sec)	95% (sec)
1	GNT PS I GN T1	17	8.1	.23	5.5	12.8
2	GNT PS I T1	15	5.2	.24	3.1	7.4
3	GNT PD GL PS I GN T1	11	11.5	.22	8.2	17.4
4	GNT PS I T1 VE	10	7.0	.34	4.5	12.1
5	GNT PD GL PS I T1	9	8.1	.17	6.0	9.8
6	GNT PD GL PD GL PS I GN T1	8	15.6	.17	11.7	20.1
7	GNT PS I GN T1 VE	7	7.7	.12	7.0	9.6
8	GNT PD GL PS I T1 VE	6	9.9	.33	7.4	14.2
9	GNT PD GL PD GL PS I T1	5	12.0	.28	9.2	17.1
10	GNT PD GL PS I GN T1 VE	5	11.5	.17	9.3	14.2
11	GNT PD GL PD GL PD GL PS I T1 VE	2	17.2	.10	15.9	18.5
12	GNT PD GL PD GL PS I T1 VE	2	12.5	.42	8.8	16.1
13	GNT PD GL PD GL PD GL PS I T1	2	17.6	.05	17.0	18.1
14	GNT PD GL PD GL PS I GN T1 VE	1	13.1	--	13.1	13.1
	Overall	100	9.5	.41	4.3	17.4

Figure 6.18. Predicted operator sequences and execution times for Task A2.

The operators have been abbreviated as follows: GNT = GET-FROM-MANU-SCRIPT (Task), GL = GET-FROM-MANUSCRIPT (VisualSearchTarget, ...), GN = GET-FROM-MANUSCRIPT (NewText, ...), PD = POINT (..., DON'T-SELECT), PS = POINT (..., SELECT), I = INSERT-TEXT, Tn = TYPE n characters, VE = VERIFY-EDIT.

6.3. CONCLUSIONS

We can state several conclusions as the result of this theoretical exploration. First, it was possible to construct a GOMS model for another and quite different text-editor, the display-oriented editor BRAVO. Both pointing at the display with the mouse and scrolling the display could be described by goals, operators, methods, and selection rules similar to those employed in the POET description.

Second, the GOMS notation was extended to a stochastic description of user behavior. Stochastic models of users could be used to get around some of the limits on predictability of sequences found in Chapter 5. They could also be used to attempt estimates of time and sequence variability.

Finally, we saw how our analysis of the task environment for editing could be extended so as to allow an explicit accounting of the information the user possesses moment-by-moment about the editing tasks on which he is working.

It is important to restate that in this chapter we have been concerned only with studying how the GOMS model could be extended. Additional empirical studies would be necessary to validate the detailed predictions of the GOMS extensions.

7. Models of Devices for Text Selection

The apparently endless options for the design of human-computer interfaces are composed from only a very few sensory-motor and cognitive operations performable by the user. These include:

(1) the perceptual operations of
visual search and
reading and
(2) the motor operations of
typing on the keyboard and
reaching with a hand to a target, including
reaching for a button and
pointing to a target on the display.

Systems can be imagined that require extensions to this list—perceptual judgments of alignment or motor drawing operations, for example—but, these operations are adequate for the models in Chapters 5 and 6 and for a great many other computer interfaces.

Each of these operations is worth studying in the context of human-computer interaction, and the results can have implications for the design of a computer interface. In this chapter, we shall focus on one issue within this realm—the implications for design of the reaching-to-target operations.

The editor BRAVO, in Chapter 6, made heavy use of the mouse pointing device for selecting text on the display screen. Other devices exist (the joystick, various key-operated devices) that might have been chosen. Which pointing device is the best choice and why? The choice of pointing device can have a significant impact on the ease with which the selections can be made. In fact, since pointing typically occurs with high frequency, ease of pointing can have a large effect on the success of the entire system.

There have been several studies of pointing devices. English, Englebart, and Berman (1967) measured mean pointing times and error rates for the mouse, lightpen, Grafacon tablet (an extendable, pivoted rod, intended originally for curve-tracing), and position and rate joysticks. They found the mouse to be the fastest of the devices. Goodwin (1975) measured pointing times for the lightpen, lightgun, and Saunders 720 step keys (RETURN, TAB, SPACE, and the reversal of these functions using the SHIFT key). She found the lightpen and the lightgun equally fast and much superior to the Saunders 720 step keys. Whereas these studies produced interesting comparative data on the devices measured, they did not simultaneously control the three variables likely to affect performance: learning, target distance, and target size. They also did not attempt to account theoretically for the results. The study that follows addresses both these issues. We consider the mouse, a rate-controlled isometric joystick, step keys, and text keys.

7.1. EXPERIMENTAL COMPARISON OF TEXT-SELECTION DEVICES (EXPERIMENT 7A)

The purpose of the experiment was to compare the relative merits for text-selection of a number of devices. To make the comparison meaningful a number of factors had to be controlled, including individual differences (controlled by using the same users on all devices); learning and asymmetrical transfer of training between devices (controlled by having each subject practice to "assymptote" before collecting comparison data); movement direction (controlled by randomizing target direction and assessing effect in separate analysis); target size, target distance (effect measured by factorially combining these variables into conditions of the experiment), users' motivation (kept high with performance feedback), and the possibility of important extraneous variables (controlled by using a realistic task and by identifying the cause of response time effects through modeling).

METHOD FOR EXPERIMENT 7A

Users. Three men and two women, all undergraduates at Stanford University, served as users in the experiment. None had ever used any of the devices previously, and all had little or no experience with computers. Subjects were paid $3 per hour with a $20 bonus for completing the experiments. One of the women was very much slower than the other users and was eliminated from the experiment leaving four users (inclusion of the eliminated user would not have changed the qualitative conclusions of the study).

Text-Selection Devices. Four pointing devices were tested (see Figure 7.1). Two were continuous devices: the mouse and the rate-controlled isometric joystick. Two were key operated: the step keys and the text keys. The devices had been optimized informally by testing them on local users, adjusting the device parameters to maximize performance.

The *mouse*, already described, was a small device which sat on the table to the right of the keyboard, connected by a thin wire. On the undercarriage were two small wheels, mounted at right angles to each other. As the mouse moved over the table one wheel coded the amount of movement in the x-direction, the other the amount of movement in the y-direction. A cursor moved simultaneously on the display, two units of screen movement for each unit of mouse movement on the table.

The *joystick* was a small strain gauge on which had been mounted a rubber knob 1.25 cm in diameter. Applying force to the joystick in any direction did not produce noticeable movement in the joystick itself, but caused the cursor to move in the appropriate direction at a rate (in cm/sec) $= (.0178)(\text{force})^2$, where force is measured in Newtons. For forces less than about 4 Newtons, the cursor did not move at all; and the equation ceased to hold in the neighborhood of 45 Newtons as the rate approached a ceiling of about 40 cm/sec.

The *step keys* were the familiar five-key cluster found on many display terminals. On the four sides of a central HOME key were keys to

Figure 7.1. Pointing devices tested in Experiment 7A.

move the cursor in each of four directions. Pressing the HOME key caused the cursor to return to the upper left corner of the text. Pressing one of the horizontal keys moved the cursor one character (.246 cm on the average) backward or forward along the line. Pressing a vertical key moved the cursor one line (.456 cm) up or down. Holding down one of the keys for more than .100 sec caused it to go into a repeating mode, producing one step in the vertical direction each .133 sec or one step in the horizontal direction each .067 sec (3.43 cm/sec vertical movement, 3.67 cm/sec horizontal movement).

The *text keys* were similar to keys appearing on several commercial "word processing" terminals. Depressing the PARAGRAPH key caused the cursor to move to the beginning of the next paragraph. Depressing the LINE key caused the cursor to move downward to the same position in the next line. The WORD key moved the cursor forward one word; the CHARACTER key moved the cursor forward one character. Holding down the REVERSE key while pressing another text key caused the cursor to move opposite the usual direction. The text keys could also be used in a repeating mode. Holding the LINE, WORD, or CHARACTER key down for longer than .100 sec caused it to repeat at .133 sec per repeat for the LINE key, .100 sec per repeat for the WORD key, or .067 sec per repeat for the CHARACTER key. Since there were .456 cm/line, 1.320 cm/word, and .246 cm/character, movement rates were 3.43 cm/sec for the LINE key, 13.2 cm/sec for the WORD key, and 3.67 cm/sec for the CHARACTER key.

Procedure. Subjects were seated in front of a computer terminal with a display, a keyboard, and one of the devices for pointing at targets on the screen. On each trial, a page of text was displayed on the screen. Within the text, a single word or phrase, the target, was highlighted by inverting the black/white values of the text and background in a rectangle surrounding the target. The user struck the space bar of the keyboard with his right hand and then, with the same hand, reached for the pointing device and directed the cursor to the target. The cursor thus positioned, the user pressed a button "selecting" the target as he would were he using the device in a text-editor. For the mouse, the selection button was located on the device itself. For the other devices, the user pressed a special key on the keyboard with his left hand.

Design. There were five different distances from starting position to target (1, 2, 4, 8, or 16 cm), and four different target sizes (1, 2, 4, or 10 characters). All targets were words or groups of words. Ten different instances of each distance × target size pair were created, with varying locations of the target on the display and angles of hand movement, giving a total of 200, randomly ordered, unique stimuli.

Each user repeated the experiment with each device. The order in which users employed the devices was randomized. At the start of each day, the users were given approximately twenty warmup trials to refresh their memory of the procedure. All other trials were recorded as data. At the end of each block of twenty trials they were given feedback on the average positioning time and average number of errors for those trials. This feedback was found to be important in maintaining users' motivations. At the end of each 200 trials they were given a rest break of about fifteen minutes. Subjects normally accomplished 600 trials/day, requiring about two to three hours of work. They each used a particular device until the positioning time was no longer significantly decreasing with practice (operationally defined as when the first 200 and last 200 trials of the last 600 trials in the day did not differ significantly in positioning time at the $p < .05$ level using a t-test). An approximation to this criterion was reached in 1200 to 1800 trials (four to six hours) on each device.   Of the 20 user $\times$ device pairs, 15 reached this criterion, three performed worse in their last trials (largely because several days elapsed between these sessions); and only two were continuing (slightly) to improve.

RESULTS FOR EXPERIMENT 7A

Improved Performance with Practice.   In order to compare the devices it is important that the effects of practice be isolated so as not to confound the analysis.   According to the Power Law of Practice in Chapter 2, practice should improve performance as given by Equation 2.5:

$$\log T_N = \log T_1 - \alpha \log N,$$ (2.5)

where

T_1 = estimated positioning time on the first block of trials,

T_N = estimated positioning time on the Nth block of trials,

N = trial block number, and

α = an empirically determined constant.

Thus, the ease of learning for each device can be described by two numbers, T_1 and α, which may be conveniently determined from a regression of (log T_N) on (log N).   Figure 7.2 shows the results of plotting the data from error-free trials according to Equation 2.5.   Each point on the graph is the average of a block N of twenty contiguous trials from which error trials have been excluded.   Only the first 60 trial blocks are shown.   Since some users reached criterion at this point, not all continued on to further trials.   The values predicted by the (fitted) equation are given as the straight line drawn through the points.   The average target size in each block was 4.23 cm (the range of the average target sizes for different trial blocks was 3.95 to 4.50 cm), and the average distance to the target was 6.13 cm (range 5.90 to 6.42 cm).

The parameters T_1 and α, as determined by the regressions, are given in Figure 7.3, along with the standard error and percentage of variance explained from the regression analysis.   Practice caused more improvement in the mouse and text keys than on the other two devices used. Use of the step keys, in particular, showed very little improvement with practice.   Equation 2.5 explains 39% of the variance in the average positioning time for a block of trials for the step keys, and 61% to 66% of

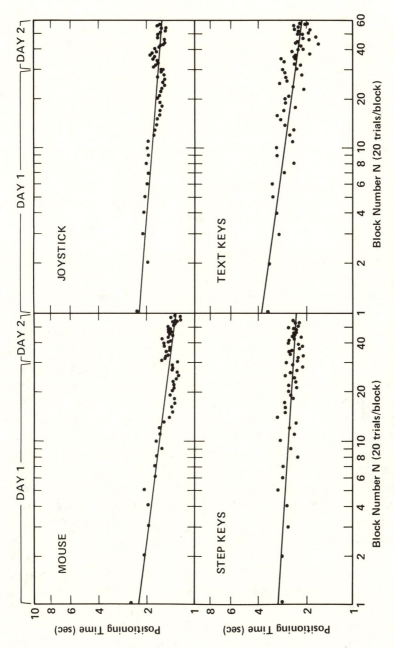

Figure 7.2. Learning curves for pointing devices.

Device	T_1 (sec)	α	Learning Curve Equation	SE (sec)	R^2
Mouse	2.20	.13	$T_{\backslash} = 2.20\, N^{-.13}$	.12	66
Joystick	2.19	.08	$T_{\backslash} = 2.19\, N^{-.08}$	.08	.62
Step Keys	3.03	.07	$T_{\backslash} = 3.03\, N^{-.07}$	.11	.39
Text Keys	3.86	.15	$T_{\backslash} = 3.86\, N^{-.15}$	.16	.61

Figure 7.3. Learning curve parameters.
N is the number of the trial block. There are 20 trials in each block. Each equation is based on 1200 trials divided into 60 trial blocks.

the variance for the other devices. The fit, at least for the mouse and the joystick, is actually better than these numbers suggest. Since users did 30 blocks of trials on a day, typically followed by a pause of a day or two before they could be rescheduled, a break in the learning curve was expected at that point; and indeed such a break is quite evident for the mouse and the joystick between the 30th and 31st blocks. Fitting Equation 2.5 to only the first day increases the percentage of variance explained to 91% for the mouse and 83% for the joystick. In the case of the step keys and text keys, there was no such obvious day effect.

Overall Speed. According to the Power Law of Practice, users' response times for text selection will continue to decrease indefinitely. But if response time were to be plotted in arithmetic coordinates as a function of number of practice trials, the plot would give the illusion of an asymptote as exponentially more trials are required for the same response-time decrease. In order to compare the devices in this region of the learning curve where response time is relatively flat (as would be the case for daily use by office workers), a sample was examined of each user's performance on each device, consisting of the last 600 trials that were not also the first 200 trials of a day (in order to diminish warmup effects). The remaining analyses are based on this subset of the data, excluding those trials on which errors occurred. Figure 7.4 gives the homing time, positioning time, and total time for each device, averaging over all the distances and target sizes. *Homing time* was measured from when the user's right hand left the space bar until when the cursor had

Device	Trials N	Movement Time for Non-Error Trials						Error Rate	
		Homing Time		Positioning Time		Total Time			
		M (sec)	SD (sec)	M (sec)	SD (sec)	M (sec)	SD (sec)	M	SD
Mouse	1973	.36	.13	1.29	.42	1.66	.48	5%	22%
Joystick	1869	.26	.11	1.57	.54	1.83	.57	11%	31%
Step Keys	1813	.21	.30	2.31	1.52	2.51	1.64	13%	33%
Text Keys	1877	.32	.61	1.95	1.30	2.26	1.70	9%	28%

Figure 7.4. Overall pointing times for all devices.
Based on data from four users x 600 trials/user, with error trials having been subtracted.

begun to move. *Positioning time* was measured from when the cursor began to move until when the selection button had been pressed. From the figure, it can be seen that homing time increases slightly with the distance of the device from the keyboard. The longest time required is to reach the mouse, the shortest to reach the step keys. Although the text keys are near the keyboard, they take almost as long to reach as the mouse. Either it is more difficult to position the hands on the text keys or, as seems likely, users often spent some of the time between hitting the space bar and beginning to press the keys in planning the strategy for their next move. Further evidence for this hypothesis comes from the relatively high standard deviation observed for the homing time of the text keys. Whereas the differences in the homing times among all device pairs except the mouse vs. the text keys are reliable statistically (at $p < .05$ or better using a t-test), the differences are actually quite small. But although the step keys can be reached .15 sec sooner than the mouse, they take 1.02 sec longer to position. Thus the differences in the homing times are insignificant compared to the differences in the positioning times.

The mouse is easily the fastest device, the step keys the slowest. As a group, the continuous devices (the mouse and the joystick) are faster than the key-operated devices (the step keys and text keys). Differences between the devices are all reliable at $p < .001$ using t-tests.

Effect of Distance and Target Size. The effect of distance on positioning time is given in Figure 7.5. At all distances greater than 1 cm, the continuous devices are faster. The positioning time for both continuous devices seems to increase approximately with the log of the distance. The time for the step keys increases rapidly as the distance increases, whereas the time for the text keys increases somewhat less than as the log of the distance, owing to the existence of keys for moving relatively large distances with a single stroke. Again, the mouse is the fastest device, and its advantage increases with distance.

Figure 7.6 shows the effect of target size on positioning time. The positioning times for both the mouse and the joystick decrease with the log of the target size. The time for the text keys is independent of target size, and the positioning time for the step keys also decreases roughly with the log of the target size. Again, the mouse is the fastest device, and again, the continuous devices as a group are faster for all target sizes.

Effect of Approach Angle. The targets in text-editing are rectangles often quite a bit wider than they are high, presenting different problems

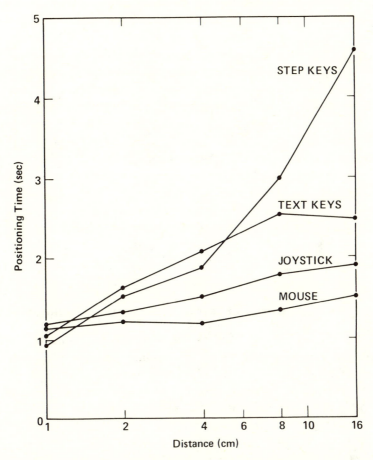

Figure 7.5. Effect of target distance on positioning time.

when approached from different angles. In addition, the step keys and text keys work differently when moving horizontally than when moving vertically. To test whether the direction of approach has an effect on positioning time, the target movements were classified according to whether they were vertical (0 to 22.5 degrees), diagonal (22.5 degrees to 67.5 degrees), or horizontal (67.5 degrees to 90 degrees). Analysis of variance shows that the angle makes a significant difference for every device except the mouse. The joystick takes slightly longer to position when the target is approached diagonally. The step keys take longer when approached horizontally than when approached vertically, a consequence probably deriving from the fact that a single keystroke

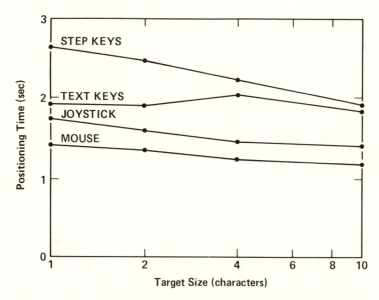

Figure 7.6. Effect of target size on positioning time.

moves the cursor almost twice as far vertically as horizontally. By contrast, the text keys take longer to position vertically, reflecting the presence of the WORD key. The differences induced by direction are not of great consequence, however. For the joystick, it amounts to 3% of the mean positioning time; for the step keys, 9%; for the text keys, 5%.

Errors. Of the four devices tested, the mouse had the lowest overall error rate, 5%; the step keys had the highest, 13%. Differences are reliable at $p < .05$ or better, using *t*-tests. The error rate increases only very slightly with distance. However, it decreases with target size for every device except the text keys (Figure 7.7). This finding replicates the result of Fitts and Radford (1966), where, in an investigation of self-initiated, discrete, pointing movements using a stylus, there was a similar marked reduction in errors as the target increased in size and a similar slight increase in error rate as the distance to the target increased.

7.2. PERFORMANCE MODELS OF TEXT-SELECTION DEVICES

Although these empirical results are of direct use in selecting a pointing device, a more useful understanding of the properties of these devices can be had in terms of the Model Human Processor in Chapter 2.

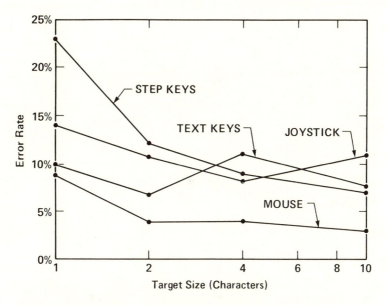

Figure 7.7. Effect of target size on error rate.

Mouse

The time to move the mouse can be analyzed in terms of the time to move the hand to a target and should therefore follow from Fitts's Law, Equation 2.3, which we may rewrite as:

$$T_{pos} = K_0 + I_M \log_2 (D/S + .5) \sec, \qquad (7.1)$$

where

$$
\begin{array}{rl}
T_{pos} & = \text{Positioning time,} \\
D & = \text{Distance to the target,} \\
S & = \text{Size of the target,} \\
I_M & = .100\,[.070 \sim .120]\ \sec/\text{bit, and} \\
K_0 & = \text{a constant.}
\end{array}
$$

The constant K_0 has been added to include the time for the hand initially to adjust its grasp on the mouse and the time to make the selection with the mouse button.

Fitts's Law predicts that plotting positioning time as a function of $\log_2 (D/S + .5)$ should give a straight line. As the solid line in Figure 7.8 shows, this prediction is confirmed. Furthermore, the prediction that the slope of the line I_M should be in the neighborhood of .100 sec/bit is

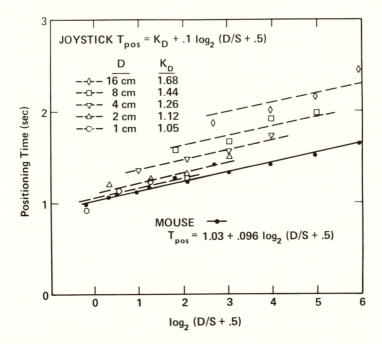

Figure 7.8. Positioning time for the continuous devices as a function of Fitts's index of difficulty.

also confirmed. The equation for the line in Figure 7.8 as determined by regression analysis is:

$$T_{pos} = 1.03 + .096 \log_2 (D/S + .5) \sec.$$ (7.2)

This equation has a standard error of .07 sec and explains 83% of the variance of the means for each condition, comparable to the percentage of variance explained by Fitts and Radford. The slope of .096 sec/bit is in the .100 sec/bit range found in other studies. Since the standard error of estimate for I_M in fitting Equation 7.2 was .008 sec/bit, the mouse would seem to be close to, but slightly slower than, the optimal rate of around .08 sec/bit observed for use of the stylus and for finger-pointing.

The values for positioning time obtained in this experiment are apparently in good agreement with those obtained by English et al. (1967). Assuming that their display characters were about the same width as ours and assuming an intermediate target distance of about 8 cm, Equation 7.2 (with the addition of the .36 sec homing time from Figure

7.4) predicts 1.87 sec for 1-character targets (English et al. measured 1.93 sec) and 1.66 sec for "word" targets of 5 characters (English et al. measured 1.68 sec).

Joystick

Although it is a rate-controlled device instead of a position device, we might wonder if the joystick follows Fitts's Law. Plotting the average time per positioning for each distance × size cell of the experiment according to Equation 7.1 shows that there is an approximate fit to the following equation:

$$T_{pos} = .99 + .220 \log_2 (D/S + .5) .$$

This equation has a standard error of .13 sec and explains 89% of the variance of the means. The size of the slope, $I_M = .220$ sec/bit, shows that information is being processed at only half the speed of the mouse, significantly below the maximum rate. Closer examination gives some insight into the difficulty. The points for the joystick in Figure 7.8 actually form a series of parallel lines, one for each distance, each with a slope of around .100 sec/bit. Setting I_M to .100 sec/bit, we can therefore write an alternative model:

$$T_{pos} = K_D + .100 \log_2 (D/S + .5) .$$

K_D is the intercept for distance D. From the figure, K_D varies from 1.05 sec for $D = 1$ cm to 1.68 sec for 16 cm. For this model, the standard error of the fit is reduced to .07 sec, the same as for the mouse. (Since the slope was not determined by the regression, a comparable R^2 cannot be computed.) Thus, the tested joystick can be thought of as a Fitts's Law device with a slope twice the .100 sec/bit slope for hand movements; or it can be thought of as a Fitts's Law device with the expected slope, but having an intercept which increases with distance. The problem with the joystick used in our experiment is probably related to the non-linearity in the control (Poulton, 1974; Craik and Vince, 1963). It should be noted that for the 1-cm distance (where the effect of non-linearity is slight) the positioning time is virtually the same as for the mouse. Thus, the possibility of designing a joystick with performance characteristics comparable to the mouse is by no means excluded.

Step Keys

The time to use the step keys should be governed by the number of keystrokes needed to move the cursor to the target. Since the keys can only move the cursor vertically or horizontally, the number of keystrokes is $D_x/.456 + D_y/.246$, where D_x and D_y are the horizontal and vertical components of distance to the target, .456 cm is the size of a vertical step, and .246 cm is the size of a horizontal step. Hence, positioning time should be

$$T_{pos} = K_0 + C(D_x/.456 + D_y/.246). \tag{7.3}$$

If the operation of the key were done with a single finger, then according to the Model Human Processor in Chapter 2, $C \simeq 2\tau_M = .140$ sec, one Motor Processor cycle would being required to cock the finger and one to press with it. But some finger-cockings could be overlapped with some key presses when the user uses two fingers, so C could be reduced somewhat. The regression to the observed data yields $K_0 = 1.20$ sec and $C = .052$ sec/keystroke (this equation has a standard error of .54 sec and explains 84% of the variance of the means). Since in the extreme case, where each cocking of the finger was completely overlapped with the keypressing by another finger, $C \simeq \tau_M = .070$ sec, the value of C obtained from these data is still a bit fast to be identified with the pressing of a key. It is also too fast to be identified with the .067 sec/keystroke automatic repetition mode. The puzzle is solved by reference to a plot of positioning time against the predicted number of keystrokes (Figure 7.9). Equation 7.3 with the above parameters (shown by the long solid line) actually confuses two phenomena. As the figure shows, positioning time is linear with the number of keystrokes until the predicted number of keystrokes becomes large (that is, until the distance to the target becomes large). In these cases, the user often has the opportunity to reduce positioning time by using the HOME key. This method change scatters the points on the right of the plot and results in a fit for Equation 7.3 with loss of physical interpretation.

Fitting Equation 7.3 to only the first part of the graph $(D_x/.456 + D_y/.246 < 40)$ gives

$$T_{pos} = .98 + .074(D_x/.456 + D_y/.246).$$

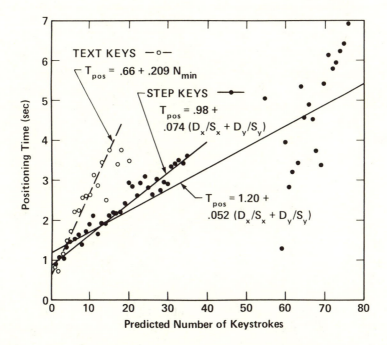

Figure 7.9.   Positioning time for the key devices as a function of the predicted number of keystrokes.

This equation, indicated by a short solid line in Figure 7.9, has a standard error of .18 sec and explains 95% of the variance in the means. The slope of $C = .074$ sec/keystroke does have a reasonable interpretation; it suggests that the .067 sec/keystroke automatic repetition feature was heavily used, and indeed, this was confirmed by observation of users.

Text Keys

The text keys present the user on most trials with a choice of methods for reaching the target. For example, he might press the PARAGRAPH key repeatedly until the cursor has moved to the paragraph containing the target. He could then press the LINE key repeatedly until it is on the target line, then use the WORD key to bring it over to the target. Or he might use the PARAGRAPH key to move to the paragraph after the target, then, holding the REVERSE key down, use the LINE key to back up to the line below the target line, and, still holding REVERSE down, use WORD to back up to the target. In fact, there are 26 different methods for moving the cursor to the target, although only a subset will be possible in a given situation. Which is the fastest method will depend on the target's location relative to the starting position and on the boundaries of surrounding lines and paragraphs.

The obvious hypothesis is that positioning time is proportional to the number of keystrokes and that, for well-practiced users, the number of keystrokes will be the minimum necessary. The constant of proportionality might be expected from the Model Human Processor to be no faster than than $\tau_M = .070$ [.030~.100] sec for multiple-finger operations and might be close to the single-finger rate of $2\tau_M = .140$ [.060~.200] sec. It is difficult to estimate how much the rate might be slowed beyond this by activities of the Perceptual and Cognitive Processors, since it is difficult to estimate the load imposed by visual search, perceptual analysis, method selection, and degree of possible overlap. To test the hypothesis that selection time is proportional to the number of keystrokes, each trial was analyzed to determine the minimum number of keystrokes N_{min} necessary to reach the target. The average positioning time as a function of N_{min} is plotted as the open circles in Figure 7.9. A least-squares fit gives

$$T_{pos} = .66 + .209 \, N_{min}.$$

The standard error is .24 sec and the equation explains 89% of the variance of the means. The keystroke rate of .209 sec/keystroke (a little higher than $2\tau_M$) is approximately equal to the typing rate for random words, Figure 2.14. Evidently, the automatic repetition mode was little used. Examination of statistics on the minimum numbers of keystrokes for each trial shows there was little need for it. For one thing, an average of only six keystrokes were necessary for the text keys to locate a target word, and ten or fewer keystrokes were sufficient to reach over 90% of the targets. For another, these keystrokes were distributed across several keys, further limiting opportunities to use the repetition mode. The PARAGRAPH key was needed on 48% of the trials, the LINE key on 85%, the WORD key on 83%, and the REVERSE key on 81%.

Comparison of Devices

Figure 7.10 summarizes the models, the standard errors of the fit, and the percentage of variance of the means explained by the model. The theory of pointing expressed in the models has strong implications for the design and selection of pointing devices. The match of the Fitts's Law slope of the mouse to the $I_M \simeq .100$ sec/bit constant observed in other hand movement and manual control studies means that positioning time is apparently limited by central information-processing capacities of the eye-hand guidance system. Taking $I_M = .08$ sec/bit as the most likely minimum value for a similar movement task and $K_0 = 1$ sec as a typical

Device	Model	SE (sec)	R^2	Notes
Mouse	$T_{pos} = 1.03 + .096 \log_2 (D/S + .5)$	.07	.83	—
Joystick	$T_{pos} = .99 + .220 \log_2 (D/S + .5)$	.13	.89	(a)
	$T_{pos} = K_D + .1 \log_2 (D/S + .5)$	.07	—	(b)
Step Keys	$T_{pos} = 1.20 + .052 (D_x/S_x + D_y/S_y)$	.54	.84	(c)
	$T_{pos} = .98 + .074 (D_x/S_x + D_y/S_y)$	.18	.95	(d)
Text Keys	$T_{pos} = .66 + .209 N_{min}$	.24	.89	—

Figure 7.10. Summary of models for positioning time.
All times in the models are in sec. Least-squares fits were performed on cell means rather than individual trials to make the results comparable to Fitts (1954). NOTES: (a) Least-squares fit to all data points; (b) Fitting a separate line with slope .1 sec/bit for each distance; (c) Least-squares fit to all data points; (d) Fit for number of keystrokes ($D_x/S_x + D_y/S_y$) < 40, where the HOME key is unlikely to be used.

value observed in this experiment, it seems unlikely that a continuous movement device could be developed whose positioning time is less than

$$T_{min} = 1 + .08 \log_2 (D/S + .5) \text{ sec},$$

unless it could somehow either reduce the information that must be centrally processed or use a different set of muscles (although something might be done to reduce the value of K_0). If this is true, then an optimal device would be expected to be no more than about 5% faster than the mouse in the extreme case of one-character targets 16 cm distant $(1 + .095 \log_2[(16/1) + .5] = 1.38$ sec vs. $1 + .08 \log_2[(16/1) + .5] = 1.32$ sec). Typical differences would be much less. By comparison to the mouse's 5% slower-than-optimal rate, the joystick (in this experiment) is 83% slower, the text keys 107% slower, and the step keys 239% slower. Even if K_0 were zero, the mouse would still be only 23% slower than the minimum. Whereas devices might be built that improve the mouse's homing time, decrease its error rate, or increase its ability for fine movement, it is unlikely their positioning times will be significantly faster.

This maximum information-processing capacity probably explains the lack of any significant difference in positioning time between the lightpen and the lightgun in Goodwin's (1975) experiment. Both are probably Fitts's Law devices, so both can be expected to have the same maximum .100 sec/bit rate as the mouse (if they are optimized with respect to control/display ratio and any other relevant variables).

In interpreting these results, highly favorable to the mouse, some qualifications are in order. Of the four devices, the mouse is clearly the most compatible for this task (cf. Poulton, 1974, Chapter 16), since less mental translation is needed to map intended motion of the cursor into motor movement of the hands than for the other devices. Thus, it would be expected to be easier to use, to put lower cognitive load on the user, and to have lower error rates. There are, however, limits to its compatibility. Inexperienced users are often bewildered about what to do when they run the mouse into the side of the keyboard while trying to move the cursor across the screen. They need to be told that picking up the mouse and setting it down at a more convenient place on the table will not affect the cursor. Even experienced users are surprised at their inability to control cursor movement when they hold the mouse backwards or sideways.

The greatest difficulty with using the mouse for text-editing occurs with selecting small targets. Punctuation marks, such as periods, are considerably smaller than an average character. The error rate for the mouse, which was already up to 9% for one-character targets, would be even higher for these sorts of targets. Yet this difficulty is even greater for many other devices, including the lightpen and the joystick.

7.3. APPLICATIONS

The theory of pointing devices developed above has immediate application in practical design and testing of commercial systems. We cite two examples from experience within our own company, Xerox: the development of a rapid test for analogue pointing devices and a computation of system throughput needed to support the mouse at maximum velocity.

Rapid Test for Analogue Pointing Devices[1]

> **Problem.** A product development group wished to pursue the development of a novel analogue pointing device. Since only subjective impressions of the performance of the device were available, and there was disagreement over these, they needed a simple test procedure which could give designers rapid, quantitative feedback about the effect of various improvements to the device.

The testing procedure described in Experiment 7A was not a practical test in this case, since it involved an expensive and time-consuming process of training users until learning was no longer a significant factor. It also required several days of trials using a computerized laboratory system capable of simulating the appearance of random targets on a display editor. Such a testing arrangement was beyond the equipment resources and time available and would not be able to give developers results quickly enough to be helpful.

---

[1] This test was developed in collaboration with Richard Sperling, Xerox Office Products Division.

OOOO
XXXX

XXXX

Figure 7.11. Four-character targets used for rapid measurement of I_M .
On each trial, the user points back and forth between the group of O's and the group of X's.

Fortunately, given the models for analogue pointing devices validated with Experiment 7A, the time-consuming procedure of Experiment 7A is now unnecessary for routine testing. Establishment of the fact that a broad class of analogue devices can be expected to follow Fitts's Law means that a simpler test, based on Fitts's (1954) dotting task, can be devised to measure the Fitts's Law slope I_M.

TEST METHOD: MEASUREMENT OF I_M FOR ANALOGUE POINTING DEVICES

Stimuli. A central target (consisting of letter O's) and two test targets (consisting completely of X's) are displayed together as shown in Figure 7.11. The midpoints of the test targets are either 1 cm or 4 cm distant from the central target.[2] There are three such displays, each consisting of only 1-, 4-, or 8-character targets, giving a total of 2 distances × 3 sizes = 6 conditions. Pure horizontal and vertical directions are avoided to minimize possible effects due to the oblong shape of the targets.

Procedure. Each user is given three blocks of trials, a different random order of target size and distance combinations occurring each block. On each trial, at a verbal signal from the experimenter, the user moves the cursor back and forth between the central O target and the appropriate X target, selecting each target in turn (the targets indicate

---

[2] An earlier version of the test also used a 16-cm distance, but it was found that users seemed to shift the set of muscles they used for movement at this distance and that the 16-cm points departed from Fitts's Law.

they have been selected by the appearance of underlining that lasts until the next target is selected). The user does this as many times as possible within a 30-sec interval. At the end of the trial, the number of selections is recorded. Users are instructed to emphasize accuracy over speed. The average time per movement is determined by dividing 30 sec by the number of selections made. The value for I_M is determined by plotting the time per movement for the third trial block as a function of $\log_2(D/S+.5)$ according to Equation 7.1, then estimating the slope of the straight line.

COMPARISON WITH EXPERIMENT 7A (EXPERIMENT 7B)

As a means of further validating this test, we ran an experiment comparing the value of I_M for the mouse obtained from the test procedure with that obtained from Experiment 7A.

Procedure. Three users were run according to the above test procedure. All users used the mouse in their daily work.

Results. The results for all three trials are given in Figure 7.12. The values for I_M are close (in fact, the third trial block value is identical) to the value of .096 sec/bit obtained in Experiment 7A. It should be noted that learning affects the value of the intercept K_0, but not the value of

	K_0 (sec)	I_M (sec/bit)	SE (sec)	R^2
Experiment 7A (from Figure 7.10)				
	1.03	.096	.07	.83
Experiment 7B (Rapid Measurement Test Procedure)				
Trial 1	.744	.102	.08	.81
Trial 2	.604	.110	.05	.94
Trial 3	.587	.096	.03	.95

Figure 7.12. Experiment 7B, rapid measurement of mouse I_M parameters for Equation 7.1 as a function of trial compared to values obtained in Experiment 7A.

Each trial lasts 30 sec. Each trial is the average of three users. Regression was performed on Target Size x Target Distance cell means.

the slope I_M (see Figure 7.12). The use of three trial blocks allows the data to stabilize on a better value, as indicated by the increase of the percentage of variance explained from 81% to 95% and the reduction in the standard error from .08 sec to .03 sec.

The test procedure described here can be run in less than a half hour using a wrist watch and requires no special computer programming (other than what would be required anyway to connect the pointing device to the display editor). It produces values for I_M practically indistinguishable from those produced by the original study, Experiment 7A. In the application, this procedure was used to quantify performance levels for the device under consideration and to identify which improvements were effective.

Maximum Mouse Velocity[3]

> **Problem.**   For technical reasons, it would have been convenient to design the mouse support hardware for the Xerox Star system in such a way that the maximum velocity with which the mouse could move the cursor across the screen would be 50 cm/sec. Is this velocity high enough not to impede user behavior?

In Chapter 2 we derived the equation for Fitts's Law from the assumption that macroscopic movements towards a target are made up of micromovements with constant error ε. The distance X_n of the cursor from the center of the target on the nth micromovement of the mouse is, according to Equation 2.1,

$$X_n = \varepsilon^n D, \tag{2.1}$$

where D is the total distance to the target from the starting point. Since $\varepsilon < 1$, the movement distance on each cycle will be less than on the previous cycle. The maximum velocity of the mouse will, therefore, be reached on the first cycle. The average velocity v_{max} on the first cycle is just:

---

[3] This analysis was performed in collaboration with Ralph Kimball, Xerox System Development Division.

$$v_{max} = (X_0 - X_I)/(\tau_P + \tau_C + \tau_M)$$
$$= (D - \varepsilon D)/(\tau_P + \tau_C + \tau_M)$$
$$= [(1 - \varepsilon)/(\tau_P + \tau_C + \tau_M)] D \qquad (7.4)$$

Using Vince's (1948, Experiment III) estimate of $\varepsilon = .07$ and our estimate from the Model Human Processor of $\tau_P + \tau_C + \tau_M = 240[105 \sim 470]$ msec to substitute into Equation 7.4 gives:

$$v_{max} = 3.9 \, D \, \text{cm/sec}, \qquad (7.5)$$

with a range of $2.0 \sim 8.9 \, D$ cm/sec. We could improve the precision of our calculation by using the experimentally measured values from Chapter 2 of $\tau_P + \tau_C + \tau_M = 190 \sim 260$ (instead of those synthesized from Figure 2.1) to reduce the range, giving:

$$v_{max} = [3.6 \sim 4.9] \, D \, \text{cm/sec}. \qquad (7.6)$$

The suggested maximum design velocity for the mouse of 50 cm/sec will be exceeded for distances greater than $50/3.9 = 13$ cm. according to Equation 7.5. The diagonal of the video display being considered at the time was about 35 cm, for which $v_{max} = (3.9)(35) = 136$ cm/sec, more than twice the suggested design velocity. The suggested maximum processing rate for mouse movement was, therefore, too low.

A good design should set the system parameters so the maximum velocity will not be exceeded by a fast user. Using the Fastman value for $\tau_P + \tau_C + \tau_M$ ($= 190$ msec), Equation 7.4 would become

$$v_{max} = 4.9 \, D \, \text{cm/sec}.$$

The Fastman calculation for maximum velocity on the longest run (the 35-cm screen diagonal) is $v_{max} = (4.9)(35) = 171$ cm/sec, more than three times the 50 cm/sec proposed. As a consequence, the Star hardware was redesigned to handle faster mouse velocities.

VERIFICATION FROM THE STEP-TRACKING LITERATURE

We have shown that mouse positioning is just an instance of hand movement and that, therefore, results from the motor movement and tracking literature should apply. In a set of experiments where subjects

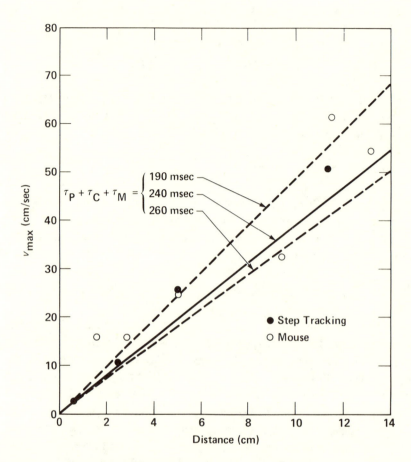

Figure 7.13. Maximum velocity of cursor as a function of distance in the step-tracking task.
Maximum velocity for step tracking derived from Poulton (1974, Figure 5.5, P. 59). Data from Craik and Vince (1963, Figure 14). Maximum velocity for mouse from Experiment 7C.

tried to keep the point of a pencil on a step track moved behind a small slit, Craik and Vince (1944, 1963, graphs reproduced in Poulton, 1974, p. 59) reported the movement velocities as a function of time for different movement distances. We have replotted in Figure 7.13 the peak velocity v_{max} read from these curves as a function of distance. The dotted lines indicate the ranges in which the points are predicted to fall by Equation 7.4. Since the points fall in this region, the maximum velocities in these experiments would seem to be well predicted.

**VERIFICATION FROM ACTUAL EDITING PERFORMANCE
(EXPERIMENT 7C)**

In order to verify that Equation 7.4 holds for actual text-editing, an experiment was run measuring v_{max} as a function of mouse movement distance during editing.   A user was given a manuscript marked with modifications and was to make these modifications on the file using the BRAVO display-oriented editor and a mouse.   The position of the cursor (moved by the mouse) was recorded 60 times per sec.   A set of 40 pointing movements were selected for analysis.   For each pointing operation, the maximum speed v_{max} of the mouse during a 1/60th sec interval was computed.   Figure 7.13 also gives the median value of v_{max} recorded for all pointings, averaged over 2-cm target distance intervals. Again the values are as predicted by Equation 7.4.

7.4.   CONCLUSIONS

Of the four devices tested, the mouse is clearly the superior device for text selection on a display:
(1) The positioning time of the mouse is significantly faster than that of the other devices.   This is true overall and at every distance and size combination save for single-character targets, for which it is roughly equal to other devices.
(2) The error rate of the mouse is significantly lower than that of the other devices.
(3) The rate of movement of the mouse is nearly maximal with respect to the information-processing capabilities of the eye-hand guidance system.
As a group, the continuous movement devices are superior in both speed and error rate.

These results can be understood in terms of the Model Human Processor.   For the continuous movement devices, positioning time is given by Fitts's Law.   For key devices, it is proportional to the number of keystrokes.

The practical use of the models derived for pointing device movement was illustrated by two applications:
(1) A modification of Fitts's dotting task can be used as a rapid method for indexing the pointing speed of an analogue device.

(2) For use in planning hardware support, the maximum movement velocity of the mouse can be taken as 4.9 D, where D is the distance to be moved in cm.

ENGINEERING MODELS

8. The Keystroke-Level Model

In this part of the book we are concerned with how our theoretical models of user behavior can be used as practical design tools. GOMS models, as we saw in Chapter 5, can be formulated at several different levels of analysis: the Unit-Task Level, the Functional Level, the Argument Level, and the Keystroke Level. We illustrate two engineering models, based on different levels of GOMS analysis. The first, called simply the Keystroke-Level Model, is similar to Model K1 of Chapter 5, but without an explicit analysis of goals and selection rules. This model is useful where it is possible to specify the user's interaction sequence in detail. The second engineering model is an application of the GOMS Model UT at the Unit-Task Level. This model is useful when it is not possible to know the details of the interaction.

We describe the Keystroke-Level Model in this chapter and take up the Unit-Task Level of analysis in Chapter 9.

8.1. THE TIME PREDICTION PROBLEM

It would be useful for a system designer to have a model that would enable him to predict how much time a user would take to accomplish a given task with a given interactive computer system. The Keystroke-Level Model addresses a restricted subpart of this general problem:

> **Given:** —A task (possibly involving several subtasks).
> —The command language of a system.
> —The motor skill parameters of the user.
> —The response time parameters of the system.
> —The method used for the task.
>
> **Predict:** —The time an expert user will take to execute
> the task using the system, providing he uses
> the method without error.

This formulation stipulates several important boundary conditions on the use of the Keystroke-Level Model. Like the GOMS models, the Keystroke-Level Model predicts only error-free expert behavior. But, unlike the GOMS models, the Keystroke-Level Model must be given the method as input. It does not predict the method—given the method, it predicts the time.

The final restriction on the Keystroke-Level Model is that it predicts only the time to execute a task, not the time to acquire it. Given a large task, such as editing a document, a user will break it into a series of unit tasks. The importance of unit tasks for our analysis is that they permit the time to do a large task to be decomposed into the sum of the times to do its constituent unit tasks.[1]

The detailed subgoal structure of the unit task was discussed in Chapter 5. For the purposes of the Keystroke-Level Model, it is only necessary to consider the top-level structure of the unit task, consisting of

---

[1] Not all tasks have a unit-task substructure. For example, inputting an entire manuscript by typing permits a continuous throughput organization. See Chapter 11 for further discussion.

two parts: (1) *acquisition* of the unit task method and (2) *execution* of the method. To acquire a unit task, the user must not only construct a mental representation of the task to be done, but also choose a method for doing it. To execute the method, the user must interact with the computer system to accomplish the task (which in text-editing includes locating the text, modifying it, and verifying the modification). The total time to do a unit task is the sum of the time for these two parts:

$$T_{unit\text{-}task} = T_{acquire} + T_{execute} .$$

Acquisition time for a unit task depends on the characteristics of the larger task situation in which it occurs. In the manuscript-editing task, where unit tasks are read from a marked-up page or from written instructions, it takes about 2~3 sec to acquire each unit task (Chapter 5). In a routine design task, where unit tasks are generated in the user's mind, it takes about 5~30 sec to acquire each unit task (Chapter 10). And in creative composition, it can take even longer.

Execution time for a unit task, though it depends on the structure of the system's command language, rarely takes over 20 sec (assuming that the system has a reasonably efficient command language). If a unit task requires a longer execution time, the user is likely to break it into smaller unit tasks.

The Keystroke-Level Model predicts only the execution time of unit tasks, and does not predict the acquisition time. Execution is the part of the task over which the system designer has most direct control (i.e., by manipulating the system's command language), so prediction of execution time suffices for many practical purposes.

Two assumptions underlie this treatment of execution time. First, execution time is assumed to be the same no matter how a task is acquired. Second, acquisition time and execution time are assumed to be independent (reducing execution time by making the command language more efficient, therefore, does not affect acquisition time).

8.2. THE KEYSTROKE-LEVEL MODEL

The Keystroke-Level Model is comprised of several primitive operators. Methods can be encoded in terms of these operators by applying a set of heuristics. In this section, we present the operators, discuss the heuristics, and demonstrate a few examples of method encoding.

Operators

The execution part of a unit task can be described in terms of four physical-motor operators, **K** (keystroking), **P** (pointing), **H** (homing) and **D** (drawing), one mental operator **M**, and a system response operator **R** (see Figure 8.1). Execution time is simply the sum of the times spent executing the different operator types:

$$T_{execute} = T_K + T_P + T_H + T_D + T_M + T_R. \tag{8.1}$$

So, for instance, the total time T_K spent in keystroking is the number of keystrokes n_K times the time per keystroke t_K, or $T_K = n_K t_K$. (Operators **D** and **R** are treated somewhat differently.)

The most frequently used operator is **K**, which represents a keystroke or a button press (on a typewriter keyboard or any other button device). **K** refers to keys, not characters (hitting the SHIFT key counts as a separate **K**). The keystroke time t_K is taken to be the standard typing rate, as determined by standard one-minute typing tests. This is an approximation in two respects. First, keying time is different for different keys and key devices (see Figures 2.14 and 2.15). Second, t_K includes the time for immediately-caught typing errors (involving BACKSPACE and rekeying). Thus t_K is computed from a typing test by dividing total test time by the total number of non-error keystrokes, giving the *effective* keying time. We accept both these approximations in the interest of simplicity.

Users can differ in their typing rate by as much as a factor of 15 (Figure 8.1). Given a population of users, an appropriate t_K can be selected. If a user population has members with large t_K differences, then the population should be partitioned into classes and the classes analyzed separately, since the different classes of users will be likely to employ different methods.

The operator **P** represents pointing to a target on a display with a mouse. In Experiment 7A, we measured the time required to point with the mouse and select a target by pressing a button, a sequence we now write as **PK**. The time required by **P** can be estimated by subtracting the time for **K** from Equation 7.2, giving:

$$t_P = .8 + .1 \log_2(D/S + .5) \sec.$$

The fastest time, according to this equation, is .8 sec, and the longest likely time ($D/S=128$) is 1.5 sec. In the interest of simplicity, we use an average value for pointing time of 1.1 sec (1.3 sec from Figure 7.4 less .2 sec for **K**).

When there are different physical devices for the user to operate, he will probably have to move his hands between them. This hand movement, including the fine positioning adjustment of the hand on the device, is represented by the **H** ("home") operator. From the studies in Chapter 5 and Chapter 7, we assign a constant t_H of .4 sec for movement between any two devices.

The **D** operator represents using the mouse to draw a set of straight-line segments. **D** takes two parameters: the number of segments (n_D) and the total length of all segments (l_D). The time $t_D(n_D, l_D)$ is a linear function of these two parameters. The coefficients of this function are different for different users; Figure 8.1 gives average values. **D** is a very specialized operator. Not only is it restricted to the mouse, it also assumes that the drawing system constrains the cursor to lie on a .56 cm grid. The **D** operator is included to indicate the wide scope of tasks potentially addressable by the model.

The user spends some time "mentally preparing" to execute the physical operators just described. Preparation can take the form of deciding how to call a command, for instance, or whether to terminate an argument string. These mental preparations are assumed to take an average of 1.35 sec each, and are represented by the **M** operator (see Section 8.3). Again, the use of a single mental operator is a deliberate simplification.

Finally, the **R** operator represents the system response time. This operator has one parameter t, a placeholder for the response time in seconds of a particular instance in which the system causes the user to wait. When, for example, a 2-sec system response is followed by a user **K** operator (on a system that does *not* allow type-ahead), the user must wait the 2 sec, which is denoted by **R(2)**. When, on the other hand, a system response is followed by an **M** operator by the user, only the system response time in excess of the overlapped 1.35 sec **M** time is counted, or $2-1.35$ sec $=$ .65 sec; and we write **R(.65)**.

Encoding Methods

Methods are represented as sequences of these operators. It is easiest to introduce the method notation with examples. Suppose that there is a

Operator	Description and Remarks	Time (sec)
K	**PRESS KEY OR BUTTON.** Pressing the SHIFT or CONTROL key counts as a separate **K** operation. Time varies with the typing skill of the user; the following shows the range of typical values:	
	Best typist (135 wpm)	.08
	Good typist (90 wpm)	.12
	Average skilled typist (55 wpm)	.20
	Average non-secretary typist (40 wpm)	.28
	Typing random letters	.50
	Typing complex codes	.75
	Worst typist (unfamiliar with keyboard)	1.20
P	**POINT WITH MOUSE TO TARGET ON A DISPLAY.** The time to point varies with distance and target size according to Fitts's Law, ranging from .8 to 1.5 sec, with 1.1 being an average. This operator does *not* include the (.2 sec) button press that often follows. Mouse pointing time is also a good estimate for other efficient analogue pointing devices, such as joysticks (see Chapter 7).	1.10
H	**HOME HAND(S) ON KEYBOARD OR OTHER DEVICE.**	.40
D(n_D,l_D)	**DRAW** n_D **STRAIGHT-LINE SEGMENTS OF TOTAL LENGTH** l_D **CM.** This is a very restricted operator; it assumes that drawing is done with the mouse on a system that constrains all lines to fall on a square .56 cm grid. Users vary in their drawing skill; the time given is an average value.	$.9n_D + .16l_D$
M	**MENTALLY PREPARE.**	1.35
R(t)	**RESPONSE BY SYSTEM.** Different commands require different response times. The response time is counted only if it causes the user to wait.	t

Figure 8.1. The operators of the Keystroke-Level Model.

The **K** times are from Figure 2.14, except the .28 sec, which is the average typing rate of the non-secretarial users in Experiment 8A.   The **P** time is from Chapter 7. The **H** time is from Chapter 5 and Chapter 7.   The **D** time function and the coefficients were derived from least-squares fits on the drawing test data from the four MARKUP users.   The time for **M** was estimated from the data in Experiment 8A.

command named PUT in some system, and that the method for calling it is to type its name followed by the RETURN key. This method would be coded by simply listing the operations in sequence: **MK**[P] **K**[U] **K**[T] **K**[RETURN], which we abbreviate as **M** 4K[P U T RETURN].   In this

Begin with a method encoding that includes all physical operations and response operations. Use Rule 0 to place candidate **M**'s, and then cycle through Rules 1 to 4 for each **M** to see whether it should be deleted.

Rule 0. Insert **M**'s in front of all **K**'s that are not part of argument strings proper (e.g., text or numbers). Place **M**'s in front of all **P**'s that select commands (not arguments).

Rule 1. If an operator following an **M** is *fully anticipated* in an operator just previous to **M**, then delete the **M** (e.g., **PMK** → **PK**).

Rule 2. If a string of **MK**s *belongs to a cognitive unit* (e.g., the name of a command), then delete all **M**'s but the first.

Rule 3. If a **K** is a *redundant terminator* (e.g., the terminator of a command immediately following the terminator of its argument), then delete the **M** in front of it.

Rule 4. If a **K** *terminates a constant string* (e.g., a command name), then delete the **M** in front of it; but if the **K** terminates a variable string (e.g., an argument string), then keep the **M** in front of it.

Figure 8.2. Heuristic rules for placing the M operations.

notation descriptive notes such as key names may be written in square brackets. If, on the other hand, the method to call the PUT command were to point to its name in a menu, then press the RED mouse button, we would write: H[mouse] **MP**[PUT] **K**[RED] H[keyboard].

As another example, consider the text-editing task (called T1) of replacing a 5-letter word with another 5-letter word, where this replacement takes place one line below the previous modification. The method for executing task T1 in the line-oriented editor POET would be described as follows:

Method for Task T1-POET:

Jump to next line	**MK**[LINEFEED]
Issue Substitute command	**MK**[S]
Type new 5-letter word	5**K**[word]
Terminate new word	**MK**[RETURN]
Type old 5-letter word	5**K**[word]
Terminate old word	**MK**[RETURN]
Terminate command	**K**[RETURN] .

Using the operator times from Figure 8.1, and assuming the user is an average skilled typist ($t_K = .2$ sec), we could predict the time it will take to execute this method:

$$T_{execute} = 4t_M + 15t_K = 8.4 \sec .$$

This method could be compared to the method for executing task T1 on the display-based system BRAVO:

Method for Task T1-BRAVO:

Reach for mouse	**H**[mouse]
Point to word	**P**[word]
Select word	**K**[YELLOW]
Home on keyboard	**H**[keyboard]
Issue Replace command	**MK**[R]
Type new 5-letter word	5**K**[word]
Terminate type-in	**MK**[ESC]

$$T_{execute} = 2t_M + 8t_K + 2t_H + t_P = 6.2 \sec .$$

Thus, we can predict that the task would take about two seconds longer using POET than using BRAVO.

The methods above are simple unconditional sequences. More complex or more general tasks are likely to have multiple methods and conditionalities within methods for accomplishing different versions of the task. For example, in BRAVO the user often has to scroll the text on the display before pointing to the desired target (see Chapter 6). In the present notation, the method would be represented as:

$$.4(\textbf{MP}[\text{SCROLL-SYMBOL}] \ \textbf{K}[\text{RED}] \ \textbf{R}(.5)) \ \textbf{P}[\text{word}] \ \textbf{K}[\text{YELLOW}] .$$

Here we have assumed the average number of scrolls per selection to be .4 and the average system response time per scroll to be .5 sec. Using these values, we would predict the average selection time:

$$T_{execute} = .4t_M + 1.4t_K + 1.4t_P + .4(.5) = 2.6 \sec .$$

When there are alternative methods for doing a specific task in a given system, we found (in Chapter 5) that expert users will, in general, use the most efficient method (i.e., the method taking the least time). Thus, in making predictions we can use the model to compute times for alternative methods and then predict that the fastest method will be used. (If the alternatives take about the same time, it does not matter which method we predict.) This optimality assumption holds, of course, only if the users are familiar with the alternatives, which, fortunately, expert users usually are. This assumption is helped by the tendency of optimal methods to also be the simplest.

Heuristics for the M Operator

It is useful to distinguish two versions of method encoding. The *physical encoding* includes only the physical operations (**K**, **P**, **H**, **D**, and **R**) required by the command language of the system. The *cognitive encoding* includes the physical encoding plus the mental (**M**) operations. The Keystroke-Level Model provides a set of heuristic rules (Figure 8.2) for placing **M**'s in a physical encoding to obtain the cognitive encoding.[2]

M operations represent acts of mental preparation for applying physical operations. Their occurrence does not follow directly from the physical encoding, but from the specific knowledge and skill of the user. The rules for placing **M**'s embody psychological assumptions about the user and are necessarily heuristic, especially given the simplicity of the model.

The rules in Figure 8.2 define a procedure that begins with a physical encoding. First, all candidate **M**'s are inserted into the encoding according to Rule 0, which is a heuristic for identifying all possible decision points in the method. Rules 1 to 4 are then applied to each candidate **M** to see if it should be deleted.

---

[2] Thus, only a physical definition of the method is required as input to the Keystroke-Level Model (see the definition of the prediction problem in Section 8.1).

A single psychological principle lies behind all the deletion heuristics. The principle is that physical operations in methods are chunked into submethods. The user cognitively organizes his methods according to these submethod chunks, which usually reflect syntactic constituents of the system's command language. Hence, the user mentally prepares for the next physical chunk, not just the next physical operation. It follows that in executing methods the user is more likely to pause between chunks than within chunks. The rules attempt to identify submethod chunks.

Rule 1 asserts that when an operation is fully anticipated in another operation, the two belong together in a chunk. A common example is pointing with the mouse and then pressing the mouse button to indicate a selection. The button press is fully anticipated during the pointing operation, and there is no pause between them (and thus **PMK** becomes **PK**, according to Rule 1). This anticipation holds even if the selection indication is done on another device (e.g., the keyboard or a foot pedal). Rule 2 asserts that an obvious syntactic unit, such as a command name, constitutes a chunk when it must be typed out in full.

The last two heuristics deal with syntactic terminators. Rule 3 asserts that the user will bundle redundant terminators into a single chunk. For example, in the POET example above, one RETURN is required to terminate the second argument and another RETURN to terminate the command; a user quickly learns simply to hit a double RETURN after the second argument (i.e., **MKMK** becomes **MKK** according to Rule 3). Rule 4 asserts that a terminator of a constant-string chunk will be assimilated into that chunk. An example is that users quickly learn to type, without pausing, a RETURN that always follows a command name.

It is clear that these heuristics do not capture the notion of method chunks precisely, but are only approximations. Further, whether something is "fully anticipated" or is a "cognitive unit" is sometimes ambiguous. Better general heuristics would help in reducing the ambiguity. However, some of the variability in what constitutes a chunk stems from a corresponding variability in expertise. Users differ widely in their behavior; their categorization into novice, casual, and expert users provides only a crude separation and leaves wide variation within each category. One way that experts differ from novices is in what chunks they have (Chase and Simon, 1973). Thus, some of the difficulties in placing **M**'s are unavoidable because not enough is known (or can be known in practical work) about the individual experts involved. Part of the variability in expertness can be represented by the Keystroke-Level Model as encodings with different placements of **M** operations.

Comparison with the GOMS Models

We are now in a position to consider the relation of the Keystroke-Level Model to the GOMS models in Chapter 5. The Keystroke-Level Model most closely corresponds to Model K1 in the GOMS family of models. Both models are at the Keystroke Level, and they both have a generic mental operator: **M** for the Keystroke-Level Model and MENTAL for Model K1.

The mental operators in the two models are not the same, as can be seen by comparing their times: **M** takes 1.35 sec, whereas MENTAL takes .62 sec (Figure 5.15). The reason for this discrepancy is that **M** is a much more aggregate operator than MENTAL: given a method, more MENTAL's will appear in a Model K1 encoding than **M**'s will appear in a Keystroke-Level Model encoding. The correspondence between **M**'s and MENTAL's can be seen by examining Model K2, which classifies the generic MENTAL into several mental operators with more specific functions (see Figure 5.15). Roughly speaking, the **M** operator corresponds to the CHOOSE operations of Model K2. Since CHOOSE operations account for less than half of the mental operations in Model K2, this explains why MENTAL takes less than half as long as **M**.

The most important difference between the Keystroke-Level Model and Model K1 has to do with method prediction. The Keystroke-Level Model does not predict methods[3] and, hence, has no goals or method selection rules (although it does predict where the mental operations occur). This difference and the slight mismatch of operators are deliberate; they represent the ways in which the GOMS description has been simplified to produce the more usable Keystroke-Level Model.

---

[3] The fact that the Keystroke-Level Model does not predict methods means that its results are more appropriately compared to the "reproduction" results than to the "prediction" results of the GOMS models, according to the distinction made in the experiment in Section 5.3.

8.3. EMPIRICAL VALIDATION OF THE MODEL (EXPERIMENT 8A)

To determine how well the Keystroke-Level Model predicts actual performance times, an experiment was run in which calculations from the model were compared against measured times for a number of different tasks, systems, and users.

Description of the Experiment

A total of 1280 user-system-task interactions were observed, comprised of various combinations of 28 users, 10 systems, and 14 tasks.

Systems. The systems were all typical application programs available locally and widely used by both technical and non-technical users. Three of the systems were text-editors, three were graphics editors, and five were executive subsystems. The systems are briefly described in Figure 8.3.

Together, these systems display a considerable diversity of user interface techniques. For example, POET, one of the text-editors, uses first-letter mnemonics to specify commands and uses search strings to locate lines. In contrast, DRAW, one of the graphics systems, presents a menu of graphic icons on the display. These icons, representing the commands, are selected by the user pointing with the mouse.

Tasks. The 14 tasks performed by the users (see Figure 8.4), though diverse, were typical. Users of the editing systems were given tasks ranging from a simple word substitution to the more difficult task of moving a sentence from the middle to the end of a paragraph. Users of the graphics systems were given tasks such as adding a box to a diagram or deleting a box (but keeping a line that overlapped the box). Users of the executive subsystems were given tasks such as transferring a file between computers or examining part of a file directory.

Task-System Methods. In all, there were 32 task-system combinations: 12 for the text editors (4 tasks × 3 systems), 15 for the graphics systems (5 tasks × 3 systems), and 5 for the executive subsystems (one task for each subsystem). For each task-system combination, the most efficient "natural" method was determined (by consulting experts) and then coded in Keystroke-Level Model operations. The encodings of the methods for all the task-system combinations are listed in the Appendix to this chapter.

System	Description
Text-Editors	
POET	Line-oriented, with relative line numbers.
SOS	Line-oriented, with absolute line-numbers.
BRAVO	Display-oriented; full-page; uses mouse for pointing.
Graphics Systems	
MARKUP	Uses mouse to draw and erase lines and areas on a display; commands selected from a hidden menu, which must be redisplayed each time.
DRAW	Lines defined by pointing with mouse to end points; commands selected with mouse from a menu.
SIL	Lines defined by pointing with mouse to end points; boxes defined by pointing to opposite vertices; commands selected by combinations of mouse buttons.
Executive Subsystems	
LOGIN	TENEX command for logging in.
FTP	Program for transferring files between computers.
CHAT	Program for establishing a "teletype" connection between two computers.
DIR	TENEX command for printing a file directory; has a subcommand mode.
DELVER	TENEX command for deleting old versions of a file.

Figure 8.3. Systems measured in Experiment 8A.

POET, described in Chapters 3 and 5, is a dialect of the QED editor (Deutsch and Lampson, 1967). For SOS see Savitsky (1969). For MARKUP see Newman and Sproull (1979), Chapter 17. For LOGIN, DIR, and DELVER, see Myer and Barnaby (1973). All the rest are experimental systems local to Xerox PARC, designed and implemented by many individuals, including: Roger Bates, Patrick Baudelaire, David Boggs, Butler Lampson, Charles Simonyi, Robert Sproull, Edward Taft, and Chuck Thacker.

Editing Tasks (used for POET, SOS, BRAVO)

T1.   Replace one 5-letter word with another (one line from previous task).

T2.   Add a fifth letter to a 4-letter word (one line from previous task).

T3.   Delete a line of text (eight lines from previous task).

T4.   Move a 52-character sentence, spread over two lines, to the end of its paragraph (eight lines from previous task).

Graphics Tasks (used for MARKUP, DRAW, SIL)

T5.   Add a rectangular box to a diagram.

T6.   Add a 5-character label to a box.

T7.   Disconnect a 2-segment line from one box and reconnect it to another box.

T8.   Delete a box, but keep an overlapped line.

T9.   Copy a box to another part of the diagram.

Executive Tasks

T10.   Phone computer and log in (4-character name, 6-character password).

T11.   Transfer a file to another computer, renaming it.

T12.   Connect to another computer.

T13.   Display a subset of the file directory and show file lengths.

T14.   Delete old versions of a file.

Figure 8.4. Tasks used in Experiment 8A.

Experimental Design. The basic design of the experiment was to have ten versions of each task on each system done by four different users, giving 40 observed instances per task-system. To avoid transfer effects, no user was observed on more than one system (except for the executive subsystems). Four tasks were observed for each of the text-editing systems, five tasks for each of the graphics systems, and one task for each of the executive subsystems.

Users. There were, in all, 28 different users (some technical, some secretarial): 12 for the editing systems, 12 for the graphics systems, and 4 for the executive subsystems. All were experts in that they had used the systems for months in their regular work and had used them recently.

Experimental Procedure. Each user was first given five one-minute typing tests to determine his keystroke time t_K. In addition, users of MARKUP (the only system to require manual drawing) were given a series of drawing tasks to determine the parameters of their drawing rate (as discussed in Section 8.2).

After the preliminary tests, each user was given a small number of practice problems of the sort to be tested and was told which method to use (see above). In most cases, the methods presented were what users claimed they would have used anyway; in the other cases, the method was easily adopted. Users practiced tasks until they were judged to be at ease with using the correct method; this was usually accomplished in three or four practice trials on each task type.

After practicing, the user proceeded to the main part of the experiment. The user was given a notebook containing several manuscript pages with the tasks to be done marked in red ink. Text-editing and graphics tasks appeared in randomized order; executive subsystem tasks were always ordered T11, T12, T13, T14; and all ten instances of task T10, logging into a computer, were done in succession.

Each experimental session, lasting approximately 40 minutes, was videotaped and the user's keystrokes were recorded automatically. Time stamps on the videotaped record and on each keystroke allowed a protocol to be constructed in which the time of each event was known to within .033 sec. This protocol is the basic data from which the results below were derived.

Results of the Experiment

Each task instance in the protocols was divided into acquisition time and execution time according to the following definitions. Acquisition time began when the user first looked over to the manuscript to get instructions for the next task and ended when the user started to perform the first operator of the method.[4] Execution time then began at that

---

[4] Technically, the boundary between acquisition and execution time was determined by taking the first recorded operator of the execution method (usually a **K**) and using it to estimate the starting time of the method's first operator.

point and ended when the user looked over to the notebook for the next task.

Those tasks on which there were significant non-typing errors or in which the user did not use the prescribed method were excluded from further consideration. After this exclusion, 855 (69%) of the task instances remained as observations to be matched against the predictions. No analysis was made of the excluded tasks.

The resulting observed times for task acquisition and execution were stable over repetition. There was no statistical evidence for times decreasing (implying learning) or increasing (implying fatigue) with repetition.

CALCULATION OF EXECUTION TIME

Execution time was calculated using the method analysis for each task-system combination together with estimates of the times required for each operator (see chapter Appendix). All times, except for mental preparation time, were taken from sources outside of the experiment. Pointing time t_P and homing time t_H were taken from Figure 8.1. Typing time t_K and drawing time $t_D(n_D, l_D)$ were estimated from the typing and drawing tests by averaging the times of the four users involved in each task-system. System response time T_R for each task-system was estimated from independent measurements of the response times for the various commands required in each method. For task T10, logging into a computer, a telephone button-press was assumed to take time t_K. Moving the telephone receiver to the computer terminal modem was estimated to take .7 sec, using the MTM system of times for industrial operations (Maynard, 1971).[5]

Mental preparation time t_M was estimated from the experimental data itself. First, the total mental time for each method was determined by removing the predicted time for all physical operations from the observed execution time. Then, t_M was estimated by a least-squares fit of the estimated mental times as a function of the predicted number of **M** operations. The result was $t_M = 1.35$ sec ($R^2 = .84$, standard error of estimate = .11 sec, standard error about the regression line = 2.48 sec). The *SD* of t_M was 1.1 sec, indicating that the **M** operator had the characteristic variability of mental operators (Section 5.5).

---

[5] One point of task T10 is to illustrate that the Keystroke-Level Model can be extended by using existing catalogues of physical operators.

Execution times for each task-system combination were calculated by Equation 8.1. The execution-time calculations are summarized in Figure 8.5, which also gives the observed execution times from the experiment for comparison.

EXECUTION TIME

The accuracy of the predictions can be seen in Figure 8.6, which plots the predicted vs. observed data from Figure 8.5. The root-mean-square (*RMS*) error is 21% of the average predicted execution time. This accuracy is about the best that can be expected from the Keystroke-Level Model, since the choice of methods used by the subjects were controlled by the experimental procedure. The 21% *RMS* error is about the same as the reproduction accuracy of Model K1 in Chapter 5 (Figure 5.16).

The distribution of relative prediction errors is evenly spread, as an analysis of Figure 8.6 shows. No particular systems or tasks make excessively large contributions. Predictions are not consistently positive or negative for systems or tasks, except that the predicted executive subsystem task times were uniformly too high. Examination of the individual observations does not reveal any small set of outliers or particular users that inflates the prediction error.

Prediction accuracy is related to the duration of the attempted prediction. Since unit tasks are essentially independent, prediction of the time to do a set of tasks will tend to be more accurate than prediction of the time to do a single unit task (see Chapter 5 for the argument). For example, using the model to predict how long it took to do *all four* text-editing tasks, the average *RMS* error is only 5% and the corresponding *RMS* error for the graphics editors over the five tasks is only 6%.

Ideally, all the parameters of the model should be determined independently of the experiment. The only parameter for which this was not possible was the mental operation time t_M, because there was no appropriate independent source of data available. The substantial variability of t_M indicates that the consequent inflation in the model's apparent accuracy is probably not too serious, since small changes in the value of t_M make little difference. For example, if a t_M as small as 1.2 sec or as large as 2.0 sec were used in the predictions, the *RMS* error for the Keystroke-Level Model would only increase from 21% to 23%. Of course, the t_M estimated from this experiment is now available as an independent estimate for use by others.

The variability in the observed task times is of interest, since user behavior is inherently variable (see Chapter 2). In our data, the average

Task-System	t_K	n_M	n_K	n_H	n_P	n_D	l_D	T_R	$T_{execute}$	$T_{execute}$ $M \pm SE(N)$	Pred. Error
	(sec)						(cm)	(sec)	(sec)	(sec)　　(sec)	
T1-Poet	.23	4	15	--	--	--	--	--	8.8	7.8 ± 0.9(27)	11%
T1-Sos	.22	4	19	--	--	--	--	--	9.6	9.6 ± 0.8(31)	1%
T1-Bravo	.23	2	8	2	1	--	--	--	6.4	5.7 ± 0.3(31)	11%
T2-Poet	.28	4	14	--	--	--	--	--	9.4	8.9 ± 0.7(17)	5%
T2-Sos	.23	4	18	--	--	--	--	--	9.5	9.7 ± 0.8(32)	- 3%
T2-Bravo	.24	2	4	2	1	--	--	--	5.6	4.1 ± 0.3(32)	26%
T3-Poet	.19	3	12	--	--	--	--	--	6.3	6.3 ± 0.4(24)	0%
T3-Sos	.23	2	7	--	--	--	--	--	4.3	4.0 ± 0.3(37)	8%
T3-Bravo	.23	1	2	1	1	--	--	--	3.3	3.5 ± 0.2(38)	- 7%
T4-Poet	.19	13	92	--	--	--	--	--	35.3	37.1 ± 4.3(20)	- 6%
T4-Sos	.23	12	47	--	--	--	--	--	26.8	32.7 ± 1.8(16)	−22%
T4-Bravo	.24	2	6	1	3	--	--	3.8	11.6	14.3 ± 1.1(33)	−23%
T5-Markup	.25	--	3.2	--	2.5	4	24.9	--	11.1	10.5 ± 1.1(27)	6%
T5-Draw	.25	7.6	12.6	--	5	--	--	--	18.9	12.5 ± 3.0(22)	34%
T5-Sil	.27	1	4	0.4	2	--	--	--	4.8	5.4 ± 0.7(32)	−12%
T6-Markup	.26	1	7	2	1	--	--	--	5.0	6.2 ± 0.4(34)	−23%
T6-Draw	.25	1	7	1	1	--	--	--	4.6	5.9 ± 0.4(34)	−29%
T6-Sil	.27	--	6	1.4	1	--	--	--	3.3	3.6 ± 0.3(19)	- 9%
T7-Markup	.24	--	8.6	--	4.8	6	13.6	--	15.1	15.0 ± 2.1(29)	2%
T7-Draw	.19	5	13	--	8	--	--	--	18.0	18.2 ± 1.9(9)	- 1%
T7-Sil	.28	1	8	--	5	--	--	--	9.1	12.3 ± 2.1(23)	−36%
T8-Markup	.26	--	8	--	8	1	4.0	--	12.3	9.3 ± 0.4(22)	24%
T8-Draw	.21	1	5	--	3	--	--	--	5.7	5.3 ± 0.3(25)	7%
T8-Sil	.27	1	5	0.7	2	--	--	--	5.2	4.1 ± 0.2(33)	20%
T9-Markup	.25	2	8	--	6.5	--	--	3.5	15.4	13.0 ± 2.5(26)	15%
T9-Draw	.22	--	5.7	--	5.7	--	--	--	7.5	10.5 ± 1.0(25)	−40% ˙˙
T9-Sil	.28	--	5	0.3	3	--	--	--	4.8	6.0 ± 1.0(28)	−24%
T10-Login	.29	2	28	--	--	--	--	15.9	27.4	25.1 ± 0.7(29)	9%
T11-Ftp	.30	5	31	--	--	--	--	10.1	26.1	19.7 ± 0.7(29)	24%
T12-Chat	.31	1	11	--	--	--	--	8.3	13.1	11.5 ± 0.6(36)	12%
T13-Dir	.30	2	20	--	--	--	--	0.5	9.2	6.6 ± 0.3(32)	28%
T14-Delver	.32	2	20	--	--	--	--	0.4	9.4	7.5 ± 0.4(33)	20%

Figure 8.5. Calculated and observed execution times in Experiment 8A.

The calculations are done according to Formula (8.1) using the operator times in Figure 8.1, except for t_K, which is the average time from the actual typing tests for the users on a given system. Each user's time is weighted by the correct number of instances for that user on a given task (column N). $SE = SD/\sqrt{N}$, which is the standard error of estimation of the population mean for samples of size N. The calculated execution time for task T10 also includes .7 sec for the operation of picking up the telephone receiver (see Section 8.3).

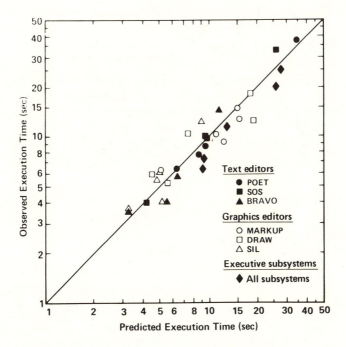

Figure 8.6. Predicted vs. observed execution times in Experiment 8A.

The plotted values are taken from Figure 8.5.

CV of the individual observations over each task is .31, which is typical of variability for behavior of this duration according to Figure 5.19. In comparing predictions by the model against any actual behavior, prediction error will always be confounded with some error from the sampling process. Sampling error for each of our observed task times is indicated in the SE column of Figure 8.5. The average standard error is 9%. The prediction error of the Keystroke-Level Model being more than two times larger than this indicates that most of the prediction error is due to a real failure in the model and not just to unreliable observations.

ACQUISITION TIME

Turning from the execution part of the task to the acquisition part, the data show that it took users 2 sec on the average to acquire the tasks from the manuscript. This number may be refined by breaking the tasks into three types: (1) those tasks that the user already had in memory (the executive subsystem tasks that were done each time in the same order);

(2) those tasks for which the user had to look at the manuscript each time (all the graphics tasks, the POET and SOS tasks, and task T11); and (3) those tasks for which the user had to look at the manuscript, then scan text on the display to locate the task. The times for these three types of acquisition are given in Figure 8.7. Users took .5 sec when the task was in memory, 1.8 sec when the task had to be retrieved from the manuscript, and 4.0 sec when the users had to get the task from the manuscript and search for it on the video display. The time for getting the task from the manuscript is similar to the results obtained in Chapter 5, where the GET-NEXT-TASK operator took 1.92 sec (Figure 5.15). It is interesting to note that although display editors are generally faster to use, they impose a 2-sec penalty by requiring the user to scan the text on the display.

Task Type	Task numbers	Acquisition Time $M \pm SE(N)$ (sec) (sec)
All tasks	T1–T14	2.0 ± 2.0 (885)
Repeated task, recalled from memory	T10, T12, T13, T14	0.5 ± 0.3 (130)
Task acquired by looking at manuscript	T1–T4 (POET, SOS), T5–T9, T11	1.8 ± 1.9 (621)
Task acquired by looking at manuscript, then scanning for task on display	T1–T4 (BRAVO)	4.0 ± 1.9 (134)

Figure 8.7. Observed acquisition times in Experiment 8A.
SE is the standard error of estimate of the population mean for samples of size N.

We can use the acquisition times in Figure 8.7, along with the calculated execution times in Figure 8.5, to predict total task times. The resulting *RMS* error of these predictions is 21%, which is just as accurate as predicting the execution times alone.

8.4. A FURTHER LOOK AT THE M OPERATOR (EXPERIMENT 8B)

We have presented evidence for the validity of the Keystroke-Level Model by showing that it predicts overall task execution times, but we have not examined unit-task executions to see if the Keystroke-Level Model predicts performance in finer detail. There is little doubt that performance of physical operations corresponds to the model. How well does the generic **M** operator predict the pause times between the observed physical operations? To investigate this issue, we consider an exploratory experiment (Moran, 1980), in which the detailed performance record of one user was compared to the model's predictions.

DESCRIPTION OF THE EXPERIMENT

Task and Method. The user was given the task of converting Sentence 8.2a to Sentence 8.2b using the BRAVO editor:

The sun shines when it rains; our weather is funny.	(8.2a)
Our weather is funny; when it rains the sun shines.	(8.2b)

In addition to switching the two outer clauses of the original, the task requires changing punctuation and capitalization. The task is complex, requiring several BRAVO commands, and there are several different ways of performing it. The timewise optimal method requires seven commands,[6] as shown in Figure 8.8. As predicted by the Keystroke-Level Model, this method requires seven **M** operations.

User. The user was an experienced BRAVO user. She had considerable technical training, including some programming; but she was not an experienced programmer.

Procedure. The optimal method was discussed verbally with the user—but without actually executing it on the system—for about 30 minutes, after which she began the experimental session. Using BRAVO, the user executed the clause-switching task 100 times. The task was exactly the same on each trial (the user edited a file with 100 copies of

---

[6] The details of how the optimal method works are not important for the present discussion. However, we take up the clause-switching task as a problem-solving task in Chapter 11.

OPTIMAL METHOD:

(C1) Delete 3rd clause and...	H[mouse] **PK PK M K**[D]
(C2) ...insert it in front of 1st clause.	**PK M K**[I] **K**[ESC]
(C3) Replace "; o" by "O".	**PK M K**[R] **K**[SHIFT]
	H[keyboard] **2K**[O ESC]
(C4) Replace "T" by "; t".	H[mouse] **PK M K**[R]
	H[keyboard] **4K**[; SPACE T ESC]
(C5) Delete 3rd clause and...	H[mouse] **PK PK M K**[D]
(C6) ...insert it in front of 2nd clause.	**PK M K**[I] **K**[ESC]
(C7) Find next task.	**M K**[F]

TIME PREDICTION:

$$T_{execute} = [24t_K + 8t_P + 5t_H] + 7t_M$$
$$= [24(.15) + 8(1.03) + 5(.57)] + 7(1.35)$$
$$= 14.7 + 9.4$$
$$= 24.1 \text{ sec}$$

Figure 8.8. Optimal method for the clause-switching task and its predicted time.

The underlined **K**'s in the method indicate command-invocation keystrokes (see Figure 8.9 and Figure 8.10). Unit operator times for the prediction were obtained by measuring the user; see the text for the rationale. Pointing time t_P decreases with practice; the value of t_P used here is the value at Trial 7 (see the caption for Figure 8.11 for an explanation).

Sentence 8.2a on it), and the user employed exactly the same method every time.[7] On subsequent days, the user repeated 100-trial sessions until she had completed eleven sessions (1100 trials) in all. Records of time-stamped keystrokes were collected for all sessions.

[7] Note that the prediction of the **M**'s in Figure 8.8 is based on the assumption that the parts of the method become fixed. For example, no **M** is predicted to occur before the terminator (ESC) of the Replace command, since the replacement string is the same every time (see Rule 4 in Figure 8.2).

ANALYSIS AND RESULTS

Analysis.   The data were analyzed first to isolate the method execution times, as was done in Experiment 8A. Execution times were then partitioned into physical and mental components using the Keystroke-Level Model. Figure 8.8 shows the model's predicted time for an error-free trial. The time for the physical component (the physical operations) was predicted to be 14.7 sec, and the time for the mental component (the **M** operations) was predicted to be 9.4 sec.

The procedure for partitioning the user's actual execution times was to estimate the physical time and then to regard the remaining time as mental time. In order to estimate the physical time as accurately as possible, the unit physical operator times (t_H, t_P, and t_K) were obtained by direct measurement of the user (rather than by taking the values from Figure 8.1). The physical component time was estimated by alotting unit operation times for the physical operations, which were inferred from the data record. For example, in executing the optimal method (Figure 8.8), the first recorded keystroke would be a mouse button-press. Because of the method analysis in Figure 8.8, this recorded keystroke is assumed to represent not just a **K** operation, but three physical operations, **HPK**; the estimated physical time is therefore $t_H + t_P + t_K = .57 + 1.03 + .15 = 1.75$ sec. If the recorded time for this keystroke were 3.0 sec, then the remaining time ($3.00 - 1.75 = 1.25$ sec) would be assumed to be mental time.

Mental Time Results.   Mental operations, as indicated by logical analysis and by pauses in the user's physical activity, appear to occur close to the locations predicted by the model; but they are less regular than idealized by the model, and they occur in more places. The user's mental time pattern is best examined by considering only error-free trials. Since all error-free trials have the same sequence of physical operations, the mental operations can be compared with each other and with the model's prediction. Figure 8.9 illustrates graphically the operation times for a subset of the error-free trials that were selected to show the user's gradual reduction in execution time. Also shown is the model's predicted time, which is within 2% of the user's overall execution time on her first error-free trial, Trial 7. Further, the user is seen to require mental operations uniformly throughout Trial 7; just as predicted, they occur in every command of the method.

The correspondence between the user's mental operations and the model's predicted **M** operations is directly exhibited in Figure 8.10. The

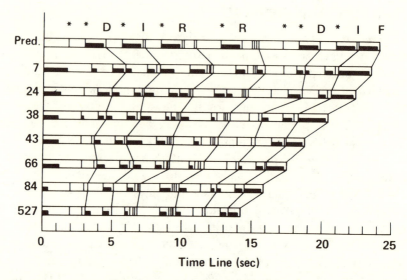

Figure 8.9. Time-line graph of keystrokes for execution of the clause-switching task in Experiment 8B.

Each of the eight horizontal bars is a time-line graph of the sequence of operations in an execution of the optimal method. The topmost time line (labeled "Pred.") represents the execution time predicted by the Keystroke-Level Model (Figure 8.8). Just above this time line are labels for the command-invocation keystrokes and the mouse button-presses (indicated by *'s); see Figure 8.8 for where they occur in the method. The remaining time lines represent several of the user's performances, labeled by their trial numbers. The vertical strokes in the time lines represent recorded keystrokes. The little black horizontal bars in the time lines represent inferred mental operators. The diagonal lines between time lines show the corresponding command boundaries (as defined in Figure 8.8) between time lines.

figure shows histograms of the user's mental time for the trials graphed in Figure 8.9. A trial consists of seven command executions. Each command has a *command-invocation keystroke*, such as D for the Delete command or R for the Replace command, indicated in Figure 8.8 by the underlined **K**'s. The model predicts an **M** operation immediately preceding each command-invocation keystroke. The command-invocation keystrokes thus provide reference points in the execution by which to compare the locations of mental time. The horizontal axis in Figure 8.10 represents the time preceding the recorded command-invocation key-strokes, normalized to be at time 0. The model's predicted **M** operations occur uniformly between the times -1.50 sec and $-.15$ sec, as is shown by the lightly-shaded histogram at the top of Figure 8.10. The histogram

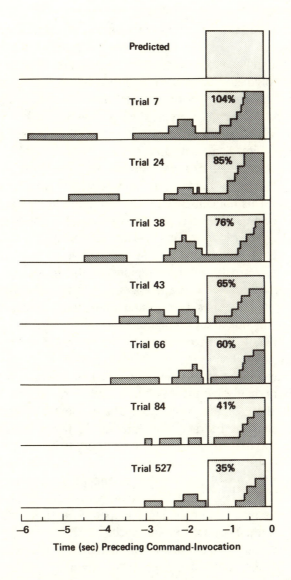

Figure 8.10. **Histograms of mental times preceding command-invocation keystrokes in Experiment 8B.**

Each histogram corresponds to one of the time-line bars in Figure 8.9. The lightly-shaded histograms represent the **M** operations predicted by the Keystroke-Level Model. The darkly-shaded histograms represent the user's actual mental times in each trial. The percentages in the lightly-shaded histograms represent the ratio of the user's mental time to the predicted mental time.

is rectangular, since it represents seven identical **M** operations stacked on top of each other. The distributions of the user's actual mental times are shown as the darkly-shaded histograms, which are superimposed over copies of the predicted mental time histogram for comparison. As predicted, most of the user's actual mental time does occur in the two or three seconds preceding the command-invocation keystrokes. The figure also gives for each trial the percentage ratio of the user's actual mental time to the predicted mental time.

Figures 8.9 and 8.10 make clear several features of the user's actual mental time. The user's mental time is more widely distributed than the model predicts. This happens mainly because pointing operations usually occur before command-invocation keystrokes, and the user requires mental time to prepare for them. Most noticeably, the user requires unpredicted mental operations at the beginning of each trial; these mental operations are the leftmost regions of the histograms in Figure 8.10. The bimodal shape of the histograms in the early trials indicates that the user often requires two mental operations corresponding to a single predicted **M** operation. For example, in Trial 7 the user engages in 13 mental operations, as compared to the predicted seven operations. (Since the total mental time in Trial 7 was equal to the predicted mental time, the average duration of the actual mental operations must have been only about half of t_M.) With repetitive practice on the clause-switching task, the user reduces the amount of mental time required by reducing both the number and the duration of her mental operations. Although the user's mental time in Trial 7 is 104% of the predicted mental time, the user reduced her mental time to 35% of the predicted time by Trial 527. Also, the number of mental operations is reduced from 13 in Trial 7 to 7 in Trial 527. However, even on Trial 527—her best trial—the user still required mental operations, although they were reduced to less than a half-second each.

Learning Results. With practice, the amount of mental preparation time spent by the user declines. We should expect the user's performance in this experiment to improve according to the Power Law of Practice (Chapter 2). The user's execution times over the 1100 trials are plotted in Figure 8.11 on log-log coordinates. The times approximately follow the power law, although the data is noisy in the later sessions. The predicted execution time is also shown. It can be seen that the user's performance corresponds to the model's prediction early in the first session, as was also evident in Figure 8.9. The user's execution time

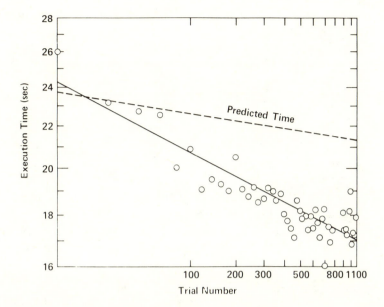

Figure 8.11. Learning curve of execution times for the clause-switching task in Experiment 8B.
Each circle represents the mean of 20 trials. The solid line is the least-squares fit to the Power Law of Practice. The dashed line shows the time predicted by the Keystroke-Level Model. The predicted time changes because the user improved her unit pointing time, t_p (from 1.31 sec in Session 1 to .86 sec in Session 11). According to the Power Law of Practice, the value of t_p for each trial was estimated by interpolating with a power function.

becomes faster than predicted by the end of the first session and continues to get faster thereafter. Most of this improvement over the predicted time is due to compression of the mental time required by the user, as just discussed.

Conclusion. In Section 8.3, we established that the Keystroke-Level Model is an accurate predictor of expert behavior under normal human-computer interaction conditions, where there is a variety of tasks. However, under the special conditions of this study, where an identical task was repeated over and over, the user's behavior became much better than the model's prediction; much of this improvement was due to compressing mental time. In particular, the user's observed mental operations were more dispersed throughout her editing activity than the description given by the **M** operator of the Keystroke-Level Model. For a more detailed account of the user's mental operations, we must turn to a GOMS model.[8]

8.5. SAMPLE APPLICATIONS

We have provided evidence for the Keystroke-Level Model in a wide range of user-computer interactions. The time required for experts to perform a unit task was predicted to within about 20% by a linear function of a small set of operators. The power of the Keystroke-Level Model lies in permitting prediction without having to do any measurements of the actual situation and in expressing the prediction as a simple algebraic expression. Its limitation lies in requiring that the physical method be specified at the Keystroke Level and in being limited to error-free expert behavior.

In this section, we illustrate how the Keystroke-Level Model can be used, both to exploit its possibilities and to work within its restrictions. The basic application—point prediction of specific interaction times—has been sufficiently illustrated in the course of the experiment, where such predictions were made for 32 different tasks involving 10 highly diverse systems. We now show three further uses: (1) calculated benchmarks; (2) parametric analysis, where predictions are expressed as functions of task variables; and (3) sensitivity analysis, where changes in the predictions are examined as a function of changes in either task or model parameters.

Calculated Benchmarks

The Keystroke-Level Model makes it possible to calculate the equivalent of a benchmark for a system and, hence, to compare systems. This has obvious cost advantages over having to obtain actual measurements. More importantly, it permits benchmarking at design time, before the system exists in a form that permits actual measurement. The experimental data from Section 8.3 can be used as a ready illustration.

Suppose we were to use the four tasks T1 to T4 as a benchmark for the three text editors, POET, SOS, and BRAVO. Without performing

---

[8] For example, the GOMS Model K2 (see Figure 5.12) has mental operators that are not only less aggregate than the **M** operator (see Section 8.2), but that also provide functional labels for each mental operation.

experiments, we could use the Keystroke-Level Model to compute the total benchmark time for each system. The computed benchmark times come directly from Figure 8.5 by summing the calculated $T_{execute}$ of T1 to T4 for each editor, giving 59.8 sec for POET, 50.2 sec for SOS, and 26.9 sec for BRAVO. Taking the POET time (the slowest) as 100, we get ratios of 100:84:45. As we might have expected, the two line-oriented editors are relatively close to each other and the display editor is substantially faster. Since we have also done the experiment, we can compare these calculated benchmarks with the observed benchmarks (by summing the observed $T_{execute}$ from Figure 8.5). This time we get 60.1 sec for POET, 56.0 sec for SOS, and 27.6 sec for BRAVO, or experimentally determined ratios of 100:93:46—essentially the same result. The agreement between the calculated and observed benchmark provides confidence only in using the calculated benchmark in place of a measured one. It does not provide evidence for the validity of the particular benchmark (tasks T1-T4) or for whether benchmarks are generally a valid way to compare editors.

A similar analysis can be performed for the three graphics systems, using tasks T5-T9 as the benchmark. The analysis predicts ratios of 100:93:46 for MARKUP, DRAW, and SIL, respectively; the observed ratios were 100:97:58. The ratio between MARKUP and DRAW is close enough to raise the question of whether the predicted difference is too small to be reliable. The calculated difference between MARKUP and DRAW on the benchmark is $59.0 - 54.7 = 4.3$ sec or 7%. The model has an *RMS* prediction error of 21% for a single unit task. Since this benchmark is essentially an independent sum of five unit tasks, the *RMS* error should theoretically be $21\%/\sqrt{5} = 9\%$.[9] Thus, predictions for the two systems are within the *RMS* error of the model, so the predicted difference between them can hardly be reliable. The fact that the model correctly predicted that DRAW was slightly faster than MARKUP was lucky—there is no reason to expect the Keystroke-Level Model to always make such fine discriminations successfully.

Parametric Analysis

We can illustrate the use of the model for parametric analysis and sensitivity analysis with the following example problem:

---

[9] Recall in Section 8.3 that the actual RMS error for the graphics systems was 6%.

Problem. A user is typing text into the BRAVO editor and detects a misspelled word n words back from the word he is currently typing. He wants to correct the misspelled word and resume typing. What methods will the user use for this task? How long will these methods take? Is it possible to design a better method for this task?

Let us compare two methods available in BRAVO for making the correction. Since the methods may behave quite differently depending on how far back the misspelled word is, we need to determine how long each method takes as a function of n.

The first method for correcting the word makes use of the Backword command (invoked by hitting the CONTROL key and then W), which erases the last typed-in word:

Method W (Backword):

Set up Backword command	**MK**[CONTROL]
Execute Backword n times	$n((1/c)\textbf{MK}[\text{W}])$
Type new word	5.5**K**[word]
Retype destroyed text	$5.5(n-1)\textbf{K}$

$$T_{execute} = (1+n/c)t_M + (1+6.5n)t_K \qquad (8.3)$$
$$= 1.6 + 2.16n \text{ sec}.$$

The execution time is a function not only of n, but also of the way the user chunks repeated keystrokes. When a user has to repeat a single-keystroke command several times, like the Backword command in the above method, he will chunk the sequence into small bursts separated by pauses (the pauses represented as **M** operations), according to Rule 2 in Figure 8.2. The average number of Backword commands chunked in a burst is represented by the parameter c. We use this parameter in the second step in the above method, where we count $1/c$ **M** operations for each use of the Backword command. Since we do not know an exact value for c, we assume the value $c = 4$ in our calculations (we return to this decision in the next section). In the calculations we also assume an average non-secretarial typist ($t_K = .28$ sec) and an average word-length of 4.5 characters (excluding associated punctuation and spaces). The second method for correcting the word is to exit type-in mode, use the Replace command to correct the word, and then re-enter type-in mode, so that type-in can be resumed:

Method R (Replace):

Terminate type-in mode	**MK**[ESC]
Point to target word and select it	**H**[mouse] **P**[word] **K**[YELLOW]
Call Replace command	**H**[keyboard] **MK**[R]
Type new word	4.5**K**[word]
Terminate Replace command	**MK**[ESC]
Point to last input word and select it	**H**[mouse] **P**[word] **K**[YELLOW]
Re-enter type-in mode	**H**[keyboard] **MK**[I]

$$T_{execute} = 4t_M + 10.5t_K + 4t_H + 2t_P$$
$$= 12.1 \text{ sec}.$$

The predicted times for the two methods as a function of n are plotted as the two solid lines in Figure 8.12a. As the figure shows, it is faster to use the Backword method up until a certain crossover point n_{WR}, after which it becomes faster to use the Replace method. Under the above assumptions, the crossover from the Backword method to the Replace method is found to be at 4.9 words.

Now, let us consider providing a new method to improve performance on this correction task. The new method will require implementing two new commands in BRAVO. We wish to determine, *before* implementing the commands, whether they are likely to be much of an improvement.

The first new command is a Backskip command (CONTROL S), which moves the text-insertion point back one word without erasing any text. The second new command is a Resume command (CONTROL R), which moves the insertion point back to where it was when the first Backskip command was invoked. These commands allow the Backskip method:

Method S (Backskip):

Set up Backskip command	**MK**[CONTROL]
Execute Backskip $n-1$ times	$(n-1)((1/c)$**MK**[S]$)$
Call Backword command	**MK**[W]
Type new word	4.5**K**[word]
Call Resume command	**M2K**[CONTROL R]

$$T_{execute} = (3+(n-1)/c)t_M + (n+7.5)t_K \qquad (8.4)$$
$$= 5.8 + .62n \text{ sec}.$$

The predicted time for the Backskip method is plotted as the dashed line in Figure 8.12a. With the addition of this method there are two

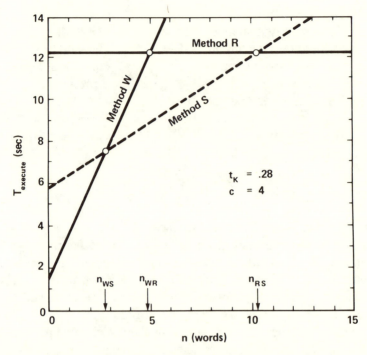

Figure 8.12a. **Execution time of three methods for the misspelled-word task as a function of** *n.*

Method W uses BRAVO's Backword command, and Method R uses the Replace command. Method S uses the proposed new Backskip command.

additional crossover points, n_{WS} and n_{RS}, between it and the other two methods. As can be seen, the Backskip method is faster than both of the other methods between n_{WS} and n_{RS}, in the range from 2.7 to 10.2 words. Thus, a brief analysis provides evidence that the proposed new feature probably will be useful in the sense that it will be the fastest method over a significant region of the task space.

Sensitivity Analysis

How sensitive are the calculations above to variations in the parameters of the methods? The question of interest is whether, over such variations, there remains a region in the task space in which the Backskip method is the fastest. An important parameter is the user's typing speed t_K. How much does the crossover between the Backward method and the Backskip method change as a function of typing speed?

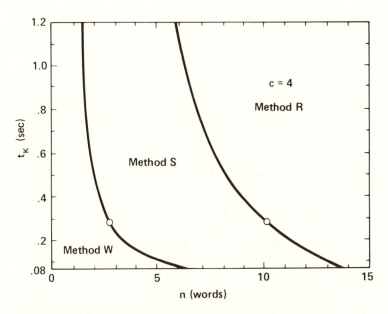

Figure 8.12b. Boundaries for the fastest method.
The space is divided into three regions; each region is labeled with the name of the fastest method over that region.

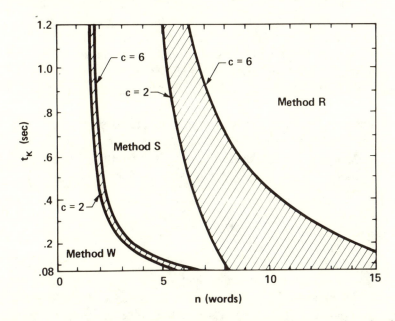

Figure 8.12c. Boundaries adjusted for different chunk sizes.
As in Figure 8.12b, each region is labeled by the fastest method over that region. The crosshatched areas indicate the variability in the boundary between regions as c is varied from $c = 2$ to $c = 6$.

291

Setting Equation 8.3 equal to Equation 8.4 and solving for n as a function of t_K gives

$$n_{WS} \equiv n = 1.2 + .43/t_K.$$

The crossover point n_{WS} increases with typing speed (decreasing t_K), rising to $n_{WS} = 6.6$ words for the fastest typist ($t_K = .08$ sec). As n increases, slow typists can be expected to switch from the old Backword method (which involves more typing) to the new Backskip method (which involves less typing, but more mental overhead) sooner than fast typists.

We can plot the crossover boundary between the two methods in the space of the two parameters: n (characterizing different tasks) and t_K (characterizing different users). The two boundaries of the new Backskip method are plotted in Figure 8.12b. These boundaries define the regions in the parameter space where each method is fastest. The circles mark the crossover points corresponding to the ones in Figure 8.12a (i.e., at $t_K = .28$ sec). There is a large region in the space that is dominated by the new method. In fact, the new method is dominant for certain values of n, no matter what t_K is. Thus our conclusion, that the new Backskip method may be a useful method, is not sensitive to assumptions about the particular typing speed of the users. (Actually, of course, the analysis should take into account the relative frequencies of various points in the parameter space; we have omitted this complication to simplify the example.)

How sensitive are these conclusions to the value assumed for c, the number of keystrokes per chunk? To find out, we re-derive the crossover between the Backword and Backskip methods by setting Equation 8.3 equal to Equation 8.4 and solving for n as a function of both c and t_K, giving

$$n_{WS} = 1.2 + .49/t_K - .24/ct_K.$$

Although we do not know an exact value for c, we can be reasonably confident that it will stay between 2 and 6. With $t_K = .28$ sec, the crossover varies between 2.5 and 2.8 words as c varies between 2 and 6; the crossoverpoint is not very sensitive to the value of c at this point.

The sensitivities of the various crossover points for other values of t_K can be assessed by replotting Figure 8.12b using the reasonable extreme values of c. The two crossover boundaries for the Backskip method are plotted in Figure 8.12c as crosshatched lines defined by setting c to 2 and to 6 in the crossover equations. The diagram shows that the value of c affects the boundary between the Backskip and Replace methods more than the boundary between the Backskip and Backword methods, since c is not involved in the Replace method. Small chunk sizes especially penalize the Backskip method. The boundary between the Backskip and Backword methods is not affected much by c, since the chunk size is involved in both methods in exactly the same way. Overall, varying c does not squeeze out the region for the new Backskip method; and our basic conclusion—that the new method is a useful addition—still holds.

The sensitivity analyses above illustrate how the Keystroke-Level Model can be used to evaluate design choices—even when many aspects of the calculation are uncertain—for the principal conclusions are often insensitive to many of the uncertainties.

8.6. SIMPLIFICATIONS OF THE MODEL

The question naturally arises as to whether further simplifications of the Keystroke-Level Model might do reasonably well at predicting execution time. One could (1) count only the number of keystrokes, (2) count just the physical operators and prorate the time for mental activity, or (3) use a single constant time for all operators. We show below that such simplifications degrade accuracy. However, they provide useful approximations where lower accuracy can be tolerated.

Keystrokes-Only Simplification

The first simplification is to consider only the keystrokes, in which execution time is proportional to the number of keystrokes:

$$T_{execute} = \kappa n_K + T_R .$$

We separate out the system response times T_R so as not to confound the comparison of the various simplifications. The constant of proportionality κ should be distinguished from the typing speed t_K. The latter, determined from standard typing tests, is the keystroke time in a copy-typing task, whereas the former is the average time per keystroke in an interaction task. Estimating the value of κ from a least-squares fit of the values of n_K and the observed $T_{execute}$ in Figure 8.5 gives $\kappa = .49$ sec/keystroke. The *RMS* error is 49% (compared to 22% for the Keystroke-Level Model). The statistics for comparing all the simplifications are presented in Figure 8.13. As can be seen, using keystrokes only is substantially less accurate than using the full Keystroke-Level Model. This simplification is inappropriate for tasks that are not dominated by keystroking. For example, it only predicts about a third of the observed time for the MARKUP tasks, which are dominated by pointing and drawing operations.

The keystrokes-only simplification is essentially the model previously introduced in Figure 3.7 (Chapter 3). The above estimate of κ is strongly influenced by one outlying point in the data, T4-POET ($n_K = 92$). Estimating κ with this one point removed gives $\kappa = .60$ sec, a value close to the .57 sec estimate obtained in Chapter 3. T4-POET is the only task that requires any input-typing of text. One obvious refinement of the keystrokes-only simplification would be to distinguish two kinds of keystrokes: mass input-typing (at t_K sec/keystroke) vs. command-language keying (at κ sec/keystroke). For command-language keying, a κ of .60 sec is the more reasonable value.

The model of Embley, Lan, Leinbaugh, and Nagy (1978; see also Embley and Nagy, 1981), though formally similar to our keystrokes-

Model Variation	Parameters	R^2	*RMS* Error
Keystrokes Only	$\kappa = .49$ sec/keystroke	.76	49%
Prorated Mental Time	$\mu = 1.67$	.66	45%
Constant Operator Time	$\tau = .43$ sec/operator	.85	34%
Keystroke-Level Model	(See Figure 8.1)	.90	22%

Figure 8.13. Comparison of the Keystroke-Level Model with simplifications of the model.

The correlations are between the execution times predicted by each of the models and the observed execution times from Figure 8.5. The *RMS* error is given as a percentage of the *observed* execution time, which was 11.0 sec. (This is why the *RMS* error for the Keystroke-Level Model is 22% here.) More useful values for the κ and τ parameters are $\kappa = .60$ sec and $\tau = .49$ sec.

only simplification, is conceptually distinct. The Keystroke-Level Model is based on the notion of a unit-task structure; the Embley et al. model is based on system commands. The Keystroke-Level Model is restricted to skilled expert behavior, whereas Embley et al. attempt to model all kinds of users (essentially, by varying their versions of the parameters $T_{acquire}$ and κ). Since they did not compare their model against empirical performance data, we cannot directly compare our results to theirs. But because of the similarities to the keystrokes-only simplification, Embley et al.'s model might be expected to have about the same accuracy.

Prorated-Mental-Time Simplification

According to the prorated-mental-time simplification, execution time is the time required for the physical operations multiplied by a factor to account for the mental time:

$$T_{execute} = \mu (T_K + T_H + T_P + T_D) + T_R .$$

The idea is that the physical operations will require a certain average overhead of mental activity. Thus, instead of trying to predict exactly how many mental operations there are, we can do fairly well by just using a multiplicative mental overhead constant, μ.

Using a least-squares analysis to determine μ from the sum of the calculated times for the physical operations and from the observed values of $T_{execute}$ in Figure 8.5 gives $\mu = 1.67$, signifying a 67% overhead for mental activity. The *RMS* error is 45%.

Like the keystrokes-only simplification, this simplification is also less accurate than the Keystroke-Level Model, as can be seen in Figure 8.13, suggesting that the extra detail in the Keystroke-Level Model, involving the explicit placements of the mental preparation operator **M**, is effective.

There is an interesting relation between these simplifications and the rules for placing occurrences of **M** in the Keystroke-Level Model (Figure 8.2). The initial placement of **M**'s (by Rule 0) near certain **K**'s and **P**'s is essentially an assumption that mental time is proportional to a subset of the physical operators. If Rule 0 had specified all physical operators, it would, by itself, have been equivalent to prorating mental time. If the other physical operators (**P**, **H**, and **D**) had been ignored, this would have been equivalent to counting keystrokes only. Thus, the deletion of the **M**'s according to Rules 1 to 4 constitutes the way in which the

Keystroke-Level Model departs from these simplifications. The evidence for the superiority of the Keystroke-Level Model presented in Figure 8.13 is also evidence that rules Rules 1 to 4 had a significant effect. The contribution of each of the rules individually is significant, in the sense that each one's removal leads to a decrease in the accuracy of the Keystroke-Level Model.

Constant-Operator-Time Simplification

According to this simplification, execution time is proportional to the *number* of Keystroke-Level Model operations:

$$T_{execute} = \tau (n_M + n_K + n_P + n_H + n_D) + T_R .$$

Support for this simplification comes from the statistical observation (Wainer, 1976; Claude, 1972) that the accuracy of linear models is not very sensitive to the differential weighting of the factors—equal weighting does nearly as well as any other weighting. Thus, we disregard the different operator times and use a single time τ for all operators. Note that the constant-operator-time simplification is formally similar to the keystrokes-only simplification; the latter can be viewed as using n_K as a crude estimate of the total number of operators.

Estimating τ by a least-squares fit of the data in Figure 8.5 gives $\tau =$ .43 sec/operator. The *RMS* error is 34%. (Removing the long typing task, T4-POET, gives $\tau = $ .49 sec/operator.)

The constant-operator-time simplification is more accurate than the keystrokes-only simplification, affirming that taking into account operators other than **K** is useful. In fact, most of the action in the constant-operator-time simplification (over the set of data in Figure 8.5, at least) comes from counting only the **K**, **P**, and **M** operators. The constant-operator-time simplification is still less accurate than the Keystroke-Level Model, showing that the use of estimates of each operator time yields yet another increment of accuracy.

In summary, all the simplifications presented in this section are less accurate than the full Keystroke-Level Model. However, these simplifications are probably good enough for many practical applications, especially for "back-of-the-envelope" calculations, where it is too much trouble to worry about the subtleties of counting the **M**'s required by the full Keystroke-Level Model.

8.7. CONCLUSIONS

The GOMS analysis at the Keystroke Level has been refined into a model of practical use, the Keystroke-Level Model.   Only a few operators—keystroke, point, home, draw, a generic mental operator, and a system response operator—are needed to describe methods in a wide range of interactive systems.   Heuristic rules are provided to predict where the mental operations are needed.

The generic mental (**M**) operator of the model appears to be more aggregate than users' actual mental operations, although there is a close correspondence between **M**'s and the actual mental time.   With highly repetitive tasks, users can reduce their mental time below the model's predictions.

The Keystroke-Level Model can be used to estimate the execution time of a method for doing a task.   In laboratory experiments, the model was accurate to a standard error of 21% over a variety of different tasks and systems.   Applications of the model include point prediction, calculated benchmarks, parametric analysis, and sensitivity analysis.

Simplifications of the Keystroke-Level Model—such as counting only keystrokes, prorating mental time, or using a constant operator time—are much less accurate at predicting execution time; but they do provide the designer with greater ease of use at the expense of accuracy.

Appendix to Chapter 8:
METHODS FOR THE TASKS IN EXPERIMENT 8A

This Appendix gives the methods and their Keystroke-Level Model encodings for all the task-system combinations used in Experiment 8A. The notation is explained in Section 8.2. The following notes elaborate on specific points:

(1) **.3P**   Fractional coefficients indicate that certain actions occurred less than 100% of the time. For example, the order in which tasks are actually done influences the necessity of certain actions.

(2) **0H**   The coefficient of zero indicates this action does not occur in this particular task, although it could in similar tasks.

(3) **R(0)**   A response time of zero indicates that the actual response time is absorbed in the beginning of the subsequent task and therefore is not added to the task time of the current task.

(4) **7K**   A search string is assumed to average 7 characters.

(5) **5K**   Line numbers in the SOS editor are 5 digits long.

(6) **M18C**   This is a special operator taken from the MTM predetermined time standards (Maynard, 1971).

(7) **5K**   A label is assumed to average 4 characters. When capitalized, its total number of keystrokes is five.

Citations to the notes appear in the extreme right-hand column of the method encodings.

Task T1: *Replace one 5-letter word with another (one line from previous task).*

Method for Task T1-POET:

Jump to next line	**M K**[LINEFEED]
Issue Substitute command	**M K**[S]
Type new word	5**K**[word]
Terminate new word	**M K**[RETURN]
Type old word	5**K**[word]
Terminate old word	**M K**[RETURN]
Terminate command	**K**[RETURN]

Method for Task T1-SOS:

Issue Substitute command	**M K**[S]	
Type old word	5**K**[word]	
Terminate old word	**M K**[ESC]	
Type new word	5**K**[word]	
Terminate new word	**M K**[ESC]	
Type line number	5**K**[number]	(5)
Terminate line number	**M K**[RETURN]	

Method for Task T1-BRAVO:

Reach for mouse	**H**[mouse]	
Point to word	**P**[word]	
Select word	**K**[YELLOW]	
Home on keyboard	**H**[keyboard]	
Issue Replace command	**M K**[R]	
Type new word	5**K**[word]	
Terminate type-in	**M K**[ESC]	
Wait for completion	**R**(0)	(3)

Task T2: *Add a fifth letter to 4-letter word (one line from previous task).*

Method for Task T2-POET:

Jump to next line	**M K**[LF]
Issue Substitute command	**M K**[S]
Type new word	5**K**[word]
Terminate new word	**M K**[RETURN]
Type old word	4**K**[word]
Terminate old word	**M K**[RETURN]
Terminate command	**K**[RETURN]

Method for Task T2-SOS:

Issue Substitute command	**M K**[S]	
Type old word	4**K**[word]	
Terminate old word	**M K**[ESC]	
Type new word	5**K**[word]	
Terminate new word	**M K**[ESC]	
Type line number	5**K**[number]	(5)
Terminate command	**M K**[RETURN]	

Method for Task T2-BRAVO:

Reach for mouse	**H**[mouse]	
Point to word	**P**[word]	
Select word	**K**[YELLOW]	
Issue Append command	**M K**[A]	
Home on keyboard	**H**[keyboard]	
Type new letter	**K**[letter]	
Terminate type-in	**M K**[ESC]	
Wait for completion	**R**(0)	(3)

Task T3: *Delete a single line of text*
(eight lines from previous task).

Method for Task T3-POET:

Indicate search string	**M K**[QUOTE]	
Type search string	7**K**[string]	(4)
Terminate search string	**M K**[QUOTE]	
Print line	**K**[SLASH]	

Issue Delete command	**M K**[D]
Terminate command	**K**[RETURN]

Method for Task T3-SOS:

Issue Delete command	**M K**[D]	
Type line number	5**K**[number]	(5)
Terminate command	**M K**[RETURN]	

Method for Task T3-BRAVO:

Reach for mouse	**H**[mouse]	
Point to line	**P**[line]	
Select line	**K**[RED]	
Issue Delete command	**M K**[D]	
Wait for completion	**R**(0)	(3)

Task T4: *Move a 52-character sentence (on two lines) to the end of its paragraph (eight lines from previous task).*

Method for Task T4-POET:

Delete sentence at current location
    Delete part of sentence on first line

Indicate search string	**M K**[QUOTE]	
Type search string	7**K**[string]	(4)
Terminate search string	**M K**[QUOTE]	
Print line	**K**[SLASH]	
Issue Edit command	**M** 2**K**[E RETURN]	
Issue Search subcommand	**M** 2**K**[CTRL S]	
Type first letter of sentence	2**K**[SHIFT letter]	
Delete rest of line	**M K**[RETURN]	

    Delete part of sentence on second line

Jump to next line	**M K**[LINEFEED]
Issue Edit command	**M** 2**K**[E RETURN]
Issue Delete subcommand	**M** 2**K**[CTRL Y]
Type first letter of sentence	2**K**[SHIFT letter]
Save rest of line	**M** 2**K**[CTRL Z]

Retype sentence at new location

Indicate search string	**M K**[QUOTE]	
Type search string	7**K**[string]	(4)
Terminate search string	**M K**[QUOTE]	
Print line	**K**[SLASH]	
Issue Append command	**M 2K**[A RETURN]	
Type sentence	52**K**[sentence]	
Terminate type-in	**M 2K**[CTRL Z]	

Method for Task T4-SOS:

Break out sentence onto its own lines

Break sentence out of first line

Issue Alter command	**M K**[A]	
Type line number	5**K**[number]	(5)
Terminate line number	**M K**[RETURN]	
Issue Search subcommand	**M K**[S]	
Type first letter of sentence	2**K**[SHIFT character]	
Issue Insert subcommand	**M K**[I]	
Type line break	**K**[RETURN]	
Terminate subcommand	**M K**[ESC]	
Terminate command	**K**[RETURN]	

Break sentence out of second line

Issue Alter command	**M K**[A]	
Type line number	5**K**[number]	(5)
Terminate line number	**M K**[RETURN]	
Issue Search subcommand	**M K**[S]	
Type first letter of next sentence	2**K**[SHIFT character]	
Issue Insert subcommand	**M K**[I]	
Type line break	**K**[RETURN]	
Terminate subcommand	**M K**[ESC]	
Terminate command	**K**[RETURN]	

Move sentence to new location

Issue Transfer command	**M K**[T]	
Specify new location	5**K**[number]	(5)
Type separator	**K**[COMMA]	
Specify first line to be moved	5**K**[number]	(5)
Type separator	**K**[COLON]	
Specify last line to be moved	5**K**[number]	(5)
Terminate command	**M K**[RETURN]	

Method for Task T4-BRAVO:

Delete sentence from current location

Reach for mouse	**H**[mouse]
Point to beginning of sentence	**P**[character]
Select beginning point	**K**[RED]
Point to end of sentence	**P**[character]
Select ending point	**K**[BLUE]
Issue Delete command	**M K**[D]
Wait for completion	**R**(3.8)

Move sentence to new location

Point to new location	**P**[character]	
Select new location	**K**[RED]	
Issue Append-deleted-text command	**M 2K**[A ESC]	
Wait for completion	**R**(0)	(3)

Task T5: Add a box (rectangle) to a diagram.

Method for Task T5-MARKUP:

Select drawing mode

Reach for mouse	0**H**[mouse]	(2)
Point to place for menu	.6**P**[display]	(1)
Display menu	.6**K**[YELLOW-DOWN]	(1)
Expand menu	.3**P**[menu]	(1)
Point to menu icon	.6**P**[icon]	(1)
Undisplay menu	.6**K**[YELLOW-UP]	(1)

Draw rectangle

Point to corner of rectangle	**P**[corner]
Begin drawing mode	**K**[RED-DOWN]
Draw rectangle	**D**(4,24.86)
Terminate drawing mode	**K**[RED-UP]

Method for Task T5-DRAW:

Get into line-drawing mode

Begin Draw mode	.6(**M K**[ESC])	(1)
Reach for mouse	0**H**[mouse]	(2)

Draw first side of rectangle
    Point to corner and select   **P**[corner] **K**[BLUE]
    Point to next corner and select   **P**[corner] **K**[BLUE]
    Draw line   **M K**[ESC]
Draw second side of rectangle
    Reselect current corner   **M K**[BLUE]
    Point to next corner and select   **P**[corner] **K**[BLUE]
    Draw line   **M K**[ESC]
Draw third side of rectangle
    Reselect current corner   **M K**[BLUE]
    Point to next corner and select   **P**[corner] **K**[BLUE]
    Draw line   **M K**[ESC]
Draw fourth side of rectangle
    Reselect current corner   **M K**[BLUE]
    Point to first corner and select   **P**[corner] **K**[BLUE]
    Draw line   **M K**[ESC]

Method for Task T5-SIL:
    Reach for mouse   .4**H**[mouse]   (1)
    Point to corner and select   **P**[corner] **K**[RED]
    Point to opposite corner and select   **P**[corner] **K**[BLUE]
    Draw rectangle   **M** 2**K**[CTRL B]

Task T6: *Add a label (5 characters, first one capitalized) to a box.*

Method for Task T6-MARKUP:
    Home on keyboard   **H**[keyboard]
    Type shift and label   5**K**[SHIFT label]   (7)
    Terminate type-in and get mouse   **M H**[mouse]
    Point to location of label   **P**[location]
    Paste label   **K**[RED]

Method for Task T6-DRAW:
    Home on keyboard   **H**[keyboard]
    Type shift and label   5**K**[SHIFT label]   (7)

Terminate type-in	**M K**[RETURN]	
Point to location of label	**P**[display]	
Paste label	**K**[RED]	

Method for Task T6-SIL:

Reach for mouse	**.4H**[mouse]	(1)
Point to location of label and select	**P**[display] **K**[RED]	
Home on keyboard	**H**[keyboard]	
Type shift and label	**5K**[SHIFT label]	(7)

Task T7: *Disconnect a 2-segment line from one box (rectangle) and connect it to a different box.*

Method for Task T7-MARKUP:

Select drawing mode

Reach for mouse	**OH**[mouse]	(2)
Point to place for menu	**.3P**[display]	(1)
Display menu	**.3K**[YELLOW-DOWN]	(1)
Expand menu	**.15P**[menu]	(1)
Point to menu icon	**.3P**[icon]	(1)
Undisplay menu	**.3K**[YELLOW-UP]	(1)

Erase two line segments

Point to end of line	**P**[end]
Enter erase mode	**K**[BLUE-DOWN]
Trace line segments	**D**(2,5.65)
Exit from erase mode	**K**[BLUE-UP]

Redraw damaged segments

Point to end of segment 1	**P**[end]
Enter drawing mode	**K**[RED-DOWN]
Redraw segment	**D**(1,1.13)
Exit from drawing mode	**K**[RED-UP]
Point to end of segment 2	**P**[end]
Enter drawing mode	**K**[RED-DOWN]
Redraw segment	**D**(1,1.13)
Exit from drawing mode	**K**[RED-UP]

Draw segments in new location
Point to end of segment	P[end]
Enter drawing mode	K[RED-DOWN]
Draw segment	D(2,5.65)
Exit from drawing mode	K[RED-UP]

Method for Task T7-DRAW:

Select Delete command mode
Reach for mouse	OH[mouse]	(2)
Point to Delete icon and select	P[icon] K[RED]	

Delete line segments
Point to line segment 1	P[line]
Delete line	K[RED]
Point to line segment 2	P[line]
Delete line	K[RED]

Redraw damaged line segments
Enter drawing mode	M K[ESC]
Point to end of line and select	P[end] K[BLUE]
Point to end of line and select	P[end] K[BLUE]
Draw line segment	M K[ESC]

Draw line segments in new location
Point to end of line and select	P[end] K[BLUE]
Point to end of line and select	P[end] K[BLUE]
Draw line segment 1	M K[ESC]
Reselect end of segment 1	M K[BLUE]
Point to end of line and select	P[end] K[BLUE]
Draw line segment 2	M K[ESC]

Method for Task T7-SIL:

Delete line segments (one by shortening)
Reach for mouse	OH[mouse]	(2)
Point to segment and select	P[segment] K[BLUE]	
Point to new endpoint	P[location]	
Shorten segment	2K[CTRL RED]	
Delete other segment	M 2K[CTRL D]	

Draw line segments in new location
Point to end of new segment	P[end]
Select	K[RED]
Point to other end of new segment	P[segment]

Draw segment	**K**[YELLOW]
Point to end of second segment	**P**[segment]
Draw segment	**K**[YELLOW]

Task T8: *Delete a box (rectangle) with an overlapped line to another part of the diagram, keeping the overlapped line.*

Method for Task T8-MARKUP:

Select area deletion mode

Reach for mouse	**OH**[mouse]	(2)
Point to place for menu	**P**[display]	
Display menu	**K**[YELLOW-DOWN]	
Expand menu	**.5P**[menu]	(1)
Point to menu icon	**P**[icon]	
Undisplay menu	**K**[YELLOW-UP]	

Erase rectangle

Point to corner of area	**P**[corner]	
Enter erase mode	**K**[BLUE-DOWN]	
Point to opposite corner of area	**P**[corner]	
Erase area	**K**[BLUE-UP]	

Select drawing mode

Point to place for menu	**P**[display]	
Display menu	**K**[YELLOW-DOWN]	
Expand menu	**.5P**[menu]	(1)
Point to menu icon	**P**[icon]	
Undisplay menu	**K**[YELLOW-UP]	

Redraw damaged line segment

Point to end of segment	**P**[end]	
Enter drawing mode	**K**[RED-DOWN]	
Draw segment	**D**(1,3.96)	
Exit from drawing mode	**K**[RED-UP]	

Method for Task T8-DRAW:

Reach for mouse	**OH**[mouse]	(2)
Point to area-select icon and select	**P**[icon] **K**[RED]	
Point to corner of area	**P**[corner]	

Enter area selection mode	**K**[RED-DOWN]
Point to opposite corner of area	**P**[corner]
Exit from area selection mode	**K**[RED-UP]
Issue Delete command	**M 2K**[CTRL D]

Method for Task T8-SIL:

Reach for mouse	.7**H**[mouse]	(1)
Point to corner of area and select	**P**[corner] **K**[RED]	
Point to opposite corner of area	**P**[corner]	
Select area	2**K**[CTRL BLUE]	
Issue Delete command	**M 2K**[CTRL D]	

Task T9: Copy a box (rectangle) to another part of the diagram.

Method for Task T9-MARKUP:

Select area deletion mode

Reach for mouse	0**H**[mouse]	(2)
Point to place for menu	**P**[display]	
Display menu	**K**[YELLOW-DOWN]	
Expand menu	.5**P**[menu]	(1)
Point to menu icon	**P**[icon]	
Undisplay menu	**K**[YELLOW-UP]	

Delete rectangle and save in buffer

Point to corner of area	**P**[corner]
Enter erase mode	**K**[BLUE-DOWN]
Point to opposite corner of area	**P**[corner]
Erase area	**K**[BLUE-UP]
Wait for deletion	**R**(1.1)

Restore deleted rectangle from buffer

Enter copy-from-buffer command	**M K**[RED-DOWN]
Point to location	**P**[display]
Copy buffer to location	**K**[RED-UP]
Wait for completion	**R**(1.2)

Copy rectangle to new location

Enter copy-from-buffer command	**M K**[RED-DOWN]
Point to new location	**P**[display]
Copy buffer to location	**K**[RED-UP]
Wait for figure to be pasted	**R**(1.2)

Method for Task T9-DRAW:

Select rectangle

Reach for mouse	**0H**[mouse]	(2)
Point to area-select icon and select	**.7**(**P**[icon] **K**[RED])	(1)
Point to corner of area	**P**[corner]	
Enter area selection mode	**K**[RED-DOWN]	
Point to opposite corner of area	**P**[corner]	
Exit from area selection mode	**K**[RED-UP]	

Execute copy command

Point to copy icon and select	**P**[icon] **K**[RED]
Point to old location and select	**P**[location] **K**[BLUE]
Point to new location and select	**P**[location] **K**[BLUE]

Method for Task T9-SIL:

Reach for mouse	**.3H**[mouse]	(1)
Point to corner of area and select	**P**[corner] **K**[RED]	
Point to opposite corner of area	**P**[corner]	
Select area	**2K**[CTRL BLUE]	
Point to new location	**P**[location]	
Issue Copy command	**2K**[CTRL YELLOW]	

Task T10: Phone the computer and login with a 4-character login name and a 6-character password.

Method for Task T10-TENEX:

Dial up computer on phone

Press 8 digits on phone	**8K**[number]	
Wait for computer tone	**R**(1.8)	
Put phone on terminal cradle	**M18C**(.7)	(6)
Wait for carrier signal light	**R**(.9)	

Login to computer

Type login prompt	**M** 2**K**[CTRL C]
Wait for system greeting	**R**(5.9)
Issue Login command	4**K**[L O G SPACE]
Type login name	4**K**[name]
Terminate name	**K**[SPACE]
Type password	6**K**[password]
Terminate password	**K**[SPACE]
Type account number	**K**[1]
Terminate login	**M K**[RETURN]
Wait for completion of login	**R**(7.3)

Task T11: *Transfer a file to a file server (5-character name), renaming the file from a 4-character filename to a 10-character filename.*

Method for Task T11-TENEX:

Connect to file server

Start up FTP program	4**K**[F T P SPACE]
Specify file server	5**K**[name]
Terminate command	**M K**[RETURN]
Wait for connection	**R**(4.1)

Transfer and rename file

Issue Store command	**M** 3**K**[S T SPACE]
Type old filename	4**K**[name]
Terminate filename	**M K**[SPACE]
Type new filename	10**K**[name]
Terminate command	**M K**[RETURN]
Wait for completion	**R**(1.4)

Close connection

Issue Quit command	**M** 2**K**[Q RETURN]
Wait for connection to close	**R**(4.6)

Task T12: *Connect from one computer to another*
computer (5-character name).

Method for Task T12-TENEX:

Start up Chat program	**5K**[C H A T SPACE]
Type computer name	**5K**[name]
Terminate command	**M K**[RETURN]
Wait for connection	**R**(8.3)

Task T13: *Display a subset of files (with a 10-character*
specification) along with their file lengths.

Method for Task T13-TENEX:

Issue Directory command	**4K**[D I R SPACE]
Specify files	**10K**[name]
Call for subcommand mode	**M K**[COMMA]
Enter subcommand mode	**K**[RETURN]
Issue Length subcommand	**2K**[L E]
Terminate subcommand	**M K**[RETURN]
Terminate command	**K**[RETURN]
Wait for completion	**R**(.5)

Task T14: *Delete all the old versions of a subset of files*
(with a 10-character specification).

Method for Task T14-TENEX:

Issue Delver command	**6K**[D E L V E R]
Terminate command	**M K**[RETURN]
Answer first system question	**K**[Y]
Answer second system question	**K**[Y]
Specify files	**10K**[name]
Terminate file specification	**M K**[RETURN]
Wait for completion	**R**(.4)

9. The Unit-Task Level of Analysis

The Keystroke-Level Model presented in Chapter 8 requires that it be possible to specify methods at the Keystroke Level of analysis. This requirement places the conceptual stages of design, where this level of detail is inappropriate, outside the range of the model's applicability. Yet it is precisely during conceptual design that important decisions on the basic configuration of a system must often be made. These decisions could be aided by approximate estimates of the time cost of various design alternatives, since the designer's concern is with a system's gross functional capabilities rather than with the details of which buttons to press. In this chapter we develop a technique of GOMS analysis at the Unit-Task Level, which is appropriate for this stage of system analysis.

The basis of the technique we describe is the unit task. We have seen that users tend to break a large task into a series of unit tasks within which behavior is highly integrated and between which dependencies are minimal. This quasi-independence of unit tasks means that their effects are approximately additive. Estimates of task time can therefore be obtained by enumerating unit tasks and estimating both their frequencies and duration. The total time for a task can be found by multiplying the total number of unit tasks by the time per unit task. This is, essentially, the GOMS Model UT.

9.1. CASE STUDY OF A PAGE-LAYOUT SYSTEM

Problem. A company is contemplating the development of a computer-based system for page layout of journal articles in the style of the journal *Cognitive Psychology*. The proposed system would assemble the elements of a document from different on-line files into a single file embodying the laid-out pages. Input files include a file of the main text (the body of the document), a file of figures, a file of figure captions, and a file of footnotes. The input text files are assumed to have been created with simple text-entry systems incapable of specifying font and format information. Thus, the task includes: (1) positioning and formatting the text, (2) setting various pieces of text into the correct fonts, and (3) numbering pages, section headings, figures, and footnotes. In order to assess the economics of the proposed system, the company's management needs to know the average time it will take to lay out a page with the proposed system. Since the system's interface has not yet been designed, the estimate cannot depend on details of its user interface.

In order to calculate the time to lay out a page by the Unit-Task-Level analysis, we need to first identify the unit tasks involved by analyzing the requirements of the task and the properties of the proposed system.

Analysis of the Task

What are the possible unit tasks? The simplest way to enumerate them is to consider the functions and objects involved in the page-layout task. The types of document objects are *text* (including both large bodies and small segments), *headings, figures,* figure *captions, footnotes,* and finally *pages.* Document objects must be *loaded* from the input files into the workspace where the pages will be laid out. Once loaded, the objects must be *positioned* on the page, the *font* must be set for various text objects, and some objects must be *numbered.* The set of all possible unit tasks can be generated by applying all the different functions to all the different object types, thus forming the array, shown in Figure 9.1.

	Loading	——— Positioning ———		Setting Fonts	Numbering
Pages	LOAD- PAGE (R)	POS-VERT- PAGE	POS-HORIZ- PAGE		NUMBER- PAGE
Headings	LOAD- HEADING (R)	POS-VERT- HEADING	POS-HORIZ- HEADING	SET-FONT- HEADING	NUMBER- HEADING
Text	LOAD- TEXT (R)	POS-VERT- TEXT	POS-HORIZ- TEXT	SET-FONT- TEXT	NUMBER- TEXT
Figures	LOAD- FIGURE (R)	POS-VERT- FIGURE (R)	POS-HORIZ- FIGURE (R)		
Captions	LOAD- CAPTION (R)	POS-VERT- CAPTION (R)	POS-HORIZ- CAPTION (R)	SET-FONT- CAPTION	NUMBER- CAPTION
Footnotes	LOAD- FOOTNOTE (R)	POS-VERT- FOOTNOTE (R)	POS-HORIZ- FOOTNOTE (R)	SET-FONT- FOOTNOTE	NUMBER- FOOTNOTE

Figure 9.1. Array of all possible unit tasks for the page-layout task.

The vertical columns of the array represent the functions involved in the page-layout task, and the horizontal rows represent the document objects to which the functions are applied. The R's indicate the unit tasks that involve a significant system response time.

It is assumed that the system has sufficient functional capability to allow each of the tasks in the array to be done by the user as a single unit task. It is assumed further that there is a simple method for loading each document object from its input file to the page-layout workspace so that **LOAD-FIGURE**, for example, is a single unit task. It is also assumed that positioning an object on a page, because it must be done vertically and horizontally, requires two unit tasks, **POS-VERT-FIGURE** and **POS-HORIZ-FIGURE**. Not all combinations of functions applied to document objects make sensible unit tasks. For example, it makes no sense to set the font of a figure, since a figure is not a piece of text; consequently, there is no **SET-FONT-FIGURE** unit task.

In the overall task of laying out a document, unit tasks are not performed in a random sequence. Rather, they are grouped together to accomplish higher-level goals, which we informally call *task groups*. For example, in laying out a figure, the figure will be loaded and positioned, its caption will be loaded and positioned, the caption font will be set, and the caption and the figure callout in the text will be numbered. All these tasks are grouped together under the task group **PROCESS-FIGURE**. Laying out a page consists of seven such task groups—setting up the new page; processing the headings, the figures, the footnotes, and the refer-

Task Groups	Sample Article				
	N_1	N_2	N_3	N_4	N_{TG}
PROCESS-NEW-PAGE	1.00	1.00	1.00	1.00	1.00
PROCESS-HEADING	.68	1.88	1.36	.58	1.12
PROCESS-FIGURE	.29	.34	.22	.41	.24
PROCESS-FOOTNOTE	.13	.18	.18	.25	.18
PROCESS-INDENTATION	2.32	2.62	3.82	3.75	3.13
PROCESS-TEXT-FONT	1.26	5.91	4.27	7.50	4.42
PROCESS-REFERENCE	3.03	1.34	1.09	.25	1.43

Figure 9.2. Frequency of the task groups per page from four sample articles.

The sample articles contained from 12 to 32 pages each. The frequencies of the task groups in the four articles are given in columns N_1 to N_4. The N_{TG} for each task group is the average frequency over all four articles.

ences; formatting indented paragraphs; and setting the font (usually italics) for various pieces of text—which are listed in Figure 9.2. It is most convenient to carry out the analysis of the page-layout task in terms of these task groups.

The average time to lay out a page depends on the frequency with which the various task groups need to be performed, that is, on the ecology of printed pages. To estimate the frequency of the task groups, a small sample of four articles was taken from the journal *Cognitive Psychology* and the number of task groups needed to lay out each article was counted. The frequencies of task groups per page are shown in Figure 9.2. The average frequencies of the four sample articles provide an estimate of the ecological frequency for each task group.

Analysis of the System

In the analysis of the task so far, only general assumptions have been made about features of the layout system, assumptions that would hold over the whole class of layout systems under consideration. In order to proceed further with the analysis, it is necessary to postulate more specific system features: Is it necessary for the user to number each page manually? Need the user manually indent each paragraph? In the present problem, we do not know the answers to these questions and therefore must make assumptions. In order to understand the effects of these assumptions, we examine two very different layout systems: a *Manual System*, in which the user has to do most of the layout steps explicitly, and an *Automatic System*, which does many of these steps for the user.   We record explicitly the assumptions in the analysis so that they can be later refined or corrected as more information about the system becomes available.

The first step in following this strategy is to define the task groups by specifying their unit-task constituents. The unit tasks for each task group in the Manual System are shown in Figure 9.3.   The decisions about what unit tasks are required for the task groups make clear many assumptions about the system.   The assumptions about the Manual System are listed in Figure 9.4, and are indexed in Figure 9.3 to the unit task decisions they affect.   For example, assumption A2, that the user must explicitly call for page, heading, figure, and footnote numbers to be placed, affects the NUMBER-PAGE, NUMBER-HEADING, NUMBER-CAPTION, NUMBER-TEXT, and NUMBER-FOOTNOTE unit tasks, as indicated by the indexing in Figure 9.3.

Unit tasks required by the Automatic System are specified in exactly the same way in Figure 9.5 and Figure 9.6.   The Automatic System shares the assumptions of the Manual System, with the exception of assumptions A2 and A5, and the addition of assumptions A9 to A12 (compare Figure 9.4 and Figure 9.6).   For example, substituting assumption A9 for A2 means that the NUMBER-PAGE unit task is not required in the PROCESS-NEW-PAGE task group for the Automatic System.   In all, many fewer unit tasks are required in the Automatic System.

We must also make assumptions about the computing technology for the layout system.   Here we assume that the layout system will reside on a small, personal computer with limited main memory and a large disk. Experiences with similar systems, such as the BRAVO editor, suggest that many of the functions of the layout system will probably require significant time for the system to carry out.   For example, locating a file on disk, loading the material from the file into main memory, and displaying the material might take a few seconds.   Since the response time of the system could be as long as a unit-task time, we need to keep track of the number of significant system responses, as well as of unit tasks, in our calculation.   In Figure 9.1 we noted (with an R) the unit tasks requiring a significant response time; those unit tasks are also marked in Figure 9.3 and Figure 9.5.

Task Groups	Unit Tasks	R's	Assumptions
PROCESS-NEW-PAGE			
	LOAD-PAGE	R	A1
	NUMBER-PAGE		A2
PROCESS-HEADING			A3
	SET-FONT-HEADING		A4
	POS-VERT-HEADING		A5
	POS-HORIZ-HEADING		A5
	NUMBER-HEADING		A2
PROCESS-FIGURE			A6
	LOAD-FIGURE	R	
	POS-VERT-FIGURE	R	
	POS-HORIZ-FIGURE	R	A7
	LOAD-CAPTION	R	
	POS-VERT-CAPTION	R	
	POS-HORIZ-CAPTION	R	A7
	SET-FONT-CAPTION		A4
	NUMBER-CAPTION		A2
	NUMBER-TEXT (callout in text)		A2
PROCESS-FOOTNOTE			A6
	LOAD-FOOTNOTE	R	
	POS-VERT-FOOTNOTE	R	
	SET-FONT-FOOTNOTE		A4
	NUMBER-FOOTNOTE		A2
	NUMBER-TEXT (callout in text)		A2
PROCESS-INDENTATION			
	POS-HORIZ-TEXT		A8
PROCESS-TEXT-FONT			
	SET-FONT-TEXT		A4
PROCESS-REFERENCE			A3
	SET-FONT-TEXT (title italic)		A4
	SET-FONT-TEXT (volume no. bold)		A4

Figure 9.3. Unit-Task-Level definitions of the task groups for the Manual System.

The R's column indicates which unit tasks involve a significant system response time (see Figure 9.1). The numbered assumptions are listed in Figure 9.4.

A1	A new page frame is initialized with margins set and just enough text loaded automatically from the text file to fill the frame.
A2	The system keeps track of all numbers (pages, headings, figures, footnotes), but the user must explicitly call for them to be placed.
A3	Headings and references are included in the text file and thus need not be loaded separately.
A4	No fonts are set in the text, footnote, caption, or reference files; they must be set explicitly during page layout.
A5	Headings must be positioned manually.
A6	When figures and footnotes are placed, the text body is automatically adjusted and the displaced text is automatically returned to the text file.
A7	Figures can go next to each other and thus need to be explicitly positioned horizontally.
A8	Indented paragraphs need to be positioned explicitly.

Figure 9.4. Specific assumptions about the Manual System.
See Figure 9.3 for which unit tasks are affected by each assumption.

Having enumerated the unit tasks and system responses, we need a reasonable estimate of the amount of time required by each of these. The unit task time is best obtained from existing data on systems as similar as possible to the proposed layout system. The system closest to the envisioned layout system is the display-based BRAVO editor, which was measured in Chapter 3. The measured error-free unit-task time from Figure 3.6 is 10.1 sec, which we round to 10 sec per unit task. (The numbers are rounded off to emphasize the rough nature of this analysis.)

The times obtained from Chapter 3 are based on tasks not requiring any significant amount of system response time. Since we assume that system response time will be significant in the page-layout environment, it is useful to partition the time for a unit task into two parts: the user's *working time* t_W and the system's response time t_R (the latter being optional for any particular unit task). (We also partitioned the unit task this way in Section 8.6.) Thus, the 10 sec estimate above is for t_W.

In order to obtain estimates of system response time, informal empirical measurements were made on several readily available display-based

Task Groups	Unit Tasks	R's	Assumptions
PROCESS-NEW-PAGE			
	LOAD-PAGE	R	A1, A9
PROCESS-HEADING			A3
	NUMBER-HEADING		A10
PROCESS-FIGURE			A6
	LOAD-FIGURE	R	
	POS-VERT-FIGURE	R	
	POS-HORIZ-FIGURE	R	A7
	LOAD-CAPTION	R	A11
	NUMBER-TEXT (callout in text)		A2
PROCESS-FOOTNOTE			A6
	LOAD-FOOTNOTE	R	A12
	NUMBER-TEXT (callout in text)		A2
PROCESS-INDENTATION			
	POS-HORIZ-TEXT		A8
PROCESS-TEXT-FONT			
	SET-FONT-TEXT		A4
PROCESS-REFERENCE			A3
	SET-FONT-TEXT (title italic)		A4
	SET-FONT-TEXT (volume no. bold)		A4

Figure 9.5. Unit-Task-Level definitions of the task groups for the Automatic System.

The R's column indicates which unit tasks involve a significant system response time (see Figure 9.1). The numbered assumptions are listed in Figure 9.6.

systems. The times to load a file of text, to move text on the screen, and to load and move various sorts of pictures were measured. Whereas the extremes of these times ranged around 2 sec an the low end and up to 425 sec on the high end, a large number clustered around 6 sec per system response. As a working estimate, we use a constant 6 sec for the system response time t_R.

A1   A new page frame is initialized with margins set and just enough text loaded automatically from the text file to fill the frame.

A3   Headings and references are included in the text file and thus need not be loaded separately.

A4   No fonts are set in the text, footnote, caption, or reference files; they must be set explicitly during page layout.

A6   When figures and footnotes are placed, the text body is automatically adjusted and the displaced text is automatically returned to the text file.

A7   Figures can go next to each other and thus need to be explicitly positioned horizontally.

A8   Indented paragraphs need to be positioned explicitly.

A9   Page numbers are automatically placed when a new page frame is loaded.

A10   A heading is automatically formatted (position and font) by just giving the type of the heading.

A11   When a caption is loaded, it is automatically positioned under the figure, its font set, and a number given.

A12   When a footnote is loaded, it is automatically positioned at the bottom of the page and numbered.

Figure 9.6.   Specific assumptions about the Automatic System.
See Figure 9.5 for which unit tasks are affected by each assumption.

Unit-Task-Level Calculation

The necessary pieces have now been gathered to do the calculation. The task group time per page is:

$$T_{TG} = N_{TG} \left(n_{UT} t_W + n_R t_R \right). \tag{9.1}$$

The number of unit tasks in the task group is n_{UT} and the number of system responses is n_R, both taken from Figure 9.3 and Figure 9.5. Thus, $n_{UT}t_W$ is the total working time in the task group, and $n_R t_R$ is the total response time. N_{TG} (taken from Figure 9.2) is the frequency with which each task group occurs per page. And T_{TG} is the total task group time per page.

The total error-free time to lay out a page is the sum of the T_{TG}'s plus a correction to account for the likely amount of errors. In Experiment 5C (Figure 5.18), we found that the user spent 25% of her time handling errors with the POET editor. Thus we charge 25% overhead to account for errors in the page-layout task.

The calculation of the page-layout time for both the Manual System and the Automatic System is given in Figure 9.7. The resulting prediction is that it takes 270 sec (about 4.5 minutes) to lay out a page with the Manual System, with time being distributed fairly evenly over all the task groups. Surprisingly, the time required by the Manual System is about half as long as it would take an average typist (60 words/minute) to type in a full page of text (550 words). The Automatic System is predicted to take 192 sec per page, about 80 sec per page faster than the Manual System, with most of this improvement lying in the PROCESS-HEADING and PROCESS-FIGURE task groups.

9.2. CHECKS ON THE UNIT-TASK-LEVEL ANALYSIS

Many assumptions about the page-layout task were made to keep the Unit-Task-Level analysis simple. Some of the assumptions and consequences of their being wrong are as follows:

(1) Performance of the page-layout task was assumed to consist of a string of independent unit tasks as we have defined them. It is possible that we have ignored some dominant global feature of the layout task. For example, the complexity of managing all the files might require significant planning time.

(2) The task analysis for the page-layout task was very approximate; it assumed that the task groups were simple linear sequences of unit tasks. Real user behavior in this task would be substantially more conditional and variable. This variabil-

MANUAL SYSTEM

Task groups	n_{UT}	n_R	N_{TG}	T_{TG}	$\%T$
PROCESS-NEW-PAGE	2	1	1.00	26.0	12%
PROCESS-HEADING	4	0	1.12	44.8	21%
PROCESS-FIGURE	9	6	.24	30.2	14%
PROCESS-FOOTNOTE	5	2	.18	11.2	5%
PROCESS-INDENTATION	1	0	3.13	31.3	15%
PROCESS-TEXT-FONT	1	0	4.42	44.2	21%
PROCESS-REFERENCE	2	0	1.43	28.6	13%

t_W = 10.0 sec   Sum of times   =   216
t_R = 6.0 sec   25% error time   =   54
                       Total time   =   270 sec

AUTOMATIC SYSTEM

Task groups	n_{UT}	n_R	N_{TG}	T_{TG}	$\%T$
PROCESS-NEW-PAGE	1	1	1.00	16.0	10%
PROCESS-HEADING	1	0	1.12	11.2	7%
PROCESS-FIGURE	5	4	.24	17.8	12%
PROCESS-FOOTNOTE	2	1	.18	4.7	3%
PROCESS-INDENTATION	1	0	3.13	31.3	20%
PROCESS-TEXT-FONT	1	0	4.42	44.2	29%
PROCESS-REFERENCE	2	0	1.43	28.6	19%

t_W = 10.0 sec   Sum of times   =   154
t_R = 6.0 sec   25% error time   =   38
                       Total time   =   192 sec

Figure 9.7. Unit-Task-Level calculation of the page-layout time for the Manual System and the Automatic System.
The n_{UT}'s and n_R's are counted from Figure 9.3 and Figure 9.5, and the N_{TG}'s are taken from Figure 9.2. T_{TG} is calculated from Formula 9.1. $\%T$ is the percentage of total time taken by each task group.

ity could require the user to spend substantial time making decisions.

(3) The Unit-Task-Level analysis does not deal with errors directly; it only uses a multiplicative factor for error. Thus, it ignores the possibility that large errors, even though they occur with relatively low frequency, could dominate the page-layout task.

(4) The analysis assumed a particular skill level for the user. From Chapter 3 we know that performance time can vary by up to a factor of three in text-editing tasks, even for expert users. We might expect even greater user variance in the page-layout task, since it is more complex.

(5) The analysis assumed a particular kind of interface design for the page-layout system. Performance can be greatly affected by the functionality and interface of the system. Although we made the specific assumptions about the system explicit, the actual layout system could be quite different.

(6) Finally, the analysis assumed a particular complexity of pages to be laid out. Again, although our assumptions were made explicit, the actual layout task could involve substantially different kinds of pages.

Many of these assumptions could be checked by sensitivity analyses on the parameters of Formula 9.1 (such analyses have been illustrated in Chapter 4 and Chapter 8). For example, the assumption about skill level of the user would mostly affect the parameter t_W.

In the remainder of this section we concentrate on the assumption of the independence of unit tasks. First, we check this assumption empirically by observing a user in a simulated layout task. (This also provides a crude check for the effect of conditional methods and large errors.) Then, we attack the independence assumption analytically by pushing the Unit-Task-Level analysis to a finer level of detail.

Experimental Check

As a check on the assumptions of the foregoing analysis, an analogue of the layout system was constructed using the BRAVO editor, in which the user performed a task approximately equivalent to the PROCESS-FIGURE task. The procedure involved loading a small text file, simulating a page, which sometimes contained a "Figure here" mark; the user was to:

—search for a "Figure here" mark and (if there was one) get the figure identifier,

—use the identifier to load a file containing the "figure" (simulated as an array of characters) and to load a caption,

—check whether the caption went with the figure,

—move the figure, if it occurred in the middle of a paragraph, to the front of the paragraph,

—make sure there were two blank lines preceding the figure, two blank lines following the caption, and one blank line between the figure and the caption, and

—edit the caption to conform to the numbering style.

Most of the steps of this procedure are conditional, so there is not a fixed sequence of unit tasks for each simulated PROCESS-FIGURE.

An expert BRAVO user performed this analogue procedure six times. Each trial consisted of "laying out" seven "pages," of which four pages had "figures" to be processed. The user's performance improved (by a factor of three) over the first three trials, but leveled off in the last three trials, indicating a certain amount of gained expertise in the analogue task.

For the page-layout problem, we are interested in the performance of an expert user. Accordingly, we need only consider data from the last trial, when the user had had enough practice to become expert in the analogue task. The user's behavior was generally in accord with the analysis given. His performance consisted entirely of unit tasks of the type we have predicted and included no non-routine errors. The user averaged 8.6 sec per unit task on the last trial—14% less than the 10 sec per unit task estimated by the Unit-Task-Level analysis. Although this result is within our expected prediction error, it may suggest that some interaction occurs among the unit tasks in this context that allows unit tasks to be done faster than in a text-editing context.

Interaction Among Unit Tasks

The Unit-Task Level of analysis assumes that the time to perform a unit task is independent of the surrounding unit tasks. There is surely some level of detail at which this is no longer the case, especially for unit tasks within a well-integrated task group. We can check how much the independence assumption affects our result by carrying the analysis down one level of detail to the Functional Level, discussed in Chapter 5.

Recall from Chapter 5 that a unit task is composed of four operations at the Functional Level, which we may for convenience denote with single letter symbols:

Acquire the unit task.	**A**
Locate the objects of the task.	**L**
Make the change specified in the task.	**C**
Verify the change.	**V**

If there were a system response **R**, it would probably occur after the **C**. But since we have already factored out the response time in the Unit-Task-Level analysis in Section 9.1, we may ignore **R** operators here.

An example is the **POS-VERT-HEADING** unit task, in which the user adjusts the vertical spacing around a heading. The Functional-Level operations for this unit task considered in isolation would be:

POS-VERT-HEADING =	
Determine the heading needs spacing.	**A**
Point to the heading.	**L**
Insert space in front of the heading.	**C**
Insert space below the heading.	**C**
Verify that the spacing is correct.	**V**

However, consider what this unit task would be like were it to occur in the middle of the highly integrated task group, **PROCESS-HEADING**:

SET-FONT-HEADING =	
Detect that the heading needs alteration.	**A**
Point to the heading.	**L**
Change the font of the heading.	**C**
POS-VERT-HEADING =	
Insert space in front of the heading.	**C**
Insert space below the heading.	**C**
POS-HORIZ-HEADING =	
Center the heading.	**C**
Verify position of heading.	**V**

NUMBER-HEADING =

Look up the heading number.	**A**
Point to the heading number location.	**L**
Replace the heading number.	**C**
Verify that the heading is correct.	**V**

Not all the Functional-Level operations are required in **POS-VERT-HEADING** in this context. The **A** (acquiring the positioning task) and the **L** (pointing to the heading) can be done in the previous unit task and the **V** (verifying that the vertical positioning is correct) in the following unit task. The vertical positioning itself, then, only requires two steps (two **C**'s). Thus, context can have a considerable effect on a unit task.

To carry out an analysis of the unit tasks in Functional-Level operations requires either that we make some more specific assumptions about the system (as in Figure 9.4 and Figure 9.6) or that we appeal to some principles of organization at the Functional Level. We take the latter course by setting out a few heuristic rules for deciding which Functional-Level operations are required for a unit task (in the spirit of the Keystroke-Level Model's rules for placing **M** operators, given in Figure 8.2). For each unit task in a task group, we assume that the unit task requires the basic set of four operations, **A**, **L**, **C**, and **V**, and use the following rules to modify the set:

(1) An **A** operation can result in more than one unit task being acquired, if they are closely related. Thus, if the current unit task is likely to have been acquired together with the previous unit task, then delete the **A** for the current one.

(2) If the current unit task deals with the same task objects as the previous unit task, then delete the **L**, since there is no need to locate them again.

(3) On a display-based system, which immediately shows the effects of changes to the user, some changes are so perceptually easy to check that a separate verification step is not required. Thus, for example, in all unit tasks concerned with font changes, delete the **V**.

(4) If two unit tasks require similar changes, then only one verification is required. Thus, for example, if the current unit task is the first of a two-part positioning task, then delete one of the **V**'s.

Task Groups	Unit Tasks	R's	Functional-Level Operations			
PROCESS-NEW-PAGE						
	LOAD-PAGE	R	A		C	
	NUMBER-PAGE		A	L	CC	V
PROCESS-HEADING						
	SET-FONT-HEADING		A	L	C	
	POS-VERT-HEADING				CC	
	POS-HORIZ-HEADING				C	V
	NUMBER-HEADING		A	L	C	V
PROCESS-FIGURE						
	LOAD-FIGURE	R	A	L	C	V
	POS-VERT-FIGURE	R	A	L	C	V
	POS-HORIZ-FIGURE	R	A		C	V
	LOAD-CAPTION	R	A	L	C	V
	POS-VERT-CAPTION	R		LL	C	
	POS-HORIZ-CAPTION	R			C	V
	SET-FONT-CAPTION				C	
	NUMBER-CAPTION		A	L	C	V
	NUMBER-TEXT		A	L	C	V
PROCESS-FOOTNOTE						
	LOAD-FOOTNOTE	R	A	L	C	V
	POS-VERT-FOOTNOTE	R	A	LL	C	V
	SET-FONT-FOOTNOTE				C	
	NUMBER-FOOTNOTE		A	L	C	V
	NUMBER-TEXT		A	L	C	V
PROCESS-INDENTATION						
	POS-HORIZ-TEXT		A	L	C	V
PROCESS-TEXT-FONT						
	SET-FONT-TEXT		A	L	C	
PROCESS-REFERENCE						
	SET-FONT-TEXT		A	L	C	
	SET-FONT-TEXT		A	L	C	

Figure 9.8. Functional-Level definitions of the task groups for the Manual System.

The R's column indicates which unit tasks involve a significant system response time (see Figure 9.1). The Functional-Level operators are: acquire a task (**A**), locate the elements of a task (**L**), make the required change (**C**), and verify the change (**V**). The double letters in the Operations column indicate that two operations are required.

328

(5) Some unit tasks require multiple operations to locate the objects or to make the changes required. If such is the case, add the requisite number of **L**'s and/or **C**'s.

This analysis is carried out for all the task groups in Figure 9.8.

This Functional-Level analysis shows that the unit tasks are, for the most part, compressed to fewer than the four functional operations per unit task assumed by the complete Unit-Task-Level analysis. To understand quantitatively how much compression there is, we define the *compression ratio* of a task group to be the ratio of the number of Functional-Level operations over the nominal four operations per unit task. For example, PROCESS-HEADING has four unit tasks and hence 16 expected operations, but there are only 11 operations in the analysis above, giving a compression ratio of $11/16 = .69$. Carrying this calculation out for all the task groups gives the following compression ratios:

PROCESS-NEW-PAGE	.88
PROCESS-HEADING	.69
PROCESS-FIGURE	.81
PROCESS-FOOTNOTE	.90
PROCESS-INDENTATION	1.00
PROCESS-TEXT-FONT	.75
PROCESS-REFERENCE	.75
Average compression ratio	*= .83*

We can now give an estimate of the error induced by ignoring interaction between unit tasks in our earlier analysis. An average task group compression ratio of .83 means that, due to interactions among the unit tasks, the average unit task time may be about 17% faster than a series of completely unrelated unit tasks, as assumed by the Unit-Task-Level analysis.

Functional-Level Calculation

Having gone to the effort to calculate the number of each of the Functional-Level operations in each task group, we are close to being able to do a Functional-Level calculation of the page-layout time. This will provide a further check on the Unit-Task-Level analysis. The task group time is:

$$T_{TG} = N_{TG} \left(n_A t_A + n_L t_L + n_C t_C + n_V t_V + n_R t_R \right). \qquad (9.2)$$

This formula is the same as Formula 9.1, except that the unit-task working time $n_{UT} t_W$ is replaced by the times for the Functional-Level operations, $n_A t_A + \ldots + n_V t_V$.

We need to estimate the unit times for the functional operators, but we do have independent data for estimating each of them. In Chapter 8 (Figure 8.7), we found that t_A was 4 sec for the BRAVO editor. (Again, we round our estimates to the nearest .5 sec to emphasize their approximate nature.) In Chapter 5 (Figure 5.15), we measured the VERIFY-EDIT operator to be 1.5 sec for the POET editor. Other experiments with BRAVO in our laboratory (not reported in this book) provided estimates of 2 sec for t_L and 2.5 sec for t_C. Note that these estimates partition the 10 sec working time t_W that we used in the Unit-Task-Level analysis, so that $t_A + t_L + t_C + t_V = t_W$. Formula 9.2 will thus yield the same time prediction as did Formula 9.1 when all unit tasks consist of the four functional operations, **ALCV**.

The Functional-Level calculation for the Manual System is shown in Figure 9.9. The Functional-Level prediction for the page-layout time is

Task groups	n_A	n_L	n_C	n_V	n_R	N_{TG}	T_{TG}	$\%T$
PROCESS-NEW-PAGE	2	1	3	1	1	1.00	25.0	14%
PROCESS-HEADING	2	2	5	2	0	1.12	30.8	17%
PROCESS-FIGURE	6	7	9	7	6	.24	25.7	14%
PROCESS-FOOTNOTE	4	5	5	4	2	.18	10.2	6%
PROCESS-INDENTATION	1	1	1	1	0	3.13	31.3	17%
PROCESS-TEXT-FONT	1	1	1	0	0	4.42	37.6	21%
PROCESS-REFERENCE	2	2	2	0	0	1.43	24.3	13%

t_A = 4.0 sec. t_L = 2.0 sec
t_C = 2.5 sec. t_V = 1.5 sec
t_R = 6.0 sec

Sum of times = 185
25% error time = 46
Total time = 231 sec

Figure 9.9. Functional-Level calculation of the total time per page for the Manual System.

The n_A's, n_L's, n_C's, n_V's, and n_R's are counted from Figure 9.8; the N_{TG}'s are taken from Figure 9.2. T_{TG} is calculated from Formula 9.2. $\%T$ is the percentage of time spent doing each task group.

231 sec, 15% faster than the 270-sec prediction of the Unit-Task-Level calculation. This reduced prediction is quite close to the 17% faster found in the compression-ratio calculation above. Both analytic calculations yield results that are quantitatively close to the experimental result (14% faster), and thus the notion of interaction among highly integrated unit tasks is sufficient explanation for the experimental results. The 14~17% differences in these predictions from the Unit-Task-Level analysis are within expected prediction errors, confirming the basic soundness of the analysis.

9.3. CONCLUSIONS

Systems in the early, conceptual stages of design can be analyzed at the Unit-Task Level. In this analysis, the unit tasks to be accomplished are enumerated, their frequencies estimated, and the time per unit task determined. Then, the total time can be found by multiplying the total number of unit tasks by the time per unit task.

To illustrate the Unit-Task-Level analysis, we have computed the estimated time per page required by two versions of a hypothetical system for laying out a scientific journal. The time per page calculated from the Unit-Task-Level analysis was close to predictions derived from a more elaborate Functional-Level analysis and close to empirical measurements on a mockup analogue of the task.

*EXTENSIONS
AND
GENERALIZATIONS*

10. An Exploration into Circuit Design

Our strategy in studying the psychology of human-computer interaction has been to focus on the specific domain of computer text-editing, then to generalize to other systems and tasks. In this concluding part of the book, we wish to place these studies of text-editing in broader perspective. First, this chapter considers to what degree our results can be further extended to a more "creative" task domain, in which the user is not given specific instructions to follow, but must use the system to solve a problem. Then, the next two chapters concentrate on understanding those general characteristics of human performance implied by our studies. Corresponding to the dual orientation put forward in Chapter 1 towards basic science on the one hand and application on the other, Chapter 11 focuses on the basic nature of cognitive skill, and Chapter 12 focuses on how our results fit into the total scheme for applying psychology to design.

The development of a theory of human-computer interaction must be based on an analysis of diverse interactive task domains. In this book, we have mostly concentrated on the task of correcting a text file from a marked-up manuscript (Chapters 3-6); but we have also studied graphics systems for creating and editing line drawings (Chapter 8), executive subsystem commands, (Chapter 8), and a page-layout system (Chapter 9). In each of the task domains, a variant of the basic GOMS model has been found applicable. Yet, all the tasks studied so far have been instruction-following tasks, in which the user follows simple directions on a set of small, independent subtasks.

The present chapter explores human-computer interaction on a task that is not an instruction-following task: the computer-aided design of a VLSI circuit-layout. In this task, there is no externally-given set of activities to be accomplished. The user is a designer who generates the tasks as he proceeds, in response to the evolving state of the design. We want to know the extent to which the GOMS account holds in this new domain.

10.1. THE ICARUS SYSTEM FOR CIRCUIT DESIGN

We report a case study of the design of an actual circuit. The design of a VLSI circuit is a complex problem, whose solution depends on finding an appropriate decomposition of a large circuit into sub-circuits (Mead and Conway, 1980). Our study will make observations of a basic element of the design process, the detailed layout of one of the sub-circuits. In this task, the user begins with a rough sketch of the circuit on paper and uses the system to produce a circuit specification that is (a) geometrically defined and dimensioned, (b) optimized to minimize the area required, and (c) contained in the memory of a computer system. In the course of producing the circuit, the user must solve several subproblems: (1) transcribe the sketch into the computer system, (2) dimension the circuit elements according to VLSI standards (called "VLSI design rules"), (3) compress the dimensioned circuit to minimize the area it occupies on a chip, and (4) define the boundaries of the compressed circuit so they will mesh with other sub-circuits. The user in our study uses a specialized VLSI circuit-layout system called ICARUS.

ICARUS is an interactive, display-based computer system for drawing and editing VLSI circuit-layouts (Fairbairn and Rowson, 1978; Mead and Conway, 1980, Ch. 4). Its user interface[1] is described in two parts (Moran, 1981a): first, in terms of the conceptual model it imposes on the circuit-layout task and, second, in terms of its command language. The reader need not assimilate all the details of ICARUS's interface, but some knowledge about how it works is necessary to understand the user behavior.

---

[1] This study was done on a very early version of the ICARUS system. The later versions of ICARUS have a somewhat different (and much improved) user interface.

ICARUS CONCEPTUAL MODEL

ICARUS processes circuit-layout descriptions stored in files. A circuit description is brought into the ICARUS *workspace* and presented graphically to the user (Figure 10.1), who can then can edit the circuit and store it on a new file.

In ICARUS, a circuit is constructed in a *circuit space*, which has virtually unlimited extent in two dimensions (i.e., in the plane) and has a limited third dimension made up of five discrete *layers*, indicated by different texture patterns on the display. (Each layer is the mask for a different step in the VLSI circuit manufacturing process.)

A circuit *layout* in ICARUS may be totally described as a set of *rectangles*, as an examination of Figure 10.1 confirms. Each rectangle has an *xy*-location in the plane and is located on a particular layer. Thus, a circuit description in ICARUS is purely geometric; there are no notions of electronic components like nodes or transistors. Circuit layouts in ICARUS are dominated by two special kind of rectangles: long, thin rectangles, called *lines* (the "wires" of the circuit), and square rectangles, called *flashes* (the connection points between layers). The system provides special command language facilities for these special rectangles. We use the term *elements* in this discussion when referring to the rectangles in a layout without regard to their shape. ICARUS allows the user to draw, delete, move, copy, mirror, and rotate elements in the circuit space.

VLSI circuits are made up of many repeated sub-circuits. ICARUS allows a sub-circuit (any contiguous rectilinear region) to be defined as a *symbol*, which can then be manipulated as a unit. We need not be concerned here with the details of the symbol facility except to note that the system provides command language facilities for defining and manipulating them.

ICARUS COMMAND LANGUAGE

The layout of the ICARUS display is shown in Figure 10.1. The display contains two windows for viewing the circuit-layout in the work-space: a *gross window* that gives an overall view of the circuit and a *fine window* that provides a close-up view of a part of the circuit. The position of the fine window's view is outlined in the gross window. Thus, the user can work on a small piece of a circuit in the fine window and still have a view of its larger spatial context. The windows can be moved around over the circuit to give views of any region of the circuit.

There is also a *parameter area* and a *layer menu* on the display. The parameter area shows the current values of various command language

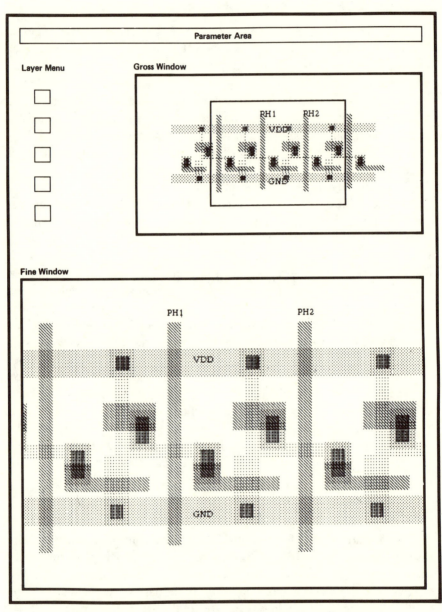

Figure 10.1. Layout of the ICARUS display.

338

parameters. For example, there are parameters for the viewing scales of the fine and gross windows, and there are parameters for the current flash size and the current line width, all of which can be reset by various commands. There are five icons in the layer menu, one for each layer of the circuit space; only one icon is highlighted at any given time to indicate the *current layer*. Creation of new elements only takes place on the current layer.

For input, ICARUS uses a mouse, a five-key keyset, and a keyboard. The mouse is used for giving commands (with its three buttons: MARK, DRAW, and DELETE) as well as for pointing to locations on the display. The keyset is used for issuing commands and specifying arguments. Only three of the five keys on the keyset need to be considered in this discussion: CENTER, FLASH, and RECTANGLE.[2] The keyboard is also used for commands and arguments. The normal (home) position for the user's hands is with his left hand on the keyset and his right hand on the mouse. He only moves his hands to the keyboard when required.

Figure 10.2 lists some of ICARUS's commands and describes the methods for executing them, using the notation from the Keystroke-Level Model (Chapter 8).

There are several commands for drawing new elements on a layout. A line, the most common element, is drawn by simply pointing to where its end points are to be located. Pressing the MARK button causes a small *mark* to be made on the display, which indicates where one end of the line will be. The line is actually drawn (from the mark to the location currently being pointed at) when the DRAW button is pressed. Sometimes the mark is already at the position of one of the end points, in which case only the other end point need be indicated (these two cases are distinguished in Figure 10.2 as two separate commands, Line2 and Line1).

A flash is drawn by holding down the FLASH key, pointing to where its center is to be located, and pressing DRAW. An arbitrary rectangle is drawn by pointing to two of its diagonally-opposite vertices while holding down the RECTANGLE key. All elements are drawn on the current layer, although the layer can be changed by pointing to one of the non-highlighted icons in the layer menu. The width of the lines is determined by the line-width parameter, which can be changed by the Width command (see Figure 10.2). Circuit elements are deleted from the layout

---

[2] The keyset keys and the mouse buttons are not actually marked with these labels, but the labels are mnemonics for purposes of this discussion.

Command	Execution Method
Drawing Commands:	
(draw) **Line2** (2 end points)	**P**[location] **K**[MARK] **P**[location] **K**[DRAW]
(draw) **Line1** (1 end point)	**P**[location] **K**[DRAW]
(draw) **Flash**	**M K**[FLASH] **P**[location] **K**[DRAW]
(draw) **Rectangle**	**M K**[RECTANGLE] **P**[location] **K**[MARK] **P**[location] **K**[DRAW]
(create) **Label**	**P**[location] **K**[MARK] **H**[keyboard] **M** 6**K**[CONTROL I string] **M K**[ESC] **H**[mouse/keyset]
Parameter Changing Commands:	
(change) **Layer**	**P**[menu] **K**[MARK]
(change) **Width** (of line)	**H**[keyboard] **M** 3**K**[CONTROL W number] **M K**[ESC] **H**[mouse/keyset]
Deletion Commands:	
Delete (element)	**P**[element] **K**[DELETE]
Undo (last deletion)	**H**[keyboard] **K**[U] **H**[mouse]
Transformation Commands:	
Move (elements)	**H**[keyboard] **M** 2**K**[M M] **H**[mouse] 4(**P**[location] **K**[MARK]) **K**[ESC] **P**[location] **K**[MARK] **P**[location] **K**[MARK] **H**[keyset]
Copy (elements)	**H**[keyboard] **M** 2**K**[M C] **H**[mouse] 4(**P**[location] **K**[MARK]) **K**[ESC] **P**[location] **K**[MARK] **P**[location] **K**[MARK] **H**[keyset]
Stretch (element)	**H**[keyboard] **M** 2**K**[M S] **H**[mouse/keyset] **P**[element] **K**[MARK] **P**[edge] **K**[MARK] **P**[location] **K**[MARK]
Display Control Commands:	
Center (drawing)	**M K**[CENTER] **P**[location] **K**[MARK]
(change) **Magnification**	**H**[keyboard] **M** 2**K**[F|G digit] **H**[mouse/keyset]
Redraw (window)	**H**[keyboard] **M K**[R] **H**[mouse/keyset]
Dimensioning Commands:	
(measure) **Distance**	2(**P**[location] **K**[MARK])
(measure) **Size**	4(**P**[location] **K**[MARK])

Figure 10.2. Methods for executing ICARUS commands.

The notation for the methods is taken from the Keystroke-Level Model (Chapter 8). The ESC and CONTROL keys are on the keyboard. FLASH, RECTANGLE, and CENTER are buttons on the keyset. MARK, DRAW, and DELETE are buttons on the mouse. The "string" in the Label command is assumed to be 4 characters long (on the average) and the "number" in the Width command is assumed to be one digit.

by simply pointing to them and pressing the DELETE button, and the most recent deletion can be undone by hitting the U key on the keyboard. There are also commands for stretching the size of existing circuit elements and for sticking labels on the layout.

The command for moving elements on the layout is rather complex. The set of elements to be moved is specified by an enclosing rectangle defined by pointing to its four corners with MARK (items that are only partially within the rectangle are included in the set). The Move vector is specified by pointing to a reference point on one element in the set and then pointing to the new position of that point. Since the Move command is so complex, ICARUS users often just delete the elements to be moved and then redraw them in the new location.

The views of the circuit-layout in the two display windows can be easily adjusted. Holding down the CENTER key and pointing with MARK to a location in either window causes the circuit to be redrawn with the specified location shifted to the center of the window. There is also a simple command for changing the scale (Magnification) of the layout in each window.

Every time MARK or DRAW is pressed, the xy coordinates of the cursor position on the circuit-layout are displayed in the parameter area, as are the differences between the current coordinates and the coordinates at the previous press of MARK or DRAW. This information allows the two successive MARK presses to be used to measure distances between points (indicated in Figure 10.2 as the Distance and Size commands).

All in all, the command language of ICARUS is not unusual. It contains both simple and complex commands; and it uses standard command language devices, such as parameters to specify default values for command arguments.

10.2. BEHAVIOR WITH ICARUS (EXPERIMENT 10A)

An experiment was run to collect a sample of actual performance with ICARUS. A single user was observed in a single session, and the behavior was recorded and analyzed.

Procedure and Data

User. The user was an experienced designer of VLSI circuits and an expert user of ICARUS (with a year's experience of regular use, roughly 300 hours of practice). The user was asked to choose a task that was "typical" of the kinds of things he did in ICARUS and that would take about a half hour to do. The task he chose was to lay out a circuit for a cell in a content-addressable memory. The function of the circuit was to store, shift, and match one bit of information. The user had previously tried to lay out this same circuit, and the experimental task was to attempt a different arrangement. Thus, the user was familiar with the experimental design task, which was of immediate interest to him.

Experimental Procedure. The user was placed in a room with a terminal connected to the ICARUS system, an arrangement similar to that in Figure 3.1. He brought a rather detailed sketch of the circuit (shown in Figure 10.3), to which he referred during the initial part of the session. He was asked to talk aloud about what he was doing as he performed the task, then he was left alone to do the task. The session was videotaped with two cameras, one viewing the user and the other viewing a monitor showing the user's display screen. The user's keystrokes were recorded

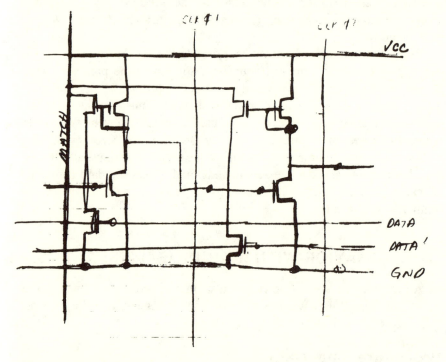

Figure 10.3. Hand-drawn sketch that the user brought to Experiment 10A.

The user's original sketch was drawn with various colored pencils, each color representing a different layer.

on a data file. The user paced himself on his task and decided when to end the session, which he did after about 40 minutes. His final circuit-layout from the experimental session is shown in Figure 10.4.

Data Calibration and Encoding. A data file recorded the name and clock time for each key and button-press plus the screen coordinates at the time of each mouse button-press. Recorded times were calibrated with the videotape to within an accuracy of about .1 sec. The recorded screen coordinates were offset from the coordinates that the ICARUS system used. It was difficult to calibrate these two, since there were no precise reference points on the screen. Although an attempt at calibration was made using the layer menu as a referent, the calibration was probably accurate to only about 1.5 cm.

Keystroke data were encoded into a sequence of ICARUS commands by a combination of heuristic programs and hand editing. All keystrokes within .3 sec of each other were grouped into *events*. A program scanned the events and attempted to recognize each of them as a command according to the command syntax in Figure 10.2. Some of the events had to be regrouped manually, and some automatic command recog-

Figure 10.4. Final ICARUS circuit layout produced by the user in Experiment 10A.

The layers on which the circuit elements lay are indicated by the texture pattern of each element.

nitions were corrected manually. The most difficult part was interpreting the recorded screen coordinates. For example, the cursor location determined whether a press of MARK was a drawing command or a selection of a layer in the menu. Most of the menu selections were identified by the program, but many had to be checked by looking at the videotape. It was not possible to recognize automatically the various circuit elements being operated upon without simulating part of the ICARUS command interpreter (since the layout display could move and change scale). The error in the coordinate calibration made this effort impractical. The circuit elements could have been recognized by scanning the videotape, but this was too expensive. Instead, only a few of the key circuit elements (about every fifth element) were extracted from the tape to identify the place in the circuit that the user was working.

Analysis of Task Structure

The user's performance during the experimental session was organized as a sequence of unit tasks; these, in turn, were grouped into major phases of the session.

MENTAL PAUSES AND UNIT-TASK STRUCTURE

The user frequently paused during the experimental session to check the work he had done and to think about what to do next. A complete protocol description of the experimental session was made, which identified and incorporated these *Pauses* between events. A threshold of 5 sec was chosen as a first cut at mechanically identifying Pauses from the inter-event intervals (which varied from .3 sec to 80 sec). This divided the session into about 100 episodes, each with an average time of 25 sec and each consisting of a Pause followed by a varying number of ICARUS commands. These episodes were then edited manually. The videotape was perused to determine which task was accomplished during each episode (drawing a transistor, making a connection, moving a structure over, etc); the user's verbalizations were very helpful in identifying these. In this process several new gaps between actions (less than 5 sec long) were found; they were identified as Pauses if there were at least 2 sec between commands. Some Pauses also occurred within Move commands. The user viewed the videotape with the experimenter and made a few minor corrections to the experimenter's interpretations.

Finally, the protocol was segmented into unit tasks (as defined in Chapters 5 and 8) by identifying many of the Pauses as task Acquisitions and the commands between them as Executions. Figure 10.5 shows an example fragment of the data record and how it was encoded into the protocol.

Data Record			Protocol Encoding	
Clock	Keys	X Y	Commands	Unit Tasks
. . .				
1:05.92	DRAW	99 300	"	"
			Pause	Acquire
1:11.70	MARK	42 49	Layer (2)	Execute
			Pause	"
1:15.77	MARK	39 466	Line2	"
1:16.75	DRAW	139 468	"	"
1:17.68	DELETE	121 469	Delete	"
1:18.68	CONTROL		Width (6)	"
1:19.22	W		"	"
1:19.92	6		"	"
1:21.93	MARK	36 478	Line2	"
1:23.00	DRAW	125 475	"	"
1:25.28	ESC		"	"
			Pause	Acquire
1:31.55	MARK	31 12	Layer (1)	Execute
. . .				

Figure 10.5. Segment of data record from Experiment 10A and its encoding.
This segment shows the third unit task in the Transcription phase (see the third task, on line 1:06, in Figure 10.7). Clock times are in minutes and seconds. X and Y are the coordinates of the cursor at the time of each mouse button-press. The names of the keys and the names of the commands are given in Figure 10.2.

The unit tasks thus identified were generated by the user to accomplish many different types of goals. The following is a useful classification of the unit tasks into four types:

Draw tasks. These create new circuit elements on the layout, using mostly Line and Flash commands.

Alter tasks. These move circuit elements or change their configuration, usually by means of Move commands, Stretch commands, or deletion and drawing commands.

Dimension tasks. These measure the dimensions of substructures, the distances between circuit elements, or the alignment of elements. This is done using the Distance and Size commands.

Check tasks. These check the circuit for connectivity, for VLSI design rule violations, or for places that can be spatially compressed. There are really many different kinds of Check tasks, all of which are characterized by long periods of thinking and little action. Since there are so few of them, they are all grouped together here as one type.

PHASE STRUCTURE

The experimental session is clearly partitioned into phases lasting several minutes each. In each phase, unit tasks are organized around one of the major subproblems of the VLSI circuit-layout problem. Phases are similar to task groups in the Chapter 9 page-layout task in that they provide a higher-level structuring of the unit tasks; they differ from the task groups in that they are more extensive and in that their unit tasks are deployed in more of a problem-solving mode.

In the experimental session, the user first input the circuit to ICARUS and then compressed it to minimize its area. His strategy for compressing the circuit was to concentrate on reducing its vertical dimension first and then its horizontal dimension. Thus, the phases in the experimental session were:[3]

Transcription phase (14 minutes). The layout of the circuit was transcribed from the hand sketch into the

---

[3] There was also a fourth phase during the experiment: a Symbol Definition phase (5 minutes), during which the circuit was checked and packaged into a unit to be used as a cell in a circuit array. In the original ICARUS study, this phase was not analyzed and so is not included here.

ICARUS system. The circuit was checked to make sure it was functionally accurate.

Vertical Compression phase (7 minutes). The circuit was compressed vertically by moving substructures around to make them fit together more closely.

Horizontal Compression phase (15 minutes). The circuit was compressed horizontally.

Figure 10.6 presents summary statistics for the three phases and for each task type within those phases. The phase structure of the layout task imposes a characteristic sequencing of task types. For example, as might be expected, most Draw tasks occurred in the Transcription phase and most Alter tasks in the Compression phases. The mean unit task time in the session was about 20 sec per task. The mean task time was about the same in each phase, but varied for different task types: 20 sec for Draw and Alter tasks, 8 sec for Dimension tasks, and 38 sec for Check tasks.

Unit-Task Type	Transcription Phase $(N) M \pm CV$ (sec)	Vertical Compression Phase $(N) M \pm CV$ (sec)	Horizontal Compression Phase $(N) M \pm CV$ (sec)	All Phases $(N) M \pm CV$ (sec)
Draw tasks	(28) 23.1±.36	(1) 19.0	(5) 11.6±.50	(34) 21.3±.42
Alter tasks	(7) 18.6±.69	(16) 20.1±.31	(31) 19.4±.72	(54) 19.5±.61
Dimension tasks	(1) 9.0	(1) 9.0	(8) 8.1±.66	(10) 8.0±.60
Check tasks	(2) 32.5	(2) 34.5	(4) 42.8±.66	(8) 38.1±.59
All Tasks	(38) 22.3±.46	(20) 21.0±.45	(48) 18.6±.86	(106) 20.4±.64

Figure 10.6. Unit-task statistics by phase and task type in Experiment 10A.

Keystroke-Level Model Analysis

To determine whether the phases of the experimental session were structured into unit tasks, it is necessary to examine the acquire-execute structure of the unit tasks and to see whether the execution parts conform to the predictions of the Keystroke-Level Model.

```
Clock  |      10        20        30        40        50        60 sec
Time   | ...:... | ...:... | ...:... | ...:... | ...:... | ...:... |
```

Transcription Phase

```
 0:18  |-------DDDD=DDD=DDDDDDDDD=DDDD=DDDDDD
 0:55  |------D=D=D
 1:06  |-----D===DDDDDDDDDD
 1:25  |-----DD==DD
 1:36  |----------------------AAAAAAAAAAAAAAAA
 2:17  |----------DDDDDD===DDDD=DDDDDDDDD
 2:50  |--------DDD=DDDDDDDDDDD===DDD=D==DDDDD
 3:30  |----DDDD=DDDDD
 3:41  |-------D---------DD
 4:01  |------------AAAAAAAA
 4:22  |------A=A
 4:31  |-------DDDDDD=DDD==DD
 4:52  |--------------------------D=DD
 5:23  |-----DD
 5:30  |----A=A
 5:38  |------DD====DDDDD=DDD
 6:00  |--------------D==DD
 6:20  |-----DDDD=DDDDDDDDDD
 6:44  |---D-D------D==DD
 6:57  |-----------DDDDDDDDD=DDDD
 7:23  |------------D---DD----D
 7:46  |---------AAAAAAAA==AAAAA
 8:11  |-------A=A=AAA=AAAAAAA
 8:33  |---------D===DDDD=DDDD==DDDD
 9:01  |---------D=DD
 9:14  |-----------DDDD==DDD=DDD==DDDDD
 9:45  |--------------D==D=DD=DDDDDDDDDDDD
10:18  |-----DDDD==DDD===DDD
10:38  |-------DD=DDDDD
10:54  |-----------D==D=DD===DDD
11:18  |-----D=DD=DDDD=DDDDD
11:44  |----MM
11:46  |----------D-DDDD----DDDD==DDDDDDDD
12:21  |C--------------------------------------------
13:05  |A=AA
13:09  |-------------DDDD=DDDDD=DDDDDDD====DDDD
13:47  |-------DDDDD=DDDDDDD
14:07  |-------------------
```

Vertical Compression Phase

```
14:26  |M=M==M==M
14:35  |----------------------------------------------C=C=C=CC
15:32  |-----AAAAAAA
15:39  |--------AAAAAAAA
15:55  |-----AAAA===AAA
16:10  |-------------AAAAAAAAAA
16:34  |----------------------A==A
17:01  |--------AAAAA+++++++A
17:26  |----A=AAA==A
17:34  |----------------C
17:51  |--------AAAAAAAAAAAAA+++++AA
18:20  |-------------AA=A====AA==AAAAA
18:49  |------AAAAAAAA
19:03  |---------AAAAAAAA
19:23  |---AA=AAA=AAA==AA==AAA=A
19:47  |---D===DDDDD=DDDDD
20:06  |-----------AAAAAAAAAAAAA
20:28  |AA--------A----AAAA==AAAAA
20:54  |-------AAAAAA
21:08  |-------AAAAAAAAAAA
```

Symbol Code:

-	Task acquisition
D	Draw task execution
A	Alter task execution
M	Dimension task execution
C	Check task execution
=	Intra-execution pause
+	Pause within Move command

(continued on next page)

Figure 10.7.   Time line representing the user's behavior sequence in Experiment 10A.

Each (single-character) symbol represents one second of behavior. The symbol sequence begins on a new line at the beginning of each unit task, and the clock time is the time at the beginning of the unit task.

348

```
Clock |        10        20        30        40        50        60 sec
Time  |....:....|....:....|....:....|....:....|....:....|....:....|
```

Horizontal Compression Phase

```
21:28  |CC=C--------------------------------------------------------//--------
22:51  |AAAAA                                                      (85 sec)
22:56  |----M=M==MM
23:07  |-------------------------------C
23:41  |--A-A-----A=AAA===AAAAAAAAA
24:05  |------AA=A==AAA=A===AAAAAAAAAAA==A
24:39  |---------AAAAAAAAAA
24:58  |-------DD=DDD
25:11  |-----AAAAAAAAAAA--AAAAAAAA
25:37  |-------------DD==DDD
25:58  |-----------------------------AAAAAAAA
26:41  |---AAAA==AAA
26:50  |------A---------------AAA
27:16  |-----A=AAA==A=AA
27:34  |--AAAAA
27:44  |----AAAAAAAAA
27:56  |---A==AAAAA
28:04  |--------AA=AAAAA==A=A===AAAA
28:32  |----------------------
28:58  |M
28:59  |----------------------------AA=AAAAAA=AAAA
29:43  |------A
29:54  |---A==AA
30:00  |--MM
30:05  |---AAA
30:11  |-M
30:13  |---A==AA=AAAAAAAAA
30:28  |--------------AAAAAAAAA
30:51  |-------M
31:00  |---------AAAAAAAAA
31:22  |---AAAA
31:26  |-------MM
31:35  |---------------------------A=AAAAA=A=AAA===AA==AAA==AA
32:34  |----AAAAAAAA
32:42  |-------A==AAAAAAAA==AAA
33:05  |-----------------
33:22  |------DDD
33:31  |---------------------------A===A==A=A====AAAA=A=AAA=AA
34:24  |------DDD
34:38  |-----AAA++++++AAA
34:53  |-----A---AAA
35:06  |----AAA
35:09  |------------MM
35:27  |--AAAAAA
35:33  |----------MM
35:46  |--------------------------
36:15  |DDDDDD
```

ACQUISITIONS AND EXECUTIONS

Let us consider more closely how each unit task was decomposed into an acquisition part and an execution part. The acquisition part of a unit task was indicated on the protocol by a Pause; the execution part consisted of a sequence of ICARUS commands, possibly including some Pauses. Sometimes the acquisition part of a task was broken by display commands, in which case the display commands were counted as part of the execution, whereas the Pauses after them were counted as part of the acquisition. The only difficulty with this decomposition was with the Check tasks, since they each consisted mostly of a Pause time followed by a few quick display-changing commands.

This decomposition of task times is graphically presented in Figure 10.7, which is a second-by-second time-line encoding of the protocol into acquisition and execution parts. Several features of the overall performance can also be seen on this figure: the long Check tasks at the end of the Transcription phase and at the beginning of both Compression phases, the relatively uniform task times in the Transcription and Vertical Compression phases, and the relatively high variance of the task times in the Horizontal Compression phase (see Figure 10.6).

Figure 10.8 shows the means and CV's of the acquisition, execution, and unit task times, grouped by different categories of tasks. Overall, acquisition and execution took about equal time, but acquisition had more variance. The greatest differences were between the different task types. Draw and Alter tasks were about the same, but the Dimension tasks were faster than either in both acquisition and execution, and the Check tasks consisted of long acquisition times with short execution times. The greater mixture of task types in the Horizontal Compression phase accounted for its greater variance and for the execution time being

Tasks	$T_{acquire}$ $M \pm CV$ (sec)	$T_{execute}$ $M \pm CV$ (sec)	T_{task} $M \pm CV$ (sec)
All tasks	10.7 ± 1.04	9.7 ± .77	20.4 ± .64
Transcription phase	10.9 ± .76	11.4 ± .69	22.3 ± .46
Vertical Compression phase	10.5 ± .91	10.5 ± .49	21.0 ± .45
Horizontal Compression phase	10.7 ± 1.28	7.9 ± .96	18.6 ± .86
Draw tasks	9.2 ± .57	12.1 ± .64	21.3 ± .42
Alter tasks	8.9 ± .86	10.6 ± .64	19.5 ± .61
Dimension tasks	5.1 ± .94	2.9 ± .95	8.0 ± .60
Check tasks	36.6 ± .57	1.5 ± 2.0	38.1 ± .59
Draw + Alter tasks	9.0 ± .75	11.2 ± .64	20.2 ± .53
Alter tasks (with Move)	9.7 ± .69	11.5 ± .44	21.2 ± .39
Alter tasks (without Move)	8.2 ± 1.03	9.8 ± .82	18.0 ± .78

Figure 10.8. Decomposition of unit-task times in into acquisition and execution parts Experiment 10A.

faster during this phase. The acquisition time remained constant between phases.

Draw and Alter tasks comprised over 80% of the tasks. They typically took about 9 sec to acquire and 11 sec to execute. The Alter tasks can be partitioned into those that used a Move command (the MOVE-COMMAND-METHOD) and those that did not, with the latter consisting of a series of deletion and (re)drawing commands to accomplish the move (the DELETE-AND-REDRAW-METHOD). The Move command itself took 9.4 sec (see below), accounting for 82% of the execution time of the Alter tasks that had a Move. The Alter tasks without a Move were 15% faster than those with a Move in both acquisition and execution time, but their task times were much more variable, since the DELETE-AND-REDRAW-METHOD was more sensitive to the number of elements being moved.

COMMAND FREQUENCY

The frequencies of use of the different ICARUS commands are listed in Figure 10.9 (in the N column). The four most frequently used commands were the Layer command, the Line commands, the Delete command, and the Flash command. Perhaps the biggest surprise was the high frequency of Layer commands—there were three Layer commands for every four drawing commands. The frequency of Delete commands was less surprising, since 90% of them occurred in Alter tasks to move items by the DELETE-AND-REDRAW-METHOD, and the rest were used to correct errors in Draw tasks (as in Figure 10.5). These four high-frequency commands—Layer, Line, Delete, and Flash—accounted for 70% of the command instances. Because they were all short commands, each requiring only one or two pointing actions, altogether they took only 48% of the total execution time (based on the calculations described below).

The fifth most frequent command type was the Move command. There were 25 Move commands in the protocol, accounting for 26% of the total execution time. The mean execution time for a Move command was 9.4 sec. (If the time for the Pauses that occur within three of the Move commands is excluded, the mean Move time becomes 8.6 sec.)

CALCULATION OF EXECUTION TIME

Using the method analysis for each command in Figure 10.2, a Keystroke-Level Model calculation of the execution time is presented in Figure 10.9. The total predicted execution time is 1192 sec, 16% longer than the total observed execution time of 1028 sec.

Command	n_P	n_K	n_H	n_M	$T_{execute}$ (sec)	N	T_{total} (sec)
Drawing Commands:							
(draw) **Line2** (2 end points)	2	2	0	0	2.8	68	188
(draw) **Line1** (1 end point)	1	1	0	0	1.4	8	11
(draw) **Flash**	1	2	0	1	3.0	51	154
(draw) **Rectangle**	2	3	0	1	4.4	1	4
(create) **Label**	1	8	2	2	6.8	6	41
Parameter Changing Commands:							
(change) **Layer**	1	1	0	0	1.4	87	120
(change) **Width** (of line)	0	4	2	2	4.6	5	23
Deletion Commands:							
Delete (element)	1	1	0	0	1.4	59	81
Undo (last deletion)	0	1	2	0	1.1	5	5
Transformation Commands:							
Move (elements)	6	9	3	1	11.7	25	292
Copy (elements)	6	9	3	1	11.7	1	12
Stretch (element)	3	5	2	1	6.9	21	144
Display Control Commands:							
Center (drawing)	1	2	0	1	3.0	15	45
(change) **Magnification**	0	2	2	1	2.7	7	19
Redraw (window)	0	1	2	1	2.4	5	12
Dimensioning Commands:							
(measure) **Distance**	2	2	0	0	2.8	11	30
(measure) **Size**	4	4	0	0	5.5	2	11

Predicted time to execute all commands	=	1192 sec
Total observed execution time	=	1028 sec
Prediction error	=	+ 16%

Figure 10.9. Execution time calculation for Experiment 10A.

This calculation is based on the Keystroke-Level Model (Chapter 8) and is similar to the calculation in Figure 8.5. The n's are obtained from the execution methods in Figure 10.2. The system response time in ICARUS is very fast and so does not need to be included. The unit operator times are taken from Figure 8.1; a typing rate of .28 sec/character is used. N is the number of instances of each command during the experimental session. $T_{total} = NT_{execute}$. Predicted total time = ΣT_{total}.

352

The 16% over-prediction is, however, more than just prediction error. For 106 unit tasks, we would expect the standard error of our prediction to be about 2% of the observed time (21%/√ 106). If we regard the Keystroke-Level Model calculation as a yardstick for expertness, then the user may be so highly practiced on the ICARUS commands that he requires fewer **M** operations than would be predicted from the rules in Chapter 8. The physical operations (**P**, **K**, and **H**) alone account for 96% of the user's execution time, which leaves little time for mental (**M**) operations. This is consistent with the idea that the user, because he is expert, has virtually eliminated his need for mental operations in executing ICARUS commands (see Section 8.3).

The Move command is the only command for which we have individual command execution times. The predicted Move time of 11.7 sec is 24% longer than the mean observed Move time of 9.4 sec. The Move commands account for almost all the long commands, with most of the remaining (non-Move) commands being much shorter. The predicted time for the non-Move commands is 13% longer than the observed time. Thus, the over-prediction is somewhat greater on the longer commands.

It is interesting to compare the tradeoff for the two methods for moving circuit elements on the layout: the **MOVE-COMMAND-METHOD** and the **DELETE-AND-REDRAW-METHOD**. For each circuit element to be moved, the **DELETE-AND-REDRAW-METHOD** requires a Delete command (1.4 sec) plus a drawing command (3.0 sec)[4] for each circuit element to be moved, which totals 4.4 sec per element. Since the Move command requires 9.4 sec, the **MOVE-COMMAND-METHOD** should be faster when there are 3 or more elements to be moved; and the **DELETE-AND-REDRAW-METHOD** should be faster when only 1 or 2 elements are to be moved. The user in the experiment moved an average of 1.8 elements with the **DELETE-AND-REDRAW-METHOD**, but on one-third of the delete-and-redraw tasks he moved 3 or 4 elements. (Data on the exact number of elements with the Move command was difficult to obtain.) Thus, the user appears to be operating near the timewise-optimal threshold in choosing between **MOVE-COMMAND-METHOD** and **DELETE-AND-REDRAW-METHOD**, with some bias towards the **DELETE-AND-REDRAW-METHOD**. This situation is comparable to the method selections observed for POET (see Section 5.2). POET users choose nearly optimally between the LF-

---

[4] The 3.0 sec is the average drawing command execution time, where the different kinds of drawing commands are weighted by their frequency in Figure 10.9.

METHOD and the QS-METHOD for locating lines, but with a bias towards the LF-METHOD.[5]

Error Data

There are 15 error tasks out of the 106 unit tasks. The errors can be grouped into four classes:

> *Selection errors.* These occurred within the Move command. Either too many circuit elements (4 cases) or too few circuit elements (1 case) were selected to be moved. New Move commands had to be performed to correct these errors.
>
> *Deletion error.* In one case, the user anticipated that a circuit element would unavoidably be selected in subsequent Move commands, even though he did not want it to be moved. To prevent this, he deleted the element with the intention of later redrawing it in its present location. After the Moves, however, he forgot, and only discovered the omission days later.
>
> *Location errors.* Because it is often difficult to specify the exact placement of elements, elements were moved to the wrong location (3 cases), flashes were drawn off center (2 cases), and a line was drawn the wrong size (1 case). There was also one case where two items at the opposite edges of the circuit were misaligned. The correction for Location errors was either to delete-and-redraw or to move the misplaced elements.
>
> *Parameter-setting errors.* Once, the user drew a line using the wrong line-width parameter; he had to reset the parameter, delete the line, and redraw it. Another

---

[5] It is interesting to speculate on the reasons for the bias. Why are the two methods (the DELETE-AND-REDRAW-METHOD in ICARUS and the LINEFEED-METHOD in POET) psychologically favored? One feature of both favored methods is that they are *incremental*: they accomplish their tasks with a series of small commands rather than by building up one big command. These methods are thus *less risky*. The user gets feedback at each increment, and the necessary corrections are small in case of an error. When using the big-command methods, however, the errors are larger and their recovery more difficult.

time the user forgot to type ESC after setting the
line-width parameter (and had to retype it later).

The most serious ICARUS error was undoubtedly the forgotten
deletion, since it remained undetected. Other errors were corrected at a
modest time cost. The time taken for the user to make all the above
corrections was a little over two minutes, about 6% of the total time spent
in the first three phases. This seems low compared to the 26% error time
observed in the earlier manuscript text-editing experiment with the POET
editor (Experiment 5C). However, 72% of the error time in the POET
session was due to three large errors in which the user lost her place in
the manuscript; the remaining errors in the POET session constituted only
about 8% of the time. These latter errors were mostly command execution errors and typing errors, comparable to the ICARUS errors above.
Perhaps an ICARUS user would have corresponding place-keeping errors
with a large circuit, but the two viewing windows provided in ICARUS
make losing one's place less severely penalizing in ICARUS than in POET.

10.3. CONCLUSIONS

The case study of computer-aided circuit design suggests that the
GOMS theory can be extended to semi-creative, human-computer
interaction tasks that are not explicitly given, but are generated by the
user. On analysis, computer-aided circuit design was seen to comprise a
creative part and a routine part not much different from the manuscript-
editing tasks studied earlier.

The routine part of the circuit-layout task had the following
similarities to the text-editing tasks studied by GOMS analysis earlier:

(1) The user's behavior was comprised of relatively independent
     unit tasks, each with a distinct acquire-execute cycle, each
     lasting about 10~30 sec.

(2) The time to execute commands was predicted by the
     Keystroke-Level Model with about 16% error.

(3) The frequency and cost of correcting execution errors was low
     (6% of execution time), as expected with skilled behavior.
     This error rate is comparable to that in the manuscript-editing
     task for the same types of errors.

(4) The user employed stereotypic methods: only six different command types were used for 85% of the command executions.

(5) Where the user had a choice of methods, he made near-optimal choices quickly.

(6) The user processed only a few elements at a time. For example, he transcribed only about three circuit elements at a time.

One important difference between circuit-layout and manuscript editing was the phase structure. The circuit-layout task was partitioned into phases of 5~15 minutes. Each phase had a distinct purpose and was governed by a loose plan that gave a distinct pattern to the unit tasks within the phase, yet that allowed the spontaneous generation of local unit tasks.

11. Cognitive Skill

Throughout this book we have been treating particular cognitive skills. The paradigm skill has been that of manuscript editing with a computer, though as Chapters 8, 9, and 10 have shown, the class of tasks to which the analysis applies is considerably broader. Our approach throughout has been to work close to the detailed structure of the tasks, drawing on the general base of modern information-processing psychology, as exemplified in Chapter 2, but not attending to how the models and results fit into a more general picture. It is this integration we now pursue.

What sets apart the human-computer interaction tasks dealt with in this book from many of the tasks that psychology has studied in detail is the combination of skilled behavior and a domain that is strongly cognitive. The primary substantive contribution of the studies of this book to basic psychology lies in helping to characterize the general nature of cognitive skill by the detailed understanding of one species of such skill.

Historically, the psychological study of skill has focused on perceptual-motor skills. Consequently, the obvious tack is to take cognitive skills to be those skills that involve cognition, as opposed to those skills that involve the motor and/or perceptual systems. However, it is not possible to distinguish the skills we have been studying from others simply by the existence of a cognitive component. As Welford says in his *Fundamentals of Skill:*

> Although a distinction is commonly drawn between sensory-motor and mental skills, it is very difficult to maintain completely. All skilled performance is mental in the sense that perception, decision, knowledge and judgment are required. At the same time all skills involve some kind of co-ordinated, overt activity by hands, organs of speech or other effectors. In sensory-motor skills the overt actions clearly form an essential part of the performance, and without them the purpose of the activity as a whole would disappear. In mental skills overt actions play a more incidental part, serving rather to give expression to the skill than forming an essential part of it. (Welford, 1968, p. 21)

Thus, all skill involves cognition. Perhaps, then, cognitive skills could be distinguished by saying that they are *primarily* cognitive. More penetrating is Welford's characterization (above) of the role of motor behavior in mental (i.e., cognitive) skill, namely, that it *expresses* the cognitive skill. Manuscript text-editing includes the skills of keystroking and viewing the manuscript and display; however, these perceptual-motor skills are not the essential activity, but the medium through which the cognitive activity gains expression.

The primacy of cognitive activity in cognitive skill does not rob the behavior of its skillful character, taking the term *skillful* to mean "competent, expert, rapid and accurate performance" (Welford, 1968, p. 12). This includes the sense of effortlessness—smoothly coordinated and patterned behavior—that is the visible hallmark of skilled performance. Our text-editing experts truly fly over the keyboard; and the contrast of their behavior with that of beginners leaves no room for doubt that skill, both perceptual-motor and cognitive, has been acquired.

We now attempt to characterize the general nature of cognitive skill to see how skill in text-editing both confirms and illuminates it. We start by presenting a view of all cognitive behavior as having a dimension of skill, so that any cognitive behavior is more or less skilled (Section 11.1). We illustrate this by showing how a task that initially requires problem solving gradually becomes skilled (Section 11.2) and how text-editing, which becomes a skill for most users, has its roots in problem solving (Section 11.3). We next examine the distinguishing features of the skill of text-editing, its most dominant feature being the unit-task structure (Section 11.4). However, skills are characterized and differentiated along many dimensions, of which the unit task is only one (Section 11.5).

Our development of the notion of cognitive skill in this chapter is consonant with current work in cognitive psychology on this topic, as exemplified by the recent collection, *Cognitive Skills and their Acquisition* (Anderson, 1981*a*). In particular, our view of cognitive skill is in substantial agreement with the picture of cognitive skill acquisition emerging from the contemporaneous work of John Anderson (1980, Chapter 8; 1981*b*).

11.1. THE SKILL DIMENSION OF COGNITIVE BEHAVIOR

Human behaviors tend to get labeled—as problem solving, skill, learning, imagining, creating, day-dreaming, etc. This leads to viewing behavior in typological terms, with many distinct species of behavior, each having its own separate characteristics. The actual situation seems to be considerably different. As epitomized by the Rationality Principle in Chapter 2, behavior is responsive to (1) the nature of the task and the human's goals, (2) the nature of the human's processing capabilities, and (3) the preparation of the human for dealing with the particular task. These factors provide different ways to classify behaviors. On the one hand, task demands and human goals are as diverse as the world itself, and human behavior reflects this diversity. However, strong communalities arise in behavior by the common involvement of the basic processing system. All behavior feeds through the perceptual, cognitive, and motor processors, whose fixed properties make all behavior similar in many respects. But also, as indicated by the third factor, humans are able to perform the same task in many ways and with many degrees of facility, depending on their state of preparation—their knowledge of facts and procedures and their internal organization to use them.

Thus, all cognitive behavior can be located in a three-dimensional space, as sketched in Figure 11.1. The task (vertical) dimension indicates the vast geography of different tasks. The processing dimension (shown as depth) reflects the perceptual, cognitive, and motor structuring of the human processor. In fact, any behavior will involve all three compo-

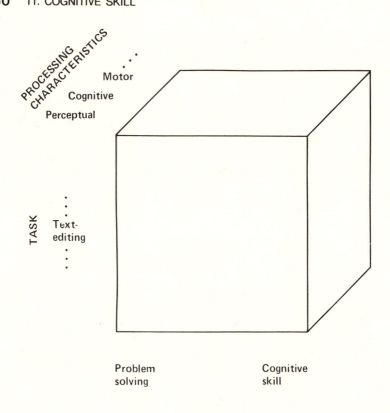

Figure 11.1. Three dimensions of cognitive behavior.
All cognitive behaviors, including both problem-solving behaviors and cognitive-skill behaviors, can be classed along three dimensions: a task dimension, a processing dimension, and a skill dimension. Any cognitive behavior on a given task will, with practice, shift to the right along the skill dimension.

nents, and so a given behavior on a task would actually be characterized by a relative emphasis on each subsystem. The skill (horizontal) dimension indicates the degree of skill with which the behavior is performed. The further to the right, the more skilled is the performance of the task; the further to the left, the more the behavior becomes that of solving a problem.

The skill dimension is the most important one for us here. All behaviors lie at some degree of skill along this dimension (although, of course, complex behaviors dealing with different subtasks of a total task can vary in their skill). Moreover, problem-solving behavior is simply the less skilled end of this dimension; it is not a separate species of behavior nor a separate class of tasks.   The notion of a skill dimension that includes both cognitive skill and problem solving is consonant with the Model Human Processor of Chapter 2, both the Problem Space Principle and the Law of Practice providing grounds for its support.   We take up each of these principles in turn.

THE PROBLEM SPACE HYPOTHESIS

The Problem Space Principle in Chapter 2 states that problem solving takes place by search in a problem space, with the partial knowledge of how to proceed—the *search control knowledge*—being used to guide the search through the space.   The principle, as stated, applies only to problem solving, where it has attained an impressive amount of support.   However, it can be extended from problem solving to cover all cognitive behavior (Newell, 1980):

> **Problem Space Hypothesis:** *The fundamental organizational unit of all human symbolic activity is the problem space.*

The status of this extended principle is much more tentative, but it can still serve our purposes.   It implies a homogeneity of structure for all cognitive behavior, from problem solving to cognitive skill.   It thus supports the notion of a continuous skill dimension.

The difficulty of performance in a problem space can be graded by how much the search control knowledge available in the problem space constrains behavior.   With little knowledge, the operators selected have small chance of being the right ones.   The paths through the problem space are bound to go astray, leading to backtracking, pauses from ignorance, trial search—the whole panoply of behaviors characteristic of problem solving.   With abundant search control knowledge, the operators selected are almost always right.   Behavior proceeds directly from the initial point in the problem space to the final goal state with only occasional error, and even then with knowledge usually available to recover quickly and get back on the track.[1]   This grading of problem

---

[1] Implicit in this account is that search control knowledge is knowledge ready to be brought immediately to bear on selection, without hesitation or puzzlement.   See Newell (1980) for elaboration.

spaces according to control knowledge provides the basis for a scale from problem solving to skilled behavior.

For behavior at the skilled end of the dimension, the sequence of operators performed are highly predictable from the structure of the task. When the task itself is only moderately complex (as are the text-editing tasks in this book), the behavior will be specifiable simply by listing the sequences of operators to be performed, explicitly conditional on the appropriate features of the task. This, of course, is just the condition under which behavior is easily specified by a standard programming language, such as Basic or Pascal. We found it convenient to use our own notational variant, the GOMS notation introduced in Chapter 5; but it comes to basically the same thing.

The GOMS scheme, however, contains one more control feature beyond operator sequence and explicit conditionals, namely goals. These help to reveal the relationship between the procedural representation of a behavior and its skill. Goals are analogous to procedure calls in standard programming languages; but they have an additional degree of conditionality in selecting the method (i.e., the procedure body) to attain the goal, depending on the characteristics of the particular environment being faced. This conditionality, embodied in the GOMS selection rules, is not provided explicitly in standard programming languages,[2] where it is assumed that sufficient analysis has been done by the programmer to directly call each procedure (via a unique name). On the other hand, artificial intelligence programming languages (e.g., Bobrow and Raphael, 1974) have goals, with methods selected by pattern-matching against the global database (so-called *pattern-evoked procedures*). This is a more open technique of method selection than the fixed selection rules of GOMS. Thus, behaviors with a greater degree of skill can be adequately represented with more rigid and explicit control structures. The rigidity of the GOMS scheme is a direct reflection of its being a model of cognitive skill.

THE POWER LAW OF PRACTICE

A second line of evidence for the continuous character of the skill dimension comes from the Power Law of Practice (Chapter 2), which states that the time to perform any task decreases with practice according

---

[2] It is not required technically; the same selection can be attained by explicitly making the requisite tests at the beginning of the procedure body. But, in representing control, the issue is always where and in what form to encode conditionality.

to a common quantitative law (a power function). This law is a quantitative reflection of the process that produces skilled behavior. The important aspect of this law in the present context is that it applies uniformly to all types of cognitive behavior, so long as the behavior is sufficiently well organized to attain the task. The law applies not only to skilled behaviors (cognitive and perceptual-motor), but also to problem-solving behaviors (Newell and Rosenbloom, 1981). The exact mechanism through which the law operates is still unclear (and under active research). For example, it cannot yet be tied firmly to the problem space organization of behavior, as implied in the Problem Space Hypothesis above. Yet, in showing that learning proceeds in a gradual way, the Power Law of Practice implies that all cognitive behavior, with practice, moves smoothly—within a homogeneous structure—along a continuous skill dimension.

THE SKILLED END OF THE DIMENSION

There is some indication that, as behavior becomes highly skilled, some of it may become organized differently, contrary to the notion of a homogeneous organization implied by the Problem Space Hypothesis. The key phenomenon is that of *automaticity*. In general, people are aware of much that goes on while performing tasks—they can comment on what they are doing, what they want, what they plan to do, what task features they have noticed, etc. But some behaviors can become highly automatic, so that the entire performance proceeds outside awareness. There may be no awareness even that the behavior occurred or no ability to recollect any intermediate aspects of internal processing. Automatic behaviors are highly skilled behaviors, and they develop gradually with extreme practice, following typical power-law practice curves, with nothing special to indicate the degree of automaticity. Thus, the natural interpretation is that automaticity is simply another attribute of skill, which increases with degree of skill, becoming highly salient at the extreme end of the skill dimension.

However, such a uniform dimension would imply that all behaviors could become fully (at least extensively) automatic; but that does not seem to be the case. The simpler and less varied a behavior, and the less cognitive (hence more perceptual-motor), the easier it appears to be for it to become automatic, though some quite complex extended tasks can become automatic. On the other hand, some rather simple (though per-ceptually varying) tasks have been shown to be very resistant to becoming automatic (see, e.g., Shiffrin and Dumais, 1981). It seems unlikely that text-editing could ever become highly automatic, however skilled it became (e.g., consider that the user in Section 8.4 still required mental preparation time even after executing the same method 1100 times). Therefore, automatic behavior could imply use of a structurally different process than cognitive skill behavior and thus a non-homogeneity in the skill dimension. This is simply another place where our simplified Model Human Processor does not yet reflect some important psychological issues, and we do not pursue it further here.

11.2. COGNITIVE SKILL FROM PROBLEM SOLVING

Problem-solving behavior will, with practice, become cognitive skill. To illustrate, we start with a classic example of problem solving, the Tower of Hanoi puzzle, whose problem space has been given ample analysis (Nilsson, 1971; Simon, 1975; Newell 1980).   We show what happens theoretically to problem solving when (1) the search control knowledge increases and (2) the problem space is altered by the construction of new operators.

Problem Space for the Tower of Hanoi Puzzle

A problem space, as outlined in Chapter 2, consists of a set of knowledge *states* plus a set of *operators* for transforming states to other states.   A *problem* within a problem space is defined by an *initial state* and a *goal* (a state or set of states).   Solving the problem consists of finding a sequence of operators (often called a *path*) to transform the initial state into a goal state.   The Tower of Hanoi puzzle consists of a set of three pegs with disks of different sizes stacked on the pegs (see Figure 11.2*a*).   The problem is to move a pyramid of disks from one peg to another peg.   A state in the problem space is a configuration of the disks on the pegs, where no disk rests on top of a smaller disk.   There is only one operator in this problem space, MOVE-DISK, and it moves a single disk from one peg to another.   There is a *path constraint* on this operator which only allows the top disk on a peg to be moved and which does not allow a disk to be placed on top of a smaller disk.   (This problem space is more carefully defined in Figure 11.2*a*.)   A problem space can be viewed as a graph, where the states are the nodes and the operators are the links between nodes. The problem space graph for a three-disk version of the Tower of Hanoi puzzle is shown in Figure 11.2b. The graph is a map of all possible paths from the initial state to the goal state.

The important psychological assertion about a problem space is that the problem solver will confine his behavior to lie within its boundaries. According to the theory of human problem solving, a person solves a problem by searching through the problem space state by state. At any point in time the problem solver resides in some state, called the *current state*; and there is a small set of previously visited states that he can still remember, called the *stock*. The units of behavior in a problem space are the successive applications of operators. For each application of an operator, there is a *control cycle* of functions to be performed (Figure 11.3). The functions involve selecting a state in the stock to work from, selecting an operator, applying the operator, and deciding whether a goal has been reached. By cycling through these functions, the problem solver will proceed from the initial state through a succession of intermediate states, perhaps reaching a goal state.

Accumulation of Search Control Knowledge

Many of the control cycle functions involve decisions, such as what operator to select. The knowledge on which to base these decisions is called *search control knowledge*. With little control knowledge, the problem solver will wander about the problem space in search of a goal state (such as is illustrated in Figure 11.4a). But with training or experience in doing the problem, the problem solver will acquire knowledge for guiding the search and making it more efficient. In the Tower of Hanoi, for example, it is quickly evident that moving back to the just-previous state is useless; other examples of control knowledge for the Tower of Hanoi are given in Figure 11.3. The problem solver eventually may build up enough search control knowledge so that he goes straight to the goal (Figure 11.4b). As his search control knowledge increases, he becomes more expert; and his behavior changes from problem solving to cognitive skill.

Thus, problem solving and cognitive skill both take place in a problem space, the main difference being the amount of search control knowledge available. In cognitive-skill behavior, decisions have to be made; but they are non-problematic, since the problem solver *knows* the

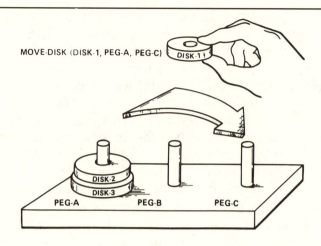

MOVE-DISK (DISK-1, PEG-A, PEG-C)

DISK-1

DISK-2
DISK-3

PEG-A   PEG-B   PEG-C

Informal Description: Three pegs—labeled PEG-A, PEG-B, PEG-C—are attached to a board as illustrated above. Disks of different diameters—labeled DISK-1 (the smallest), DISK-2, DISK-3, etc.—are stacked on PEG-A in a pyramid. The goal of the puzzle is to move the pyramid of disks to PEG-C by moving the disks from peg to peg, one at a time. A disk may be moved from any peg to another providing that it is the top disk on its peg and that it is not moved on top of a smaller disk.

States: Any configuration of the disks on the pegs such that a larger disk is not on top of a smaller disk.

Operator: MOVE-DISK (Disk, FromPeg, ToPeg)
                    causes Disk to be moved from FromPeg to ToPeg.

Path Constraint: On each MOVE-DISK, Disk must be the top disk on FromPeg and Disk must be smaller than the top disk on ToPeg.

Problem: All disks on PEG-A. ⇒ All disks on PEG-C.

Figure 11.2*a*. Problem space definition for the Tower of Hanoi puzzle.

The description above completely defines the state space, which is laid out explicitly in Figure 11.2*b*.

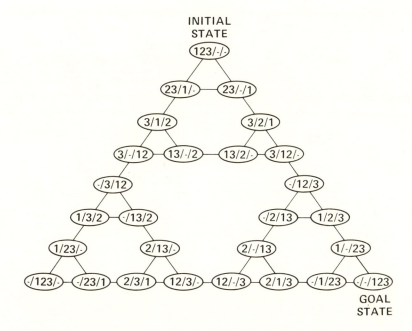

Figure 11.2*b*. Problem space graph for the three-disk Tower of Hanoi puzzle.

Each state is described as "Disks-on-PEG-A / Disks-on-PEG-B / Disks-on-PEG-C." For example, the state 23/1/ – indicates that DISK-1 is on PEG-B and DISK-2 and DISK-3 are on PEG-A; this state comes about by applying the operator MOVE-DISK (DISK-1, PEG-A, PEG-B) to the initial state, 123/ – / –. Each link in the graph represents an application of the MOVE-DISK operator.

appropriate actions to take. It is as though the flexible control structure of problem solving were frozen into specific procedures that are available at performance time. We have called these procedures *methods* in previous chapters. What distinguishes cognitive skill from problem solving is the packaging of operator sequences into integrated methods. As we have seen in previous chapters, methods are not simply uncon-ditional sequences. They also have conditional actions, although the conditionality is limited to prepared alternatives. Method selection in the GOMS model does not lead to search, but to the selection of whole methods from a fixed repertoire. Furthermore, the method selection process occurs rapidly and without any external signs of decision making. These conditionalities and method selections are part of search control knowledge.

As an example of the packaging of search control knowledge into methods for cognitive skill, consider the method of solving the Tower of

Control Cycle		Examples of Control Knowledge in the Tower of Hanoi Puzzle
Step No.	**Control Function**	
(1) (a)	Select a state from the stock, making it the current state.	– *Make the new state the current state.*
(b)	Select an operator.	– *Move a disk to the peg specified by a current goal.* – *Move an obstructing disk to the non-target peg.* – *Do not move back to the just-previous state.* – *Do not move a disk twice in a row.*
(2)	Apply the operator to the current state, producing a new state.	
(3) (a)	Decide whether the new state is a goal state.	– *A state is a goal state if it exactly matches the goal state pattern.*
(b)	Decide whether to quit.	– *Quit if successful.* – *Quit when told by the experimenter.*
(c)	Decide whether to add the new state to the stock.	– *Add the new state to the stock.*
(4)	Go back to step (1), unless decision is to quit.	

Figure 11.3. Control cycle for searching in a problem space.
This scheme is adapted from Newell (1980). Note that the lettered functions within numbered functions of the control cycle can be executed in any order.

Hanoi puzzle by simply memorizing all the moves. This method, the **MEMORIZED-MOVE-METHOD**, is described at the top of Figure 11.5 in the GOMS notation. The search control knowledge implicit in this method can be categorized by the search control decisions (identified by the number-letter labels on the functions in Figure 11.3) necessary for problem solving:

    (1a) Proceed from the current state.
    (1b) Select the next operator in the method.
    (3a) The goal state is reached after executing the method.
    (3b) Quit after applying the last operator in the method.
    (3c) Make the new state the current state.

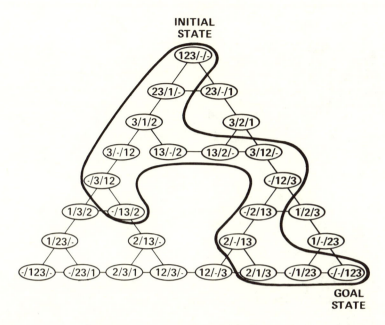

Figure 11.4a. Solution path in the Tower of Hanoi problem space for a hypothetical novice.

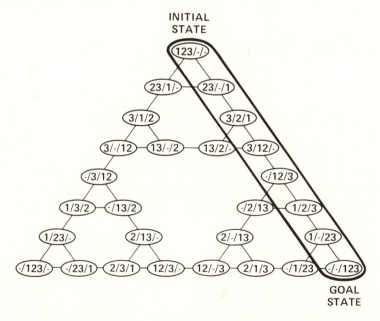

Figure 11.4b. Solution path in the Tower of Hanoi problem space for an expert.

MEMORIZED-MOVE-METHOD:

> GOAL: MOVE-PYRAMID (Pyramid(3), PEG-A, PEG-C)
> . MOVE-DISK (DISK-1, PEG-A, PEG-C)
> . MOVE-DISK (DISK-2, PEG-A, PEG-B)
> . MOVE-DISK (DISK-1, PEG-C, PEG-B)
> . MOVE-DISK (DISK-3, PEG-A, PEG-C)
> . MOVE-DISK (DISK-1, PEG-B, PEG-A)
> . MOVE-DISK (DISK-2, PEG-B, PEG-C)
> . MOVE-DISK (DISK-1, PEG-A, PEG-C)

GOAL-RECURSION-METHOD:

> GOAL: MOVE-PYRAMID (Pyramid(n), StartPeg, TargetPeg)
> . FIND-OTHER-PEG (StartPeg, TargetPeg) → OtherPeg   *if n > 1*
> . GOAL: MOVE-PYRAMID (Pyramid(n – 1), StartPeg, OtherPeg)   *if n > 1*
> . MOVE-DISK (Disk(n), StartPeg, TargetPeg)
> . GOAL: MOVE-PYRAMID (Pyramid (n – 1), OtherPeg, TargetPeg)   *if n > 1*

PERCEPTUAL-MOVE-PATTERN-METHOD:

> GOAL: MOVE-PYRAMID (Pyramid(n), StartPeg, TargetPeg)
> . GOAL: DEFINE-PEG-ORDER
> . . FIND-OTHER-PEG (StartPeg, TargetPeg) → OtherPeg
> . . PegOrder = {StartPeg, TargetPeg, OtherPeg}   *if n is odd*
> . . PegOrder = {StartPeg, OtherPeg, TargetPeg}   *if n is even*
> . GOAL: DO-MOVE-PYRAMID   *repeat until satisfied*
> . . GOAL: MOVE-SMALLEST-DISK
> . . . FIND-SMALLEST-DISK → DISK-1, PegOfDisk1
> . . . FIND-PEG-AFTER (PegOfDisk1, PegOrder) → NextPeg
> . . . MOVE-DISK (DISK-1, PegOfDisk1, NextPeg)
> . . GOAL: MOVE-SECOND-SMALLEST-TOP-DISK
> . . . FIND-SMALLEST-DISK → DISK-1, PegOfDisk1
> . . . FIND-SECOND-SMALLEST-TOP-DISK → Smallest2, PegOfSmallest2
> . . . FIND-OTHER-PEG (PegOfDisk1, PegOfSmallest2) → OtherPeg2
> . . . MOVE-DISK (Smallest2, PegOfSmallest2, OtherPeg2)

Figure 11.5. Three methods for the Tower of Hanoi puzzle.
The methods are described using the GOMS notation (from Chapter 5), which has been augmented to include variables (names in combined upper and lower case) and variable assignment (→).

There can be several methods for solving a problem, and different methods may place quite different demands on the problem solver. Simon (1975) has analyzed several methods (he called them strategies) for the Tower of Hanoi. The three methods described in Figure 11.5 (the MEMORIZED-MOVES-METHOD, the GOAL-RECURSION-METHOD, and the

PERCEPTUAL-MOVE-PATTERN-METHOD) are taken from from Simon's analysis.

The MEMORIZED-MOVES-METHOD we have just seen. The principal difficulty with this method is not only the large number of steps that must be memorized, but the fact that the steps are very similar and hence susceptible to interference. Also, because the steps are different depending on the number of disks in the puzzle, this method has no generality.

The GOAL-RECURSION-METHOD is based on a few observations about the structure of the Tower of Hanoi puzzle. Note that a pyramid of disks contains subpyramids—a three-disk pyramid consists of DISK-3 plus a two-disk pyramid. Moving a pyramid can be broken into moving its largest disk plus moving its subpyramid. Moving a subpyramid is structurally similar to moving the pyramid (which can be seen as the nested triangular regions in the state space graph in Figure 11.2b). Hence, the following recursive procedure will move a pyramid to a target peg:

—Move the subpyramid to the non-target peg.
—Move the largest disk to the target peg.
—Move the subpyramid to the target peg.

Moving a subpyramid is a subgoal of the goal of moving the pyramid, and moving the largest disk can be done with a MOVE-DISK operation. Each subgoal generates a further subgoal, until one works down to the subgoal of moving a one-disk pyramid, which can be done with a single MOVE-DISK operation. This method is very elegant, all of the operations falling into place within the subgoal structure—provided that the problem solver keeps track of the entire analysis. (In fact, if he could remember the dynamic goal structure, he would not even have to look at the puzzle's state.) The main problem with this method, of course, is the large amount of Working Memory required. Again, not only are there a large number of subgoals, but they are very similar, causing interference.

In contrast, the PERCEPTUAL-MOVE-PATTERN-METHOD uses the visible state of the puzzle to determine the next move. The first subgoal in this method is to determine which direction the smallest disk will cycle, which depends on whether the number of disks in the puzzle is odd or even. Then the disks can be moved according to a simple pattern: The smallest disk is moved to the next peg in the cycle, and the next-smallest exposed disk is moved to the one peg it can go to. This pair of moves is repeated until the puzzle is solved. Thus, to determine each move, one only need examine the visible state of the puzzle, which is accomplished by the perceptual (FIND) operators in Figure 11.5. In this method there is no dynamic goal structure to keep track of in Working Memory.

A person using any of these methods to solve the Tower of Hanoi puzzle is engaged not in problem solving, but in cognitive skill. Although the search control knowledge implicit in each of these methods could be represented as individual items of control knowledge (as in Figure 11.3), we have cast the methods in GOMS notation (Figure 11.5) to emphasize that the problem solver's search control knowledge is compiled into integrated procedures for efficient performance. However, a skilled expert is not restricted to executing precompiled methods only. Since the methods are embedded in a problem space, the expert can often revert to problem solving when necessary.

Consider what happens when an expert inadvertently makes an error while executing a method. In Chapter 5 we showed that most (but not all) error correction in text-editing just involves the execution of more GOMS methods, but we had to strain the GOMS control structure to make this scheme work. The problem space provides a better way to understand errors. While executing a method, the user is moving along a path of states in the problem space. The occurrence of an error throws the user off the path into another state, *but one that is still a state in the problem space.* He then has to formulate another goal to get back onto the intended path. This goal may be reachable by an available error-correction method; but, if there are no methods available, the user can search for the solution in the problem space (which is what happened in the three large errors observed in Experiment 5C in Chapter 5). This ability to revert gracefully to problem solving allows the expert user to deal with new, unfamiliar tasks.

Construction of Problem Space Operators

As a problem solver accumulates search control knowledge, he becomes more skilled. The process is gradual, so that a problem solver has only partially integrated methods, and his behavior is a mixture of problem solving and cognitive skill. This process can be illustrated with the Tower of Hanoi puzzle. The most frequently moved disks are the two smallest disks, DISK-1 and DISK-2; a problem solver quickly learns a special method[3] for moving them:

---

[3] This method is, of course, a special case of the **GOAL-RECURSION-METHOD** (Figure 11.5). However, the problem solver in this case does not think of it as such, but as a specific method that only applies to the two smallest disks.

—Move DISK-1 to the non-target peg.
—Move DISK-2 to the target peg.
—Move DISK-1 to the target peg.

With this method, the subgoal of moving the two smallest disks is no longer a problem. The effect is to give the problem solver the equivalent of a new operator, MOVE-DISKS-12, which moves these two disks. The addition of this new operator reduces the size of the problem space—from 27 states (Figure 11.2*b*) to only 9 states (Figure 11.6*a*)—and the number of operations to the goal has been reduced from seven to three.

Thus, the acquisition of enough search control knowledge to define a partial method results in the construction of a new operator that restructures the problem space into two nested problem spaces: a *reduced problem space* (Figure 11.6*a*) and a *skill space* (Figure 11.6*b*). The reduced problem space will still occasion problem-solving behavior, though there will be fewer states through which to search. When the problem solver executes the new operator, he descends into the skill space, where the execution takes place. There will be little or no search in the skill space, for the problem solver has methods for guiding him through.

Levels of Cognitive Behavior

Complex and extended cognitive behavior is organized hierarchically into many levels. This is reflected in the hierarchical goal structures and the various models at different levels in the GOMS framework in Chapter 5. This is also reflected in the problem space framework, where complex cognitive behavior is organized into a nested hierarchy of problem spaces. The various parts of a behavior (i.e., the various problem spaces) are differentially skilled. Although in the Tower of Hanoi example it is the lower levels of behavior that are the more skilled (implying a strictly bottom-up growth of skill), it is also possible for the higher levels of behavior to be more skilled than the lower levels (imagine, for example, a computer-game version of the Tower of Hanoi being played for the first time by an expert Tower of Hanoi player who does not understand how to specify moves to the computer). Thus, a complex cognitive behavior is a medley of varying degrees of skill. When we speak of a given degree of skill for a behavior, we are really referring to some sort of average skill over the various levels.

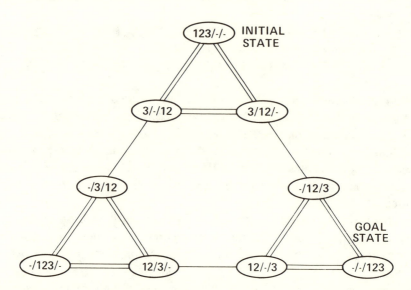

Figure 11.6a. Reduced problem space for the three-disk Tower of Hanoi puzzle.

The single-line links represent MOVE-DISK operations, just as in Figure 11.2b. The double-line links represent MOVE-DISKS-12 operations.

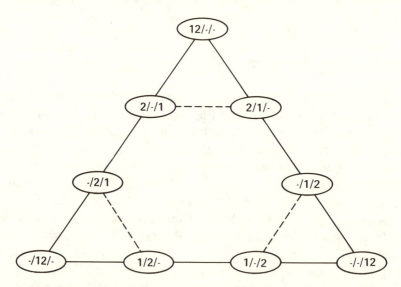

Figure 11.6b. Skill space for the two-disk subproblem of the Tower of Hanoi puzzle.

The solid links represent MOVE-DISK operations. The dashed links represent potential MOVE-DISK operations that the user does not perform, since they are not needed in the skill space.

Another good example of the composite nature of cognitive behavior is the mixture of problem solving and cognitive skill observed in the circuit-layout task studied in Chapter 10.   During the phase of compressing the circuit, the user was observed to be problem solving, yet he was obviously highly skilled at using the layout system.   The user was clearly skilled in executing the various kinds of unit-tasks, since he had methods for accomplishing them.   The unit-task executions were available as operators in the problem space where the user was searching for a more compressed circuit configuration.   Thus, the user's problem-solving behavior was taking place in a reduced problem space for circuit compression—reduced by the unit task operators—and his skilled behavior was taking place in skill spaces within the unit-task executions.

11.3.   PROBLEM SOLVING PRECEDING COGNITIVE SKILL

We have illustrated how a familiar problem-solving behavior will, with practice, become skilled.   We now wish to look at a familiar cognitive skill and show how it arose from problem solving with practice. The obvious illustrative example for us is text-editing, the paradigm cognitive skill of the book.   Hence, let us consider the clause-switching task introduced in Chapter 8 (Experiment 8B).   The task is to change Sentence 11.1a to 11.1b using the editing facilities of BRAVO:

The sun shines when it rains; our weather is funny.   (11.1a)
Our weather is funny; when it rains the sun shines.   (11.1b)

The task involves switching the two outer clauses and adjusting the punctuation and capitalization of the sentence.   In this task there is a restriction that the user is not to retype any of the text within the clauses. With this restriction, it is more convenient to consider the task in the more abstract form:

$$[A\text{—} \ b\text{—}; \ c\text{—}.] \ \Rightarrow \ [C\text{—}; \ b\text{—} \ a\text{—}.] \ .$$

Each clause is represented by a letter followed by a dash.   The letter labels the clause and indicates whether the first letter of the clause is upper or lower case; the dash represents the rest of the text in the clause, which is treated as an indivisible substring in this task.

Problem Space for BRAVO Text-Editing

Our first job is to lay out the problem space for this task. For the most part, the problem space is determined by the structure of the BRAVO editor and the further restrictions of the clause-switching task. It is simplest to see how BRAVO shapes the problem space by first considering a subtask of the clause-switching task—moving a substring of text to a new location.

PROBLEM SPACE FOR THE MOVE-TEXT TASK

The problem space for the move-text task is defined in Figure 11.7. Each state in the problem space has three parts: (1) the current configuration of text in the workspace, (2) the substring of the text that is currently selected, and (3) the contents of the deletion buffer. BRAVO's commands provide the operators to change state and allow the user to set and alter the selected text by pointing (see Figure 6.1). This is represented in the problem space by the SELECT-TEXT operator. This operator has one argument, the text to be selected (which is designated by pointing). The Delete command in BRAVO, represented by the DELETE-TEXT operator, deletes the selected text from the workspace (and makes the current selection be the character following the deleted text). The Insert command in BRAVO has several variants, which differ so much that they are represented by different operators. The simplest variant of the Insert command is to type in new text (from the keyboard) in front of the current selection; this is represented by the INSERT-NEW-TEXT operator.

These three operators, SELECT-TEXT, DELETE-TEXT, and INSERT-NEW-TEXT, are sufficient to allow text to be moved. A method for doing this, the RETYPE-AND-DELETE-METHOD, is shown in Figure 11.7. Text to be moved is (re)typed in its new location and then deleted from its old location. Although this method is easily derived from means-ends analysis, even moderately competent users of BRAVO have assimilated it as an integrated method. The method is not allowed in the clause-switching task because of the restriction against retyping the clauses. However, there is an allowable method, the COPY-AND-DELETE-METHOD, which is very similar, the only difference being that text is copied instead of retyped. This method requires the use of another variant of the Insert command, which inserts a copy of a piece of text from another location. This variant of the Insert command is represented by two operators, one

Informal Description: Move a substring of text to another location using the commands available in BRAVO.

States: [Text with <u>selection</u> underlined] {Deletion Buffer}

Operators:
SELECT-TEXT (TextToSelect)
DELETE-TEXT
INSERT-NEW-TEXT (NewText)
SELECT-COPY-TEXT (TextToBeCopied)
INSERT-COPIED-TEXT
INSERT-DELETED-TEXT
etc.

Problem: [A— b—; c—.] ⇒ [b—; c—A— .]

Methods: **RETYPE-AND-DELETE-METHOD**

	[A— b—; c—.]	{}
SELECT-TEXT	→ [A— b—; c—<u>.</u>]	{}
INSERT-NEW-TEXT	→ [A— b—; c—<u>A—</u> .]	{}
SELECT-TEXT	→ [<u>A—</u> b—; c—A— .]	{}
DELETE-TEXT	→ [<u>b</u>—; c—A— .]	{A— }

COPY-AND-DELETE-METHOD

	[A— b—; c—.]	{}
SELECT-TEXT	→ [A— b—; c—<u>.</u>]	{}
SELECT-COPY-TEXT	→ [<u>A—</u> b—; c—.]	{}
INSERT-COPIED-TEXT	→ [A— b—; c—<u>A—</u> .]	{}
SELECT-TEXT	→ [<u>A—</u> b—; c—A— .]	{}
DELETE-TEXT	→ [<u>b</u>—; c—A— .]	{A— }

DELETE-AND-INSERT-METHOD

	[A— b—; c—.]	{}
SELECT-TEXT	→ [<u>A—</u> b—; c—.]	{}
DELETE-TEXT	→ [<u>b</u>—; c—.]	{A— }
SELECT-TEXT	→ [b—; c—<u>.</u>]	{A— }
INSERT-DELETED-TEXT	→ [b—; c—<u>A—</u> .]	{A— }

Figure 11.7. Problem space and three methods for the move-text task.

The methods are shown as paths in the problem space, i.e., as sequences of states and the operators that change the states.

377

for selecting the text to be copied (SELECT-COPY-TEXT) and one for making the copy and insertion (INSERT-COPIED-TEXT). The COPY-AND-DELETE-METHOD is known and used by all expert BRAVO users.

A third method for moving text, the DELETE-AND-INSERT-METHOD, requires not only another operator, but also knowledge of BRAVO's deletion buffer. Whenever a piece of text is deleted (DELETE-TEXT), it is saved in a deletion buffer. The text in the buffer may be accessed by the third variant of the Insert command, INSERT-DELETED-TEXT, which inserts a copy of the buffer at the selected location. The DELETE-AND-INSERT-METHOD, shown in Figure 11.7, is thus to delete the text to be moved and to copy it from the buffer to the new location.[4] This method is riskier than the first two methods (the wrong text could be deleted, the buffer contents could be lost, etc.), but is faster to execute (according to the Keystroke-Level Model). Almost all BRAVO experts know this method, and most use it at least occasionally despite the risk.

For an expert BRAVO user, moving text is not a problem; it is a routine unit task, for which he knows several methods (of which the three methods presented above are only a sample). The problem space for moving text (Figure 11.7) is thus a skill space, and the activity of moving text is available to the expert as a unitary operator in a larger problem space context.

PROBLEM SPACE FOR THE CLAUSE-SWITCHING TASK

We now describe a problem space for the clause-switching task in which the MOVE operator is used. We could extend the problem space in Figure 11.7 (by adding more operators to represent other commands) to encompass the clause-switching task; but that space would be at too fine a level of detail to represent expert behavior. That this is so can be seen by considering the SELECT-TEXT operator. The expert user never has to decide explicitly to use SELECT-TEXT, for he has control knowledge telling exactly when to use it. He knows that SELECT-TEXT is always used just before other operators that effect the selection. For example, it is used before DELETE-TEXT to specify the text to be deleted; to the expert, these two operators are bound together as a unit.

Thus, there should be a reduced problem space for the clause-switching task in which selections are incorporated within the operators.

---

[4] This method is described at the Keystroke Level as task T4-BRAVO in the Appendix to Chapter 8.

Informal Description: Starting with a three-clause sentence, switch the outer clauses and adjust the punctuation and capitalization, using the commands available in the BRAVO editor. There is a restriction that the text within the clauses is not to be retyped, except for the first letters of the clauses.

States: [Text]

Operators: MOVE (ToLocation, TextToBeMoved)
            COPY (ToLocation, TextToBeCopied)
            DELETE (TextToBeDeleted)
            INSERT (ToLocation, NewText)
            REPLACE (TextToBeReplaced, NewText)
            etc.

Path Constraint: No text within any clause (the "—" part of the clause) is to be retyped.

Problem: [A— b—; c—.] ⟹ [C—; b— a—.]

Optimal Method:

		[A— b—; c—.]
MOVE	→	[; c—A— b—.]
REPLACE	→	[C—A— b—.]
REPLACE	→	[C—; a— b—.]
MOVE	→	[C—; b— a—.]

Figure 11.8. Problem space and optimal method for the clause-switching task.

The optimal method is the same one described at the Keystroke Level in Figure 8.8. Note that the state description does not include a selection or a buffer, as does the state description in Figure 11.7. The underlined portions of text thus do not indicate the selections, but simply show the parts of the text affected by the operators.

For example, SELECT-TEXT followed by DELETE-TEXT is represented by the single operator DELETE, which takes the text to be deleted as an argument. Also, moving text is the single operator MOVE, which takes as arguments the text to be moved and the new location. The other relevant operators in this reduced problem space are listed in Figure 11.8. Note that the state descriptions in this reduced problem space are simpler than in the problem space of Figure 11.7. There is no longer any need to keep track of the current selection or the deletion buffer, since these are managed *within* each operator.[5]   In this reduced space, the optimal method for doing the clause-switching task (see Figure 8.8) takes only four operations (see Figure 11.8): two MOVE operations to switch the clauses and two REPLACE operations to clean up the punctuation and capitalization.

Problem Solving Behavior in Text-Editing (Experiment 11A)

We have characterized a problem space for the clause-switching task; the problem space is for a user who is expert in BRAVO, but who is not (yet) expert in the clause-switching task. We now consider whether such a user actually exhibits problem-solving behavior, in addition to the usual cognitive-skill behavior of an expert. We expect problem-solving behavior to take place within the reduced problem space (Figure 11.8) and skilled behavior within the operators of that space (i.e., within the skill spaces of the operators, such as in Figure 11.7). We consider two types of evidence for problem-solving behavior. First, we expect to see an inefficient use of the operators for the task at hand, possibly including some backtracking. Second, we expect the time required to do the task to be considerably longer than for skilled behavior, including long pauses for deciding which operators to apply. We illustrate these phenomena with some actual behavior.

Experimental Procedure. A pilot experiment, Experiment 11A, was run with an expert BRAVO user, a secretary with considerable technical ability, performing the clause-switching task repeatedly. The procedure

---

[5] The current selection and the buffer contents are, of course, aspects of BRAVO's state that the user may be aware of. But this reduced problem space makes the substantive psychological assertion that these aspects of state are only considered locally by the user. A user in this problem space cannot, for example, deliberately make use, in a subsequent operator, of an operator's side-effect on the buffer.

was the same as in Experiment 8B, with one important exception. The user in Experiment 8B was told a method for the task and then proceeded to execute that method as quickly as possible. In the present experiment, the user was given no information about how to do the task. Since the user did not have a method, it was for her a problem-solving task. She eventually acquired a method by repeatedly performing the task. (The instructions to the user were to repeat the task over and over. She was not to stop and attempt an elaborate analysis, although the latter is a reasonable strategy.) There were two experimental sessions. The user performed the task 60 times in the first session and 50 more times in the second session (two weeks later). As in Experiment 8B, time-stamped keystrokes were recorded.

Operator Sequence Results. The user's behavior for the first three trials, as a sequence of states and operators in the problem space, is shown in Figure 11.9. The behavior in these trials is representative of the behavior up to about Trial 33, when the user began to settle on a single method. The figure also shows the best method found by the user, which she first executed without error on Trial 35. Although the user's best method was not quite optimal (requiring one more MOVE operator than the optimal method shown in Figure 11.8), it was a considerable improvement over the first few trials (which required 3-4 more operators than the optimal method).

There are several indications of the problem-solving nature of the user's behavior in the early trials. First is the use of COPY operators instead of MOVE operators, which are inefficient because they require the use of extra DELETE operators (e.g., operators 1.1, 1.4, and 1.7 of Trial 1 in Figure 11.9). A second indication of problem solving is the overt correction of previous operators, such as doing a DELETE (3.5) to modify a previous COPY (3.4). A third indication is the user's failure to structure the sequence of operators; even small local consolidations of operators are missed. For example, the INSERT (1.3) and REPLACE (1.6) in Trial 1 could have been accomplished by a single REPLACE operator, since the text being edited was contiguous at the time of the INSERT. Any of these features of the behavior might occur occasionally in skilled behavior, but they occur very frequently in these trials. The asterisks in Figure 11.9 mark the operators that could have been avoided by one of the above considerations, that is, by acquiring specific search control knowledge about this task.

The user's lack of a stable method in the early trials is indicated by the radically different operator sequences from trial to trial. On each

Trial 1:   [A— b—; c—.]

1.1	COPY	→	[c—A— b—; c—.]
1.2	REPLACE	→	[C—A— b—; c—.]
1.3	INSERT	→	[C—; A— b—; c—.]
1.4	COPY	→	[C—; b—A— b—; c—.]
1.5	INSERT	→	[C—; b— A— b—; c—.]
1.6*	REPLACE	→	[C—; b— a— b—; c—.]
1.7*	DELETE	→	[C—; b— a—.]

Trial 2:   [A— b—; c—.]

2.1	MOVE	→	[; c—A— b—.]
2.2	INSERT	→	[; c—; A— b—.]
2.3	REPLACE	→	[C—; A— b—.]
2.4*	REPLACE	→	[C—; a— b—.]
2.5	COPY	→	[C—; b—a— b—.]
2.6	INSERT	→	[C—; b— a— b—.]
2.7*	DELETE	→	[C—; b— a—.]

Trial 3:   [A— b—; c—.]

3.1	MOVE	→	[c—A— b—; .]
3.2	MOVE	→	[c—; A— b—.]
3.3	REPLACE	→	[C—; A— b—.]
3.4	COPY	→	[C—; b—A— b—.]
3.5*	DELETE	→	[C—; b—A— b—.]
3.6	INSERT	→	[C—; b— A— b—.]
3.7*	REPLACE	→	[C—; b— a— b—.]
3.8*	DELETE	→	[C—; b— a—.]

Trial 35 (User's Best Method):   [A— b—; c—.]

35.1	MOVE	→	[A—; b— c—.]
35.2	MOVE	→	[; b— c—A—.]
35.3	MOVE	→	[c—; b— A—.]
35.4	REPLACE	→	[c—; b— a—.]
35.5	REPLACE	→	[C—; b— a—.]

Figure 11.9. **User's behavior on selected trials in the clause-switching task in Experiment 11A.**

The behavior is described in the problem space of Figure 11.8. The asterisks after the operator numbers mark the operators that could have been eliminated by simple local planning; see text.

trial the user was trying a different way to improve by reacting to local aspects of the task. It was not until Trial 35 that the user settled on a stable method, after which she always used this method (except for the first four trials of the second session, during which she was trying to recall this method). The user's performance became more and more skilled after Trial 35, just as did the performance of the user in Experiment 8B.

Performance Time Results. The most dramatic indicator of the character of the user's behavior throughout the experiment is the performance time curve, plotted in Figure 11.10. The top line in the figure, representing the total performance time per trial, gives the overall story of the behavior. The user took about 50 sec per trial for the early trials, dropping sharply to about 32 sec per trial at about Trial 35. Performance fell back in the first few trials of the second session as the user was getting reoriented to the task, but quickly improved and leveled off at about 22 sec per trial. It is instructive to compare this performance against a standard of skilled performance. One obvious standard, the calculated time to execute the user's best method (the sequence in Trial 35) according to the Keystroke-Level Model, is plotted as the dashed line in Figure 11.10. This clearly shows that the user had reached a skilled level of performance on the last 40 trials.

More light can be shed on the user's performance by decomposing the performance time into its physical and mental components. (The method for making this decomposition was explained in Section 8.4.) The mental time component is further decomposed into large and small mental time components, with the *large mental time* consisting of all pauses over 3 sec and the *small mental time* consisting of the pauses under 3 sec. A mental pause over 3 sec long must be more than simply a preparation for skilled performance, since it is significantly longer than skilled mental preparation (the **M** operator of the Keystroke-Level Model).[6] Thus, in the present context, we can interpret the large mental time as the time needed for planning the operator sequence. This decomposition is represented in Figure 11.10 by the variously shaded regions under the execution time curve. The execution time is composed of the physical time (unshaded region), the small mental time (lightly

---

[6] The rationale for choosing 3 sec as the threshold for large mental time is as follows. In Chapter 8, the mean **M** operator time was estimated to be 1.35 sec, and its standard deviation was estimated to be 1.1 sec. 3 sec is 1.5 standard deviations from the mean **M** time. Thus, assuming a normal distribution for **M** times, only about 10% of **M** times will be 3 sec or greater.

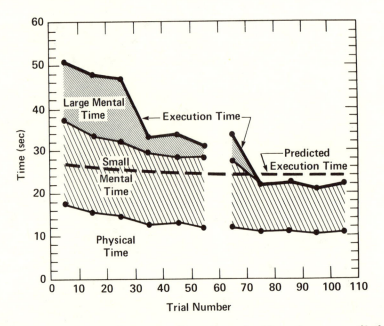

Figure 11.10. User's execution time in the clause-switching task in Experiment 11A.

The times are averaged in 10-trial blocks. The top solid line shows the user's actual execution times; and the dashed line shows the predicted time to execute the user's best method, as calculated with the Keystroke-Level Model. The user's time is decomposed into three components: the physical time (the unshaded area on the bottom), the small mental time (the lightly shaded area in the middle), and the large mental time (the darkly shaded area on the top). See the text for how these time components are defined and interpreted.

shaded), and the large mental time (darkly shaded). The figure clearly shows that the dramatic changes in overall performance time are not in the physical time, but in the mental time, especially the large mental time. It is the large mental time that drops around Trial 35, when the user finally settles on a stable method. And further, the large mental time is eliminated altogether around Trial 70, just as the user attains a completely skilled performance level, i.e., betters the predicted time of the Keystroke-Level Model.

Summary. We have observed a user who has attained a skilled level of behavior in a specific task, switching two sentence clauses with a text-editor; and we have shown that this skill starts out as problem solving and becomes skill after practice on the task. We have illustrated a problem space for the task and have described the user's problem-solving behavior in terms of this problem space. Finally, we have shown that the observed skill could have emerged within the problem space by the acquisition of specific search control knowledge about the task.

11.4. THE UNIT TASK

So far in this chapter we have been characterizing the general nature of cognitive skill. We now return to the particular skill studied in the book, text-editing. As shown by the preceding section, the text-editing task fits well enough into the general picture, eliciting behavior that moves along the skill dimension with practice from problem solving to skilled performance. But text-editing has a striking feature—its unit task structure—which sheds some light on the organization of cognitive skills. The organization of behavior into a sequence of short quasi-independent tasks is pervasive in the studies in this book. It originated in our analysis of the manuscript-editing task (Chapter 5) as an obvious way to divide the total task into smaller parts, based on the localized nature of the corrections on the manuscript. In the Keystroke-Level Model (Chapter 8), it provided a useful way to separate task acquisition from task execution. It also proved an easy way to analyze other interactive tasks, one involving page layout (Chapter 9) and another involving computer-aided design (Chapter 10). The unit task partitions the behavior stream, thus providing the basic structural foundation on which the detailed models can be erected. We review here the basic nature and function of the unit task and the determinants of unit task structure in more fundamental psychological factors.

The Nature and Function of Unit Tasks

We consider several aspects of unit tasks: their well-defined internal structure, their basic function as a control construct for the user, their characteristic durations, and their relationship to problem solving.

STRUCTURE

In each of the cases of unit-task behavior studied, there has been a structured cycle of repeated actions. The user first acquires a task and a method for doing it (reads or decides what to do and how to do it) and then executes the method:

$$\text{Unit Task} = \text{Acquire} + \text{Execute}.$$

The important point about acquisition is that more than just a task (goal) is acquired—a method for accomplishing the task is also acquired. Without a repertoire of readily available methods, behavior cannot be structured into unit tasks.

The execution of a method involves locating the necessary data, acting on the data, and then (optionally) checking to see if the action was correct:

$$\text{Execute} = \text{Locate} + \text{Act} + (\text{Verify}).$$

The Act operation is the main purpose of the unit task, locating the data being just a preliminary step enabling direct access to materials necessary for carrying out the action. In text-editing, Locate and Act were specific operations: locating a piece of text in a file and modifying the text. However, Locate and Act (and, of course, Verify) are general functions that apply to a wide variety of tasks. For example, the task of checking a canceled check received from the bank involves locating the check and the checkbook entry and then making a detailed comparison. Thus, one of the properties of the unit task is its characteristic functional structure.

CONTROL FUNCTION

The most important point to understand about the unit task is that *the unit task is fundamentally a control construct, not a task construct.* This distinction can be made clear by considering the manuscript text-editing task. The manuscript contains a set of spatially separated marks, each denoting a different modification to be made. Thus, the task is structured as a set of separate modifications. However, it is up to the user to decide how to organize these modifications into a series of unit tasks. Usually, as we have seen, the user makes each modification as a unit task. But the user may perform two nearby modifications within a single unit task, or he may sometimes find a way to make a whole series of modifications at once (e.g., using a global substitution command). In the other direction, a single complex modification may be broken into several unit tasks.   Inexperienced users may even as a general rule separate the location of the site of the modification from the execution of the modification, making separate excursions to the manuscript to seek the information for these two aspects. Although the frequency of such behavior in text-editing is low enough not to make them a prominent part of our analysis, we have seen enough examples of such behavior to make clear the basic point: that the unit task is not given by the task environment, but results from the interaction of the task structure with the control problems faced by the user.

The reason a user imposes unit-task structure on his behavior is because each unit task can be kept within his performance limitations. In a skilled behavior streams of inputs and outputs must be managed in Working Memory if performance is not to be degraded. If the input and output streams can be managed in a parallel (pipelined) fashion, then the behavior can have a continuous structure rather than a unit-task structure. But when conditions on the inputs and outputs do not allow this pipelined processing (e.g., when the output process cannot keep up with the input stream), then behavior must be structured into a series of unit tasks.

Let us examine a particular task in more detail to see how the user controls his behavior in response to his resource limitations. Consider the skill of touch-typing, which does not have a unit-task structure, but rather a continuous structure. The reason why it is continuous can be seen by observing how behavior in this task is shifted to a unit-task structure when the task is modified slightly so as to exceed the user's processing limits. In normal touch-typing, memory load remains low and within Working Memory limits. The behavior is continuous—while the user types one word, he reads the next, in accord with the pipelined parallel model of Chapter 2. There is no unit-task structure.

Norman and Bobrow (1975) have suggested that a useful way to depict the relationship between performance on a task and the processing resources available is to plot the task's performance-resource function as in Figure 11.11. The figure shows the idealized relationship between a resource (in this case, the amount of Working Memory available) and a measure of performance (in this case, accuracy). Consider the resource-performance curve labeled "Touch-Typing." If the user had no Working Memory at all, he could not remember what he had read long enough to type it; and the accuracy would be zero (the curve would begin at the

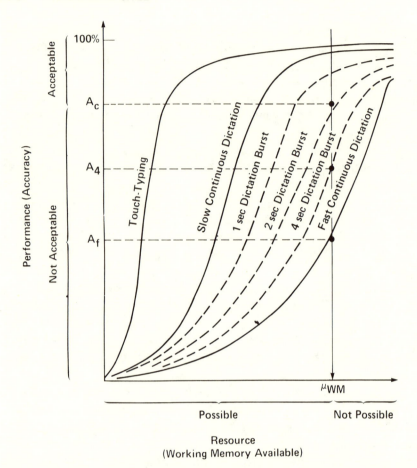

Figure 11.11. Idealized performance-resource functions for touch-typing and dictation.

Touch-typing is a continuous task (does not have unit-task structure) because acceptable accuracy can be obtained with available resources. For fast dictation, however, either the user must lower his minimum acceptable accuracy, or the task must be broken into unit tasks, each of which can be done at acceptable accuracy. μ_{WM} is the user's Working Memory capacity, A_c is the user's minimum acceptable accuracy, A_f is the user's accuracy for fast dictation, A_4 is the user's accuracy for 4 sec bursts of fast dictation.

origin). By devoting a small amount of Working Memory to the task, however, typing accuracy increases to some asymptotic value. Performance, as measured by accuracy, is "resource-limited" (more available resource leads to better performance) up to about one word's worth of Working Memory and then is "data-limited" (more available resource does not lead to improved performance, only better data can improve performance).

Touch-typing is user-paced, with the user varying his reading speed to maintain input and output in balance. If the task is changed to be machine-paced, as in the similar task of transcribing dictation, where the user plays out speech at its recorded speed, then segments of speech may flow into Working Memory faster than they can be processed; and the user needs more Working Memory to buffer the input until he can get caught up during the pauses. In Figure 11.11, this need for more resource (to obtain the same level of performance) is represented by a leaning of the resource-performance function a little to the right, as in the curve labeled "Slow Continuous Dictation," and further to the right, as in the curve labeled "Fast Continuous Dictation."

Now, according to the Model Human Processor, each user has a Working Memory capacity μ_{WM}, so performance requiring greater than μ_{WM} amount of Working Memory is not possible. Let A_c be the minimum performance accuracy acceptable to the user. The user's maximum possible accuracy (using μ_{WM} amount of Working Memory) for Fast Continuous Dictation is A_f, which is lower than A_c and therefore not acceptable. For speech this fast, the user, if he is to avoid unacceptable performance, must stop treating the task as a continuous task; instead, he must break it into a unit-task pattern by listening to a bit of tape at a time and then typing it. As an idealization, let us suppose he listens to a 4-sec burst of tape at a time. The resource-performance function is then shifted back to the left, and the user's accuracy improves to A_4—better, but still below the acceptable level A_c. But if he listens to small enough bursts of, say, 2 sec, he will finally be able to shift the resource-performance curve far enough to the left to meet his accuracy requirements. Of course, he could make the bursts smaller yet, achieving acceptable accuracy using less than his full Working Memory capacity. But the smaller the bursts become, the longer the total time required to transcribe the tape, a force that will tend to keep him from being any more accurate than necessary.

To summarize, we have seen how a continuous task can be shifted by degrees into a task where the user's performance is no longer acceptable and how a unit-task control structure comes about as a method for the user to shift performance back into an acceptable level.

DURATION

The unit task is an internally determined control construct, and its duration is limited. In fact, we have seen very few error-free unit tasks in this book that lasted longer than 30 sec.[7] Most of the text-editing unit tasks lasted 10~15 sec (Chapters 5 and 8), and the routine circuit-layout unit tasks lasted 10~50 sec (Chapter 10). The duration of the acquisition phase of a unit task is determined by how well the task is specified and presented to the user, and the duration of the execution phase is determined by the length of the method used. These two phases independently determine the duration of the whole unit task (e.g., a long acquisition time does not necessarily produce a long method). To acquire unit tasks from a marked-up manuscript requires 2 sec (Chapters 5 and 8);[8] to generate unit tasks mentally in a routine design task takes about 10 sec (Chapter 10); and to generate unit tasks in a creative composition task would take even longer.

The execution phase of unit tasks averages about 10 sec in all the interaction tasks we have studied (Chapters 5, 8, and 10). Execution can be as short as a second or two (for a one-key command); we have observed only two editing-task executions lasting over 30 sec.[9] How long an execution can be depends on how complex a method the user is willing to spend the time and effort to assimilate. The most complex methods we have studied are for the clause-switching task (Sections 8.4 and 11.3), which require about 25 sec to execute. The user in Experiment 8B required a half-hour of discussion plus about 5 trials to assimilate the optimal method. In Experiment 11A, the user required 35 trials to formulate a method for the task, but the method was not yet well-enough assimilated be treated as a single unit task; about 35 more trials were required for that to happen. Given the considerable effort

---

[7] Unit tasks containing errors do, of course, take longer, but errors often introduce new unit tasks for error correction. Thus, the time per unit task may not be much different for error unit tasks.

[8] In Chapter 8, (Figure 8.7) we found that an added 2 sec is needed for acquisition in a display-based editor, since the display has to be scanned for the text. However, we believe that the scanning operation actually belongs to the execution phase, as part of the Locate function.

[9] Tasks T4-POET and T4-SOS took 37 sec and 33 sec, respectively, to execute (Figure 8.5). However, the T4-POET execution included about 10 sec of continuous copy-typing behavior.

needed to assimilate complex methods, it is not surprising that we only observe short and simple methods in human-computer interaction tasks.

RELATION TO PROBLEM SOLVING

Since unit tasks are a predominant feature of some cognitive skills and since cognitive skill emerges from problem solving, we would expect unit tasks to play some role in problem solving. Unit-task-structured cognitive skill occurs in problem solving, with the unit tasks functioning as operators in problem spaces. We pointed out earlier in this chapter that the unit task of locally altering the parts of a circuit configuration served as an operator for the problem of compressing a large circuit layout (Chapter 10). Actually, only the execution part of a unit task serves as the operator, for the execution part is based on a method—a well-integrated, purposeful unit of behavior. Thus, problem-solving operators derive from the existence of integrated methods. This was illustrated in the text-editing example in Section 11.3, where the existence of methods for performing the move-text task provided a MOVE operator in the problem space for the clause-switching task.

The structure of unit tasks can be related to the search control cycle of problem solving (Figure 11.3). The execution part of a unit task can be identified with the control function of applying the operator (step 2); and the acquisition part of a unit task can be identified with all the rest of the control cycle, that is, with all the decisions about what to do next. In manuscript editing, operators (unit tasks) can be acquired by a straightforward interpretation of the markings on the manuscript. When there is a strict method for acquiring operators—such as taking them in the order they occur on the manuscript—then no problem solving is required at all, since the choice of operators is not problematic. But when there is no well-defined method for acquiring operators—as in a complex rearrangement of text where individual modifications interact with each other—then problem solving is required to decide the order in which to make the modifications; and operator (unit-task) acquisition becomes the search control cycle.

Determinants of Unit-Task Structure

There are constraints, both internal and external, on the user in structuring his behavior to accomplish a task. The unit task is a control construct available to the user in meeting some of these constraints:

> *Working Memory Capacity.* The performance of tasks
> requires the maintenance of temporary data, which
> must be kept within the limits of the user's Working
> Memory.
> *Information Horizons.* The performance of tasks must
> remain within certain information limits: the data
> and task limits imposed by the task environment and
> the repertoire of methods known to the user.
> *Error Control.* The performance of tasks must control the
> probability of errors and control the damage done by
> the inevitable occurrences of errors.

These constraints tend to shape the user's behavior into a series of unit
tasks. Let us consider them each in turn.

WORKING MEMORY CAPACITY

Probably the most severe constraint on behavior arises from the
limited capacity of Working Memory and the need to keep working data
within this limit. As we have discussed in Chapter 2, the capacity of
Working Memory involves both the number of chunks and interference
between chunks. In this section we limit our consideration of Working
Memory capacity to the former.

Smaller tasks generally require less Working Memory for their
performance than do larger tasks. This is one reason why a user, when
confronted with a large task, will break it into smaller tasks, which we
have called unit tasks. In text-editing from a marked-up manuscript, we
have observed that the user decomposes the overall editing task into a
series of small edits. This is not just a result of the task structure.
Although the manuscript contains a series of marks denoting
modifications, these modifications are not necessarily identical to the unit
tasks that the user generates, as we have already noted. A single
modification mark can result in multiple unit tasks (e.g., a mark
indicating the alignment of items in a table can result in several unit
tasks), and multiple marks can be handled in a single unit-task (e.g., a
pair of marks to put quotes around a phrase). The user will, of course,
try to take advantage of the structure of a large task in deciding how to
decompose it into unit tasks.

There is a characteristic pattern of Working Memory load for
behavior structured in unit tasks. Consider a text-editing example, based
on the simulation program described in Chapter 6. Figure 11.12 shows

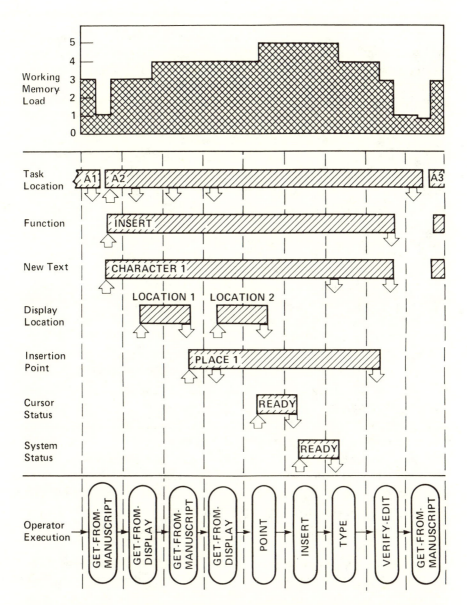

Figure 11.12. Data in Working Memory during a unit task.
This figure is a hypothetical trace of the performance of one unit, the same unit task that is traced in Figure 6.17. Time runs to the right on the horizontal axis. The bars indicate the time during which each slot (according to the scheme in Chapter 6) in Working Memory is filled with data. The arrows indicate writing into and reading from Working Memory slots. The histogram on the top plots the total Working Memory load over time, showing how the load peaks within a unit task and dips between unit tasks.

393

the data which must be kept in Working Memory while the user executes the unit task of inserting a character. The figure also plots the Working Memory load during the execution. The memory load can be seen to rise from a low of one chunk between unit tasks to a high of five chunks within the task. Thus, the number of chunks in Working Memory rises and peaks within unit tasks and dips between unit tasks. The frequent dips in Working Memory are the important benefit of unit-task structure, for these localize the use of the data. This pattern of Working Memory usage makes behavior more robust: it helps to reduce the number of Working Memory errors; it limits the scope of those errors; and it makes the behavior more interruptable.

INFORMATION HORIZONS

A user often finds himself running out of information, forced to generate new unit tasks to continue his performance. There are three kinds of information limits: (1) He can run out of data to work with—a *data horizon*. (2) He can run out of tasks to do—a *task horizon*. (3) He can run out of method to execute—a *method horizon*.

Data Horizons. The execution of a task usually requires information from the task environment. In text-editing, for example, the user needs to know both the location of the change and the details of the text to be changed. Missing pieces of data must be retrieved. If the retrieval is simple, accomplished by a glance at the screen or at the manuscript, it is a routine part of the execution of the unit task. But if the retrieval is difficult, then the retrieval itself will require one or more unit tasks. This is a data horizon: the user runs out of immediately accessible data needed to complete a task. A data horizon is not (necessarily) an issue of Working Memory capacity; the user could remember the data—it simply is not available. Thus, data horizons force the generation of data-gathering unit tasks.

Task Horizons. The user may not only run out of data, but may also run out of tasks. For example, a manuscript to be edited may contain a single short edit on each page forcing the user to take each one as a separate unit-task. Again, the horizon does not involve Working Memory capacity—the user could possibly remember two or three of the edits as a single unit task if they were close to each other on the same page—but the task environment supplies the tasks in small chunks that force a particular unit task-structure.

Method Horizons. The user may not know the method for doing a whole task, but he may know methods for doing some parts of it. This circumstance will force him to break the task into unit tasks corresponding to the methods he knows. For example, the task of changing thirteen instances of "Alan" to "Allen" on a page of text could require up to thirteen unit tasks for a user who must make each change individually, whereas a user who knows the multiple-substitution command could do it in a single unit task.

ERROR CONTROL

A final determinant of unit-task structure is the need for error control. By breaking down larger tasks into smaller unit tasks, a user can verify the correctness of each unit task, thus localizing the effect of errors. An example is the way in which users of the POET editor employ POET's Transfer command to move lines of text. Suppose the task is to find the paragraph beginning with Alpha and ending with beta and to move it to follow the paragraph ending with gamma. The command syntax of POET allows this to be accomplished in a single command:

"Alpha","beta" Transfer "gamma" .

This command causes POET to (1) search for the line containing Alpha, (2) search for the line containing beta, (3) search for the line containing gamma, and finally (4) move the lines. Few experienced users would have the temerity to do the task this way, however, for errors are probable (the wrong lines might be found in steps 1 to 3) and the consequences severe (the text would be scrambled). What users actually do in this task is to break it into four unit tasks. Each of the first three unit tasks locates a line and produces its unique line number:

"Alpha" / = .

This command causes POET to search for the line containing Alpha, display that line (the /) so the user can verify that it is the correct one, and print the line number (the =). After obtaining the three line numbers (which are, say, 11, 22, and 33), the user issues the Transfer command with them:

11,22 Transfer 33 .

This method is relatively safe from damaging errors, since the user has checked that each line found was correct and since the line numbers are unambiguous. Thus, the user breaks the large task into a series of individually verifiable unit tasks. Again, the reason has nothing to do with Working Memory capacity (although this could also be a factor), but only with controlling potential errors.

Summary. There are three classes of constraints on user behavior—Working Memory capacity, information horizons, and error control—that tend to give the behavior a unit-task structure as it develops from problem solving into a cognitive skill.

11.5. TEXT-EDITING WITHIN COGNITIVE SKILL

Manuscript text-editing is a paradigm for many similar tasks. We saw several of these in various chapters of the book: use of drawing programs (Chapter 8), elementary command language interactions (Chapter 8), page formatting (Chapter 9), and integrated circuit layout (Chapter 10). The cognitive skill involved in all these tasks has many of the same characteristics as does text-editing. On the other hand, text-editing is quite different from many other cognitive skills. It is important to appreciate this diversity of cognitive skills, so as not to over-generalize the characteristics of text-editing. Unfortunately, there is no basis for constructing a general taxonomy of cognitive skills. Cognitive skills exist for all cognitive tasks (i.e., all situations that permit problem solving), provided that practice on them is possible. Hence, as indicated in Figure 11.1, the taxonomy of all cognitive skills is an image of the taxonomy of all possible tasks—hardly something to be taxonomized easily. Furthermore, there does not at present exist a population of studies of other cognitive skills that have been analyzed in ways that would permit deep comparison, either with text-editing or with each other (though, as we noted, some appropriate analyses are beginning to emerge). Thus, we simply present a large handful of cognitive skills that differ on a number of dimensions, providing an informal context within which to locate text-editing.

The dimensions of cognitive skill we consider come from the nature of the Model Human Processor and from the demands of the task. They fall into four groups that address, respectively, the character of the skill, the demands on Working Memory, the demands on Long-Term Memory, and the external task demands. Figure 11.13 lists the dimensions and

CODE: ● HIGH ◑ INTERMEDIATE ○ LOW TASKS	COGNITIVE SKILL DIMENSIONS								
	SKILL CHARACTER			WM		LTM		TASK DEMANDS	
	PROBLEM SOLVING	PERCEPTUAL-MOTOR	PLANNING	UNIT TASK STRUCTURE	MEMORY LOAD	INPUT TO LTM	RETRIEVAL FROM LTM	PACING	ACCURACY
CPA DOING INCOME TAX	●	○	●	●	◑	◑	●	○	●
ROUTINE MEDICAL DIAGNOSIS	●	○	●	●	●	●	●	◑	●
PLAYING BRIDGE	●	○	●	●	●	◑	◑	●	●
WRITING BUSINESS LETTER	●	○	●	●	◑	◑	◑	○	◑
BALANCING CHECKBOOK	●	○	◑	●	●	◑	◑	○	●
AIR TRAFFIC CONTROL	◑	◑	●	●	●	◑	○	●	●
MENTAL MULTIPLICATION	◑	○	○	●	●	○	◑	○	●
MANUSCRIPT TEXT-EDITING	◑	◑	◑	●	◑	○	○	○	●
TYPING	○	●	○	○	○	○	○	○	◑
SORTING MAIL	○	●	○	●	○	○	○	○	●
FOOTBALL COMMENTARY	○	●	○	◑	●	●	●	●	○
TENNIS	○	●	○	●	○	○	○	●	●
SHORT-ORDER COOK	○	●	◑	◑	●	○	◑	●	●
DRIVING CAR	○	●	◑	○	◑	○	○	●	●
ASSEMBLY TASK	○	●	○	○	○	○	●	●	●

Figure 11.13. Dimensions of cognitive skill.
Several generic tasks are rated approximately on dimensions of cognitive skill. This collection of tasks shows the variety of cognitive skills and how the manuscript text-editing task compares with them.

several tasks and locates the cognitive skills associated with the tasks on these dimensions. Since this discussion is only illustrative, we consider the tasks and their associated cognitive skills at a generic level. Each of these generic tasks has many variants with different properties; a serious analysis would have to consider task specifics.

SKILL CHARACTER

Problem-Solving Embedding. As we have discussed, cognitive skill is often embedded in a context of problem solving. For example, an accountant may perform skilled computations in the course of searching for ways to solve an income tax problem, but a short-order cook simply prepares orders as they arrive. We have already seen that a task will demand more problem solving behavior from a novice than from an expert. A new short-order cook must figure out the best way to manage multiple orders. But even with expertise held constant, problem solving varies with task. Manuscript text-editing, as we have discussed, usually involves little problem solving.

Perceptual-Motor Involvement. At the high end of this dimension are tasks in which perceptual-motor involvement is crucial, such as driving a car or playing tennis; at the low end are purely mental tasks, such as the mental multiplication of multi-digit numbers. The presence of a strong perceptual or motor component presents possibilities for overlapping the operation of the perceptual, cognitive, and motor processors (as discussed in Chapter 2), allowing for a considerable gain in speed and the external appearance of coordinated motion so characteristic of skill. Along this dimension, manuscript text-editing occupies an intermediate point.

Requirement for Planning. Writing a business letter and playing bridge are tasks in which people usually plan ahead; assembling electronic parts is a task in which they do not. Planning, in problem space terms, refers to taking steps in a simplified, abstracted problem space (the planning space), and using the results to guide steps in a more detailed problem space. Planning is unnecessary for simple text-editing tasks, which is mostly what we have studied in the book; but more complex text-editing tasks, such as the clause-switching task discussed in Section 11.3, often require planning.

WORKING MEMORY DEMANDS

Unit-Task Structure. Some tasks, such as text-editing, have a strong unit task character, whereas others, such as touch typing, are continuous.

Working Memory Load. Some tasks put higher demands on a user's Working Memory than do others. High on this dimension is mental multiplication, where the limits on performance are due to the limits on keeping track of intermediate results. A task with low Working Memory demands is freeway driving. Text-editing is intermediate.

LONG-TERM MEMORY USAGE

Long-Term Memory contains knowledge of the cognitive skill itself—the methods—and knowledge about the objects being processed by the skill—the data. The former is, of course, involved in all cognitive skill execution. We are here concerned with the latter.

Input to Long-Term Memory. There is little long-term, task-specific information to remember in manuscript text-editing (once the basic skills of the editor have been learned). In the game of bridge, on the other hand, there is a large amount of task-specific data that players must retain in Long-Term Memory—all the bidding and every card played.

Retrieval from Long-Term Memory. Some tasks, such as sorting the mail or routine diagnosis of disease, make impressive demands on knowledge stored in Long-Term Memory. This is quite different from manuscript text-editing, which only involves a small amount of transient data.

EXTERNAL TASK DEMANDS

Pacing. The extent to which a task is externally paced may make the task substantially more difficult than it would otherwise have been, because it may demand processing that exceeds the resources of the user. Driving a car and certain assembly-line work are both tasks in which external pacing is important; the text-editing tasks we have studied are unpaced. Closely related to the rate of pacing is the input load. For example, although the task of an air traffic controller may be paced, this only becomes an issue when the number of planes he must control per unit time reaches a certain level.

Accuracy (vs. Speed). A calculation, such as balancing a bank statement, where the task is not finished until a balance is obtained, is different from producing a running commentary on a football game, where only a few descriptive features of the action need to be reported. The latter example also illustrates that accuracy is traded off against speed—the accuracy of the commentary is limited by the demand that it be produced quickly. Manuscript text-editing has high accuracy demands; and, since it is user-paced, the necessary time and precautions are taken to insure accuracy.

The main purpose of Figure 11.13 is to emphasize the vast variety of cognitive skills, of which manuscript text-editing and its variants are only one small group. Perhaps the most important additional facts that can be read from the figure are about the unit-task structure: (1) it is simply one among many independent dimensions, and (2) it is not unique to text-editing.

11.6. CONCLUSIONS

The human-computer interaction that we have studied is a cognitive skill with modest perceptual-motor involvement. We have attempted to characterize the general nature of cognitive skill in a broader psychological framework.

All cognitive behavior can be characterized along a continuous skill dimension that includes both problem-solving behavior (at the low-skill end of the dimension) and cognitive-skill behavior (at the high-skill end). This view is consistent with two principles of the Model Human Processor in Chapter 2. The Problem Space Hypothesis states that all cognitive behavior has a homogeneous structure—that of problem solving in problem spaces. The Power Law of Practice shows the continuous character of changes in cognitive behavior with practice.

Problem-solving behavior, such as is exhibited by a novice attempting the Tower of Hanoi puzzle, will, with practice, become a cognitive skill that can be characterized by a GOMS-like model. Examples of the mechanisms by which problem solving evolves into cognitive skill are the accumulation of search control knowledge (which eventually becomes skilled methods) and the construction of new operators (which effectively reduce the problem space to be searched).

Cognitive-skill behavior, such as the text-editing behavior we have analyzed in this book, has its roots in problem solving. Text-editing by a novice can be characterized as search in a problem space. To become skilled in text-editing requires the acquisition of editing-specific search control knowledge and more powerful editing operators. We observed a specific learning sequence in which a user clearly began with problem solving on a complex editing task and proceeded to become a skilled expert at the task (i.e., performed the task at the level predicted by the Keystroke-Level Model).

The most striking feature of text-editing skill is its unit-task structure. We have shown that unit tasks are primarily control (not task) constructs, that they seem to have characteristic durations, and that they emerge from the structure of problem solving in problem spaces. Further, several constraints on skilled behavior—Working Memory limits, information limits, and the need for error control—tend to shape skill as a sequence of unit tasks.

Text-editing is but one skill in a vast population of cognitive skills, varying among themselves in many ways (other than unit task structure): in their basic skill/problem-solving character, in their demands on Working Memory and Long-Term Memory, and in their external task demands (such as pacing and accuracy). Although unit-task structure is characteristic of text-editing, it is not unique to it.

12. Applying Psychology to Design

In this chapter, we return to the theme of an *applied* psychology introduced in Chapter 1 and attempt to tie our studies together from the vantage point of how such knowledge might be used in design. We do this first by presenting a framework showing the way in which psychological results can be applied to design and then by mapping our studies into this framework. We also summarize an application-oriented extension of our work by Roberts. Finally, we list some general system design principles suggested by the studies.

12.1. A FRAMEWORK FOR APPLYING PSYCHOLOGY

In Chapter 1 we proposed that an applied psychology of human-computer interaction should be relevant to the system design process itself (not just to after-the-fact evaluation) and that the designer himself should do the actual application. Such an applied science must be based on information-processing models, whose applicability to design depends on three critical features: task analysis, calculation, and approximation. This proposal rests on a view (until now implicit) of how psychology can be applied to system design. We now present our view by briefly sketching a framework for application. This framework includes (1) the structure and performance of the human-computer system, (2) performance models for predicting the performance of the human-computer system, and (3) design functions for using the performance models in the design process.

THE HUMAN-COMPUTER SYSTEM

Our present object of study is the *human-computer system*, which consists of a human user interacting with a computer to accomplish a task. The user, the computer, and the task are the structural components of the system. Human-computer systems vary in many different respects, called *structural variables*, in each of the components. Systems address different task domains, and they have different models of the tasks in any given domain. Users vary widely in general intellectual ability, experience with computers, specific knowledge of the task, specific knowledge of the computer, cognitive style, and perceptual-motor skills. User-interface aspects of computers vary in system architecture, dialogue style, command syntax, input devices, and so on.[1] The combination of all these variables produces a vast space of possible human-computer systems.

The ultimate concern of an applied psychology is not so much with the structure of the human-computer system per se, as with its *performance*. There are many different aspects to performance, which we call *performance variables*. The basic performance variables of a human-computer system are concerned with what tasks the system can do (functionality), how long it takes to acquire the functionality (learning), how long it takes to accomplish tasks (time), how frequently errors occur and how consequential they are, how well tasks are done (quality), and how robust the system is in the face of unexpected conditions. Other performance measures are possible, such as performance under extreme conditions (fatigue and stress) and the performance demands on the user's memories (Working Memory and Long-Term Memory). Finally, there are variables concerning the user's subjective feeling about the system. All these performance variables are potential areas of concern to the system designer.

The performance variables of a human-computer system are determined by its structural variables. This can be summarized in a formula analogous to the Rationality Principle (Chapter 2):

$$Task + User + Computer \rightarrow System\ Performance. \qquad (12.1)$$

---

[1] A systematic analysis of the structure of the user-interface aspects of interactive computer systems is a difficult undertaking. For some attempts, see Moran (1981*a*); Young (1981); Newman and Sproull (1979, Ch. 28); and Ramsey and Atwood (1979).

It is the task of an applied psychology to discover the specific relationships between the structural and performance variables of human-computer systems.

PERFORMANCE MODELS

The design of a human-computer system begins with a set of requirements, which includes both structural constraints and performance goals. The designer's job is to specify a human-computer system satisfying the requirements. But while specifications of a system are readily checked against the structural constraints, the performance aspects of a system are not derivable from a descriptive specification. A special kind of representation of the human-computer system is needed for this, which we call a *performance model.* To predict the performance of a system, the designer must construct a specific performance model from the system's structural specifications and then use the model to generate a prediction:

$$Model\,(Task,\,User,\,Computer)$$
$$\rightarrow Performance\,Prediction\,. \qquad (12.2)$$

The concept of a performance model is the key notion in this framework. It is useful to construe this notion functionally, i.e., as any model or description that can be used to predict system performance. Performance models can be roughly categorized as experimental models, symbolic models, and database models. Experimental models consist of actual human users with actual running programs or physical mock-ups. Such models are *run,* and performance variables are *measured.* Symbolic models are calculational, algebraic, or simulation models. They are represented on paper or in a computer and have no actual human component (although, of course, they model the user). Performance values are obtained by *computation* (by hand or computer). Database models are stores of pre-measured or pre-calculated data. Performance values are obtained simply by *look-up.* Each of these different kinds of performance models has its place in the system design process.

DESIGN FUNCTIONS

The predictive function of performance models is primarily *evaluative*: given a structure, predict performance. The designer's problem, however, is *generative*: given performance requirements, design the structure.

Although it is possible to invert, partially at least, some performance models to generate design ideas, it is not possible to invert Formula 12.2 in a general way. For any interesting real-world domain of design, there cannot be any global synthesis function that maps requirements into a structure. How, then, can performance models be useful in design? To answer this question we must consider the nature of the design process.

Design, as all designers know, is not a simple top-down or bottom-up process of synthesizing a design solution from requirements. Design is an open process, in the sense that the design problem is constantly being redefined. Many requirements can emerge only in the course of the design process, when partial design solutions provide enough context to realize which issues are really important. Thus, design proceeds in a complex, iterative fashion in which various parts of the design are incrementally generated, evaluated, and integrated. At the risk of being over-simplistic, we characterize the complex process of design as consisting of a set of different kinds of *design functions*, each attending to a specific design subproblem:

$$Design\ Process\ =\ a\ set\ of\ Design\ Functions. \tag{12.3}$$

Although we do not pretend to have even a crude taxonomy of design functions, we can list some examples to make the notion more concrete. We can group design functions into three broad categories: evaluation, parametric design, and structural design. Evaluation, as we just noted, refers to the situation in which the structure of the system (or of part of the system) has been specified and its performance needs to be understood. Parametric design refers to the situation in which the structure of the system is relatively fixed and there are a set of quantitative parameters of the structure to be determined. (What makes parametric design tractable for analysis is the assumption that the remaining structure of the system will not change in the range of parameter values under consideration.) Structural design is where a part of the system is configured or restructured to satisfy specific requirements. There are several functions in structural design, such as to identify an opportunity for a change, to diagnose a problem, to generate an improvement, and to synthesize a new structure.

Design functions require the use of performance models to solve particular design subproblems:

Design Function (*Design Subproblem, Model*) → *Solution*.

The dependence on performance models is clear in evaluation and parametric design. Formal performance models are seldom used in structural design, although there are usually implicit, informal performance assumptions underlying the design functions, which can be viewed as vague, informal performance models. But partial inversions of more formal performance models can also be used (e.g., to diagnose the causes of performance deficits).[2]

This framework for applying psychology to design emphasizes the pivotal role of performance models. Without models, the designer cannot predict the performance of the system he is designing. If he cannot predict performance, he will not be able to come to grips with performance requirements. And if he does not deal with performance requirements, then other requirements will dominate the design of the system—and the user will be neglected. Thus, in our view it is clear that *the role of an applied psychology is to supply performance models for the designer.*

12.2. CONTRIBUTIONS TO APPLICATION

We now have a framework for considering how the studies in this book address the issues of applying psychology to system design. We proceed by enumerating the principal aspects of the application framework—the human-computer system, performance models, and design functions—and by showing how far we have progressed and where needs exist for future research.

THE HUMAN-COMPUTER SYSTEM

According to Formula 12.1, the structural variables of the human-computer system—the task, the user, and the computer—determine its performance variables. Figure 12.1 lists a set of structural variables for characterizing the variety of possible human-computer systems, and

---

[2] Some of the design principles to be presented in Section 12.4 can be viewed as inversions of our models of cognitive skill in human-computer interaction. For example, Principle 8 takes the GOMS model of method-selection in Chapter 5 and, instead of using the model to predict performance, observes that performance will be better if method alternatives are designed so they can be selected with a simple set of method-selection rules.

Structural Variables	Studies (Chapters/Sections)
TASK VARIABLES	
Task domain	Text-editing (3-6, 8)
	Graphics (8)
	Page layout (9)
	Circuit design (10)
Task model	POET editing analysis (5.1)
	BRAVO editing analysis (6.1)
USER VARIABLES	
INTELLECTUAL ABILITIES	
General intelligence	——
Technical ability	Individual differences (3.3)
COGNITIVE STYLE	
Risk preference	Selection rules (5.2)
	Error rates (12.3)
Curiosity	——
Persistence	——
EXPERIENCE	
Experience on system	Individual differences (3.3)
Frequency of system use	Individual differences (3.3)
KNOWLEDGE	
Method knowledge	Selection rules (5.2)
Conceptual knowledge	——
Task expertise	——
PERCEPTUAL-MOTOR SKILL	
Typing rate	Individual differences (3.3)
	Model validation (8.4)
Manual skill	——
COMPUTER VARIABLES	
Dialogue style	Compare editors (3.2, 12.3)
	Editor vs typewriter (4)
	Interactive systems (8.3)
Command syntax	Interactive systems (8.3)
Naming conventions	——
Display layout	——
Input devices	Pointing devices (7)
Response time	——

Figure 12.1. Studies classified by structural variables.

Performance Variables		Studies (Chapters/Sections)
BASIC PERFORMANCE MEASURES		
Functionality	What tasks can the user accomplish with the system?	Editing task population (12.3)
Learning	How does his performance improve over time?	Pointing devices (7.3) Text editing (12.3)
Time	How long does it take the user to do a task with the system?	Editing benchmarks (3.2) Individual differences (3.3) Editor vs. typewriter (4) POET editing (5) BRAVO editing (6) Pointing devices (7) Interactive systems (8) Page layout (9) Circuit design (10) Text editing (12.3)
Error	What errors are made, how frequently, and how consequential are they?	POET editing (5.4) Pointing devices (7.3) Circuit design (10.3) Text editing (12.3)
Quality	How good is the output?	———
Robustness	How does performance adapt to unexpected conditions or to new tasks?	———
SUBJECTIVE MEASURES		
Acceptability	How does the user subjectively rate the system?	———
Enjoyableness	How much fun is it to use?	———
EXTREME CONDITIONS		
Fatigue	How does performance degrade over time?	———
Stress	How does performance degrade under adverse conditions?	———
MEMORY VARIABLES		
WM Load	How much immediate information does the user have to keep in Working Memory?	———
LTM Recall	How easy is it for the user to recall information needed to accomplish a task?	———

Figure 12.2.   Studies classified by performance variables.

Figure 12.2 lists a set of performance variables for characterizing the behavior of these systems.

Task Variables. Our strategy has been to focus on a single task domain and then to try to generalize to other domains. We have therefore been largely concerned with text-editing. We have generalized the results to other human-computer interaction task domains in the Keystroke-Level Model (Chapter 8), in the page-layout analysis (Chapter 9), and in the study of a circuit-layout system (Chapter 10). Within the domain of text-editing, we have analyzed two types of task models for text-editing—the line-structure model of text in POET and the two-dimensionally displayed character-stream model of text in BRAVO.

User Variables. We have not attempted to explore user variables systematically, except for the preliminary individual-differences study in Chapter 3. Instead, we have focused on expert users (who are best characterized by the knowledge and experience variables in Figure 12.1). Our strategy was to build a solid theoretical and empirical characterization of the expert user before attending to novice and casual users. However, we have seen some variations within experts, such as their knowledge of methods in the method selection study in Section 5.2.

Computer Variables. We have not attempted to explore computer variables systematically. Rather, our focus has been on how the user adapts to a given computer system structure. However, we have studied a variety of computer systems interfaces, from 1960's-style teletypewriter-oriented systems to state-of-the-art display-based systems. In some cases, we have directly compared behavior on alternative styles of system, such as in the studies of Chapters 3 and 8.

Performance Variables. Figure 12.2 clearly reveals our deliberate emphasis on performance time, which goes hand-in-hand with our emphasis on expert users. We have also presented a few modest accounts of the errors made by expert users. Our focus on performance time does not imply that we think the other performance variables are less important; indeed, they may be more important in many human-computer interaction contexts.

PERFORMANCE MODELS

Performance models, according to Formula 12.2, predict the performance of the human-computer system from a specification of its structure. Figure 12.3 lists several kinds of performance models. As can be seen in the figure, almost every kind of performance model has been

Performance Models		Studies (Chapters/Sections)
EXPERIMENTAL MODELS		
Running system	Use actual system.	Benchmark comparison (3.2) Editor evaluation (12.3)
Analogue system	Use another similar running system.	Layout test (9.2)
Mock-up system	Use a physical mock-up.	———
SYMBOLIC MODELS		
Calculational model	Code performance as a set of operations.	Manuscript editing (5) Keystroke-Level Model (8) Unit task analysis (9)
Simulation model	Code performance in a runable program.	BRAVO simulation (6)
Algebraic model	Represent relationships between variables and parameters as equations.	Editor vs. typewriter (4) Fitts's Law (7) New method analysis (8.4)
DATABASE MODELS		
Data table	Look up a pre-measured or pre-calculated value.	Model Human Processor (2) Benchmark data (3.2) Individual differences (3.3) Manuscript editing (5) Pointing devices (7) Keystroke-level operators (8) Editing data (12.3)
Checklist	Check design against principles or guidelines.	Design principles (12.4)

Figure 12.3. Studies classified by performance models.

presented. Our main emphasis, of course, has been on the development of symbolic models, especially calculational models: the GOMS family of models (Chapter 5), the Keystroke-Level Model (Chapter 8), and the Unit-Task-Level Model (Chapter 9). We have also presented a simulation model (Chapter 6) and several algebraic models, such as Fitts's Law (Chapter 7). We used running systems for the benchmark studies in Chapter 3. Finally, we have tabulated data that are useful databases (e.g., Figures 2.1, 2.2, 5.15, 7.4, 8.1, and 8.2).

DESIGN FUNCTIONS

The process of system design, according to Formula 12.3, consists of a set of design functions, which address design subproblems and which use performance models in finding solutions. We classified the design

Design Function		Studies (Chapters/Sections)
EVALUATION		
Compare systems	Compare on given performance variables.	Benchmark comparison (3.2) Device evaluations (7) Computed benchmark (8.4) Editor comparison (12.3)
Evaluate system	Compare against some standard.	Layout system (9) Editor evaluation (12.3)
PARAMETRIC DESIGN		
Optimize parameter	Find best value on given performance variables.	Editor vs. typewriter (4.2)
Analyze sensitivity	Relate parameter value to performance.	Editor vs. typewriter (4.4) New method analysis (8.4) Layout calculation (9.3)
STRUCTURAL DESIGN		
Identify opportunity	Find place where system can be improved.	Crossover point (4.2) Information rate limit (7.3) New method (8.4) Icarus Move command (10)
Diagnose problem	Pinpoint structural component causing problem.	Crossover point (4) New method (8.4)
Generate improvement	Find structural change.	New method (8.4)
Synthesize structure	Create new structure.	———

Figure 12.4. Studies classified by design functions.

functions as evaluation, parametric design, and structural design. Figure 12.4 lists several design functions, along with the studies illustrating them. The coverage is heaviest in evaluation and parametric design, where performance models are most clearly useful. System comparison usually involves experimentation (such as the benchmark study in Chapter 3 and the comparison of pointing devices in Chapter 7), but we have proposed the notion of a calculated benchmark (Section 8.4) for making comparisons analytically. The typewriter-versus-editor analysis in Chapter 4 illustrates both parameter optimization and sensitivity analysis. We have had less to say about structural design, especially the synthesis of a new design, for which we have no examples. Perhaps the best illustration of structural design functions is the analysis of alternative methods in Section 8.4, where an opportunity was identified (the task), the problem diagnosed (the awkwardness of the existing methods), and an improvement generated (the new method).

APPLICATION SUMMARY

The principal contribution of our studies to application is a set of specific performance models. This is consonant with the view, sketched in Section 12.1, that performance models are the keystones in the application of psychology to system design. The main limitations are that the models are restricted to predicting the error-free performance time of expert users. The models have been validated in a variety of human-computer systems, which has also produced a useful database of empirical performance data. We believe that these models may be useful in design, in the style exhibited in Chapters 4 and 9 and in the example of Section 8.4.

However, as of yet we have only small bits of evidence for the usefulness of the models in actual design situations. Let us cite one interesting application by a product testing group within our own company, Xerox. The group was testing alternative command schemes for a particular set of routine tasks. They taught the schemes to novice users in order to evaluate how easy they were to learn. However, they did not have enough time to train the users to become experts and so could not measure expert performance. Instead, they used the Keystroke-Level Model to calculate the expert performance time. That is, the initial part of the learning curve was measured experimentally, while an asymptote was calculated from the model. Thus, they were able to put together, within their constrained time limits, a fairly complete picture of behavior with the alternative command schemes using both experimental and calculational performance models.

12.3. EXTENSION: AN EVALUATION METHODOLOGY

A study that builds on and extends the work in the present book towards practical application was conducted in our laboratory by Teresa Roberts (for her Ph.D thesis in computer science at Stanford University). The goal of her study was to develop a practical methodology for evaluating computer text-editors. In this section, we briefly describe Roberts's methodology for evaluating text editors, her empirical results, and how these extend the results so far reported (see Roberts, 1979, for the original technical report and Roberts and Moran, 1982, for additional data and analysis).

METHODOLOGY

Roberts began by enumerating a population of 212 text-editing tasks. Each task was expressed in a way neutral with respect to any particular type of text-editor. From this population she selected a set of 32 *core tasks*, which included the basic editing tasks that any text-editor can be expected to perform. This set of core tasks provides a common basis for comparing the performance of different editing systems.

Roberts's evaluation methodology covers four performance variables: functionality, time, learning, and errors. The latter three are measured over the core tasks, whereas functionality measures how well an editor extends beyond the core tasks.

Functionality. In Roberts's methodology, the functionality of an editor is measured by having expert users rate whether each task in her task population can be accomplished with the editor. (The rating levels are: "can't be done," "can be done at manual speed," "can be done clumsily," "can be done efficiently.") Scores are summed up to give each editor an overall functionality rating. The scores can be partitioned into different task categories to show the strengths and weaknesses of the editor.

Learning. Learning is measured experimentally in Roberts's methodology by teaching a novice with *no* computer experience how to do the core tasks with the editor. The experimental learning session is made up of five cycles, each consisting of a teaching part, followed by a quiz to measure what the novice knows how to do (a learning session usually takes from two to five hours). Learning is scored by taking the total time in the session and dividing by the total number of tasks that the quizzes reveal the novice has learned, i.e., the learning time per task. The overall learning score for an editor is the average learning time for the four novices.

Time. The time it takes experts to perform core tasks is also measured experimentally. An expert user of the editor is clocked while performing a benchmark set of about 60 editing tasks (usually taking about 30 minutes). Note that this experiment is similar to our experiment in Chapter 3, except that the times are measured with a stopwatch, rather than with an on-line data collection facility. In addition to the overall time, the time the expert spends correcting large errors (i.e., large enough to be timed with a stopwatch) is also noted. The time score is the error-free time (total time minus the error-correcting time) to edit the benchmark tasks. The overall time score for the editor is the average error-free time for the four expert users.

Errors. Errors are difficult to measure in a simple experiment. Large damaging errors are rare enough so that it takes a long time to collect a reasonable sample. Further, there are large individual differences in error rates, even for routine errors. Roberts explored several methods of assessing errors, none of which seemed satisfactory enough to be used in practice. A modest indication of error effects, however, is the the percentage of time spent correcting errors in the core benchmark experiment above. Thus, the error score for an editor is the average error time, as a percentage of error-free time, for the four expert users.

Cost of Evaluation. An important constraint on this methodology is that it must be relatively easy to use. This is why only manual (stopwatch) measurements are required and why the minimal number of users are measured in the time and learning experiments. The time required to do a complete evaluation of a single editor depends on many factors—the evaluator's familiarity with the methodology, the effort required to prepare the materials for the specific editor, and the difficulty of recruiting users for the experiments. We have found that an editor evaluation takes roughly a week to do for an experienced evaluator.

VALIDATION AND EMPIRICAL RESULTS

Roberts tested her methodology by evaluating four widely-used editors: TECO (BBN, 1973), WYLBUR (Stanford, 1975), NLS (Englebart and English, 1968), and a display-based WANG word processor. We also include here an evaluation of BRAVO, BRAVOX (an extended version of BRAVO), GYPSY (another experimental editor developed at Xerox), and EMACS (Stallman, 1981).

The methodology provides a multidimensional evaluation of the editors. Each editor can be characterized by a 4-tuple of numbers. This summary evaluation is presented in Figure 12.5 for the eight editors, which shows the performance tradeoffs between these editors. The major differences are between the non-display editors (TECO, WYLBUR) and the display editors (all the others). With the exception of NLS on error time and GYPSY on functionality, the display editors are better on all performance dimensions. The display editors are up to twice as fast to use and have about 50% more functionality. On learning, TECO stands out as taking nearly three times as long to learn as the others. One surprising result is the high correlation ($R = .80$) between the time and learning scores. It is usually thought that systems that are highly efficient

Editor	Evaluation Scores			
	Functionality (% tasks)	Learning $M \pm CV$ (min/task)	Time $M \pm CV$ (sec/task)	Errors $M \pm CV$ (% time)
TECO	39%	19.5 ± .29	49 ± .17	15% ± .70
WYLBUR	42%	8.2 ± .24	42 ± .15	18% ± .85
EMACS	49%	6.6 ± .22	37 ± .15	6% ± 1.2
NLS	77%	7.7 ± .26	29 ± .15	22% ± .71
BRAVOX	70%	5.4 ± .08	29 ± .29	8% ± 1.0
WANG	50%	6.2 ± .45	26 ± .21	11% ± 1.1
BRAVO	59%	7.3 ± .14	26 ± .32	8% ± .75
GYPSY	37%	4.3 ± .26	19 ± .11	4% ± 2.1

Figure 12.5. Evaluation summary of eight text-editors.
The Functionality score is the percent of the 212 tasks in Roberts's task population that can be accomplished with each editor. The Learning score is the average learning time per task for four novices. The Time score is the average error-free time per task for four expert users on the benchmark set of tasks. (The time scores are large, because many of the tasks on the benchmark required many unit tasks to perform.) The Error score is the average percentage of time the four expert users spent correcting errors; the score is given as a percentage of the error-free time. The *CV*'s show the amount of between-user variance. The evaluation results for TECO, WYLBUR, NLS, and WANG are from Roberts (1979).

to use by experts take longer for novices to learn. This is not the case in this set of editors; the faster editors to use are also faster to learn.

The experimental results of the time dimension were compared against the predictions of the Keystroke-Level Model. The model predicted over 75% of the error-free benchmark time for most of the editors. For TECO, however, the model only predicted 50% of the time. The problem here was that the methods actually used by the expert users were not predicted correctly; the users were much more cautious than predicted in using TECO. When the model's prediction for TECO was adjusted for the actual methods used by the test users, then it accounted for 87% of their error-free time. These predictions are quite reasonable, given the differences between the assumptions of the model and the conditions of Roberts's experiment.

Finally, since these evaluation experiments were run on 32 expert users and on 32 novice learners, they provide us with some useful empirical results on individual differences. On expert performance, there was a factor of 1.5 to 2 between the fastest and slowest users within each editor, which is consistent with the results of Chapter 3. It is interesting and somewhat surprising that there was not a great deal more variation among the novice learners than among the experts, i.e., there was about the same range of ratio between the fastest and slowest learners as between the fastest and slowest experts. By far the greatest individual differences occurred with the error times. Expert users spent from as little as 0% to as much as 28% of their time in errors, averaging 10% error time. Roberts measured error time in real time with a stopwatch, and she had to ignore the small errors. A more careful measurement of errors (on videotape, say, as was done with all the error measurements we have reported) would yield somewhat higher percentages.

APPLICATION

The places where Roberts's study contributes to application have been shown in Figures 12.1 to 12.4. Her study is mainly oriented to the design function of system comparison. Now that her data can be used as a standard of comparison, her methodology also enables the system evaluation of individual editors. The most important aspect of Roberts's work, in the context of this chapter, is that it extends the scope of our studies on two performance variables—functionality and learning.

12.4. ADVICE TO THE DESIGNER

We have presented an approach to applying psychology to design that centers around the notion of performance models. Our implicit advice to the system designer has been to use these models in design. We now present this advice more explicitly in the form of a set of system design principles (listed briefly in Figure 12.6) derived directly from the main results of our studies. Since, as we have seen in Section 12.2, the studies are highly skewed towards certain issues, the principles do not cover the whole spectrum of design concerns. Nor do we attempt to exhaust all the principles implicit in the models; we only present some of the more important and fundamental principles.

1. <u>Early</u> in the system design process, consider the psychology of the user and the design of the user interface.

2. Specify the <u>performance requirements</u>.

3. Specify the <u>user population</u>.

4. Specify the <u>tasks</u>.

5. Specify the <u>methods</u> to do the tasks.

6. Match the <u>method analysis</u> to the level of commitment in the design process.

7. To <u>reduce the performance time</u> of a task by an expert, eliminate operators from the method for doing the task. This can be done at any level of analysis.

8. Design the set of <u>alternative methods</u> for a task so that the rule for selecting each alternative is clear to the user and easy to apply.

9. Design a set of <u>error-recovery methods</u>.

10. Analyze the <u>sensitivity</u> of performance predictions to assumptions.

Figure 12.6. Some principles for user-interface design.

The first few principles summarize some high-level concerns and attitudes about design.

Principle 1: Early in the system design process, consider the psychology of the user and the design of the user interface.

This may seem too obvious to mention, but it is fundamental and often stated (e.g., Hansen, 1971). If consideration of the human-computer interaction is put off until the computer system is designed, then the psychology of the user will not have any weight among the variety of concerns that face the designer. This principle does not itself tell the designer what to do; the next few principles spell out some concrete actions.

According to Formula 12.1, the human-computer system consists of the task, the user, and the computer, which together determine the system's performance. The designer's job is to specify the total human-computer system. The designer does not have to be told to specify the computer; but he may need to be reminded of the performance requirements, the user, and the task.

Principle 2: *Specify the performance requirements.*

There are many performance variables—functionality, time, errors, learning, etc. Designing to improve performance on one dimension does not necessarily help performance on other dimensions. For example, optimizing the performance time of a system does not improve its learnability (in fact, high concentration on time optimization may make a system harder to learn). There are tradeoffs to be made in performance. For example, using the models we have presented to calculate performance time and using Roberts's methodology for measuring learning, one can quantitatively compare the tradeoffs between ease of learning and speed of execution in a system. Thus, it is important that the designer be clear about his priorities on the performance variables.

Principle 3: *Specify the user population.*

In Chapter 3 we have seen that there is about a factor of three in performance time among expert users—about the same range as the performance among different editing systems. Considering non-expert users, the range of user performance is much greater. Thus, in order to predict the performance of the human-computer system, the designer must know the important characteristics of the user population. If the target population of users is highly varied, it is important to characterize the different kinds of users, for their performances will be quite different. Much of this characterization can be done quantitatively. For example, the Keystroke-Level Model (Chapter 8) shows how the user's typing speed affects his performance time.

Principle 4: *Specify the tasks.*

Performance can only be assessed relative to the set of tasks that must be done. It is not possible to specify all the tasks that the user will want to do. However, specifying a reasonable benchmark sample of tasks is infinitely better than just listing gross task characteristics. The benchmark sample should include representatives of the qualitatively different kinds of tasks the user will face. An example of task generation is given in Chapter 9. The different types of tasks occur with unequal frequency—most of the user's time will be spent doing a very few task types. It is important to specify these high-frequency tasks. The user will become highly skilled on these tasks, and they should be made easy and efficient to do.

Task analysis can be done at different levels of detail, for any task can be decomposed into a task-subtask hierarchy (as was done in Chapter 5). What is the appropriate level of task analysis? In order to keep the range of design possibilities open, tasks should be specified in a way that makes minimal assumptions about the structure of the computer system, except for the structure that is fixed a priori as part of the design requirements. The Unit-Task Level of task analysis, as illustrated in Chapter 9, is the most detailed level of task specification that is practical early in design.

As analysis becomes more and more dependent on system structure, task analysis turns into method analysis. Task analysis reflects more the demands of the external environment, whereas method analysis reflects more the demands of the computer system and the ways in which the user adapts to them. There is, of course, no sharp line between task analysis and method analysis.

Principle 5: *Specify the methods to do the tasks.*

It is important to grasp the central role that the methods play in determining the level of performance. Skilled human-computer interaction consists of execution of assimilated methods. What makes a user skilled is his highly integrated knowledge of tasks, methods, and the connections between them. System designers tend to concentrate on the commands of the computer system (just look at the documentation for almost any system, which is usually a catalogue of commands). Yet it is *how the commands are used*—the methods—that is most important to the user. Once methods are laid out explicitly, many of the gross aspects of performance can be seen by inspection, even without formal models. For example, particularly long or awkward methods will stand out. Also, it is possible to assess informally the consistency between different methods.[3]

> *Principle 6: Match the method analysis to the level of commitment in the design process.*

As with tasks, methods can be specified at different levels of detail. In order to predict performance from a method specification, a performance model is required, which in turn determines the method description. There is no single best model; different models are appropriate at different stages of design—depending on the amount of detail known about the system under design.

Several levels of method analysis were introduced in Chapter 5. The Unit-Task Level requires only a modest commitment to the structure of the computer system. This level of analysis is appropriate early in design to assess the task domain by getting a rough picture of the total system performance. It is also useful when trying to decide on major components of the computer system, as was illustrated in Chapter 9. At the Functional Level of analysis, the unit tasks are decomposed into their four functional components—Acquire, Locate, Change, and Verify. This level begins to show how the unit tasks interact, as illustrated in Chapter 9.

The next lower level of analysis is the Argument Level, in which there is commitment to the set of commands and the arguments they take. This level is appropriate while the command set is being designed, but where the small details of the command syntax are ignored. Although we have not given any illustrations, the Argument Level is actually a quite useful level of analysis. For example, this is the level at which the scheme for defaulting arguments can be considered. And finally, at the Keystroke Level there is commitment to the actual keystrokes and other physical operations for executing commands. This level of detail is not appropriate until fairly late in design. But once this level of detail is reached, it is possible to do considerable quantitative analysis of performance time, as we have shown with the Keystroke-Level Model in Chapter 8.

---

[3] Various kinds of rule-based descriptions of the methods can be used to assess consistency more precisely (e.g., Moran, 1981*a*; Reisner, 1981), although existing rule-system proposals are not yet developed enough to be performance models.

> *Principle 7: To reduce the performance time of a task by an expert, eliminate operators from the method for doing the task. This can be done at any level of analysis.*[4]

Once methods are laid out, at whatever level, and the performance time calculated, the designer may then want to make the performance more efficient for expert users. All the performance models we have presented suggest that expert performance is composed of a sequence of operators and that the performance time is the sum of time for each of the operators.

Which operators can be eliminated depends on the stage of design and the level of analysis. For example, the performance time for a job can be reduced either by reducing the number of unit tasks or by reducing the time per unit task; however, only the former is possible early in design at the Unit-Task Level of analysis. At the Unit-Task Level, the most likely way to reduce unit tasks is to extend the functionality of the computer system to, in effect, combine unit tasks (e.g., a text-editing function for inserting a pair of parentheses around a piece of text can combine what would otherwise be two unit tasks into one). At the Argument Level, the most obvious way to reduce time is to devise appropriate default values for arguments and even for commands (e.g., with a Redo command). At the Keystroke Level, the way to efficiency is to devise short codes to specify commands and arguments and to eliminate redundant terminators.

> *Principle 8: Design the set of alternative methods for a task so that the rule for selecting each alternative is clear to the user and easy to apply.*

Alternative methods can be provided to do a task. This allows the different methods to better take advantage of the specific structural features of the different task instances. From our study of method selection in Chapter 5, we characterized the expert user as having simple decision rules for selecting an appropriate method in each task instance.

---

[4] This principle might well be called the "Gilbreth Principle," for Gilbreth (1911) was one of the first to systematically code behavior into a sequence of physical movements (which he called "therbligs," but which we would call "operators") and to optimize performance by eliminating unnecessary movements. Gilbreth, however, did not have any notion of levels of analysis; all his analyses were at the same level of physical movement.

Another useful notion from the method selection study is the notion of the "default method," the method that the user selects by default in preference to other alternatives. The default method is not the most efficient method, but it is a general method; other alternative methods are more specialized, but more efficient. Categorizing methods this way provides a good strategy for designing method alternatives: provide a general-purpose method plus a set of efficient special-purpose methods. A similar strategy is to design alternative methods for specifying commands, easy-to-remember but slow methods (such as typing out the command names) and fast but harder-to-remember methods (such as special single-key codes). These strategies allow incremental learning. The novice user need only learn the general-purpose, easy-to-remember methods at first; he can acquire the more efficient methods one by one as he becomes more expert.

Principle 9: *Design a set of error-recovery methods.*

Another aspect of expert performance is errors. We have seen that expert users adopt strategies that permit up to about 30% of their time to be spent correcting errors. Error-correction is also highly skilled behavior. Thus, error-recovery methods should also be designed for learnability and efficiency. This suggests, for example, that an Undo command would be worthwhile. In designing the Undo command, consider carefully how it will be used to help the user recover from specific kinds of errors, for all errors are not the same in the scope or severity of their effects.

In the analysis of errors, it is useful to separate the occurrence of errors from the treatment of errors once they occur. We do not yet have any models to help predict errors, although common sense can suggest a few sources of errors, such as the user accidentally hitting an adjacent key on the keyboard.[5] But given that an error has occurred, the expert user's handling of the error is a skilled activity and is thus amenable to quantitative analysis by the performance models of the sort we have described.

---

[5] Such an accident would be a motor "slip." See Norman (1981) for a categorization of action slips, including cognitive slips. For some evidence that error occurrences can be modeled, at least in closed task domains, see Brown and VanLehn's (1980) Repair Theory model of the sources of "bugs" in arithmetic procedures.

> **Principle 10:**   *Analyze the sensitivity of performance predictions to assumptions.*

In carrying out any kind of performance analysis, the designer must make assumptions about the user (psychology does not have all the answers), about the computer system (the design is not fully specified until the end), and about the task environment (which cannot be fully anticipated). Thus, any predictions of human-computer performance should be checked for their sensitivity to these assumptions.

One of the main advantages of symbolic models, as we have emphasized, is that they allow unknowns to be parameterized and hence to be analyzed for their effects on performance. This use of parametric and sensitivity analysis was illustrated in the examples in Chapter 4 and in Section 8.4. Although performance usually does change, it often does not change in ways that affect the design decisions that motivated the performance analysis. Even if sensitivities are found, it is much better to make design decisions knowing what factors critically effect the decision and what factors do not matter. With such knowledge, it is possible to know which previous design decisions must be re-evaluated as new knowledge develops during the design process.

12.5. CONCLUSIONS

We have proposed that an applied psychology should take a particular form in order to be of use in the design of interactive human-computer systems. The central feature of this applied psychology is the packaging of psychological knowledge into performance models that can predict the performance of the human-computer system from specifications of its structure. The design process can be decomposed into several different kinds of design functions, most of which require the use of performance models.

In this book we have concentrated most heavily on performance models for calculating expert performance time.

Roberts (1979) has extended the use of our models by developing a practical methodology for evaluating text-editing systems along four performance dimensions—functionality, learning, time, and errors.

We have expressed some of the results in this book as a set of design principles to aid in the design of systems for human-computer interaction.

13. Reprise

In this book, we have reported on a program of research directed towards understanding human-computer interaction. Let us briefly summarize the extent of our progress.

The flow chart in Figure 13.1 gives the argument of the book in terms of research questions addressed and the results obtained. Starting at the very top of the figure, from the proposition argued in Chapter 1 that current techniques of human-computer interaction can be improved upon, we addressed four basic questions (numbered Q1–Q4 in the figure):

Q1. How can the science base be built up for supporting the design of human-computer interfaces?

Q2. What is the nature of user behavior and what are the consequent user performance characteristics for a specific human-computer interaction task (we chose to study text-editing)?

Q3. How can our results be cast as practical engineering models to aid in design?

Q4. What principal generalizations arise from the specific studies, models, and applications of Q2 and Q3?

Let us trace through the figure, considering each of these in turn.

SCIENCE BASE

If human-computer interaction is to be improved, there needs to be a science base of knowledge about human performance on which designers can draw for actual design. Recent advances in cognitive psychology and allied sciences can aid us in building this science base.

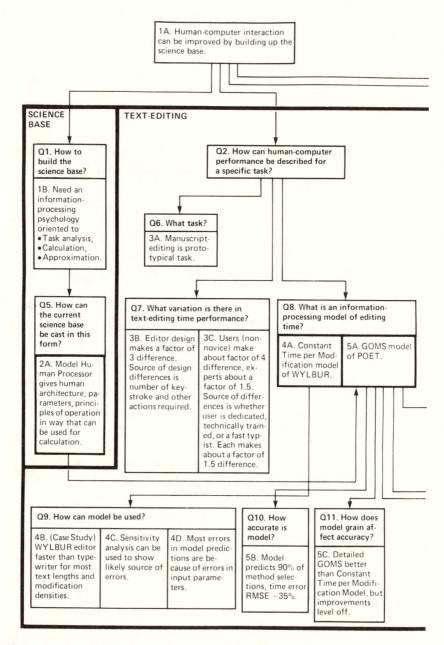

Figure 13.1. Reprise.

Principal research questions addressed are numbered Q1, Q2, etc. Answers to the research questions in the form of empirical facts established, propositions argued,

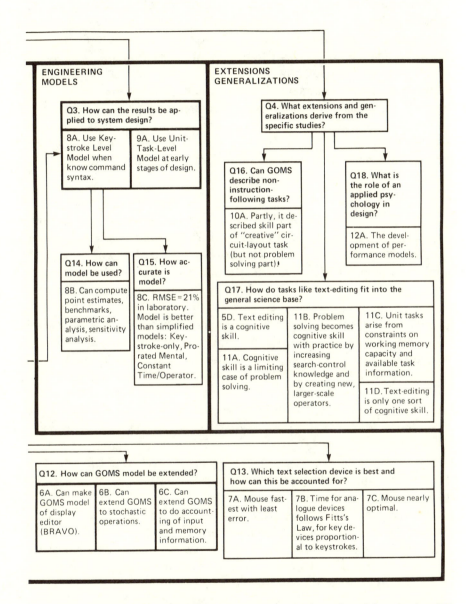

ENGINEERING MODELS

Q3. How can the results be applied to system design?

8A. Use Keystroke Level Model when know command syntax.	9A. Use Unit-Task-Level Model at early stages of design.

Q14. How can model be used?	Q15. How accurate is model?
8B. Can compute point estimates, benchmarks, parametric analysis, sensitivity analysis.	8C. RMSE = 21% in laboratory. Model is better than simplified models: Keystroke-only, Prorated Mental, Constant Time/Operator.

EXTENSIONS GENERALIZATIONS

Q4. What extensions and generalizations derive from the specific studies?

Q16. Can GOMS describe non-instruction-following tasks?	Q18. What is the role of an applied psychology in design?
10A. Partly, it described skill part of "creative" circuit-layout task (but not problem solving part)ꜝ	12A. The development of performance models.

Q17. How do tasks like text-editing fit into the general science base?

5D. Text editing is a cognitive skill. 11A. Cognitive skill is a limiting case of problem solving.	11B. Problem solving becomes cognitive skill with practice by increasing search-control knowledge and by creating new, larger-scale operators.	11C. Unit tasks arise from constraints on working memory capacity and available task information. 11D. Text-editing is only one sort of cognitive skill.

Q12. How can GOMS model be extended?

6A. Can make GOMS model of display editor (BRAVO).	6B. Can extend GOMS to stochastic operations.	6C. Can extend GOMS to do accounting of input and memory information.

Q13. Which text selection device is best and how can this be accounted for?

7A. Mouse fastest with least error.	7B. Time for analogue devices follows Fitts's Law, for key devices proportional to keystrokes.	7C. Mouse nearly optimal.

or models developed are attached to the question and numbered by chapter (3B means the second result listed in the figure from Chapter 3). Many of the results lead in turn to more detailed research questions as indicated by the arrows.

427

We have argued that a successful applied cognitive psychology of human-computer interaction requires an approach based on three tenets: (1) Primary emphasis should be placed on task analysis, calculation, and approximation. (2) The approach should be based on a theory of the user as an information-processor. (3) It should be relevant to design, that is, to the analysis of systems before they have been built. These tenets lead to the question of how the current science base of psychological knowledge can be summarized in this form (Q5)? One answer is to organize the science base into the architecture, parameters, and principles of operation of a compact, engineering-oriented model such as the Model Human Processor. This, in turn, leads us to consider human-computer interaction in terms of human information-processing operations (5A).

TEXT-EDITING

Since new knowledge and insight are often achieved by first focusing on concrete cases and then generalizing, it is necessary to select a task for detailed study (Q6), and the task we have selected is computer text-editing. Computer text-editing is a prototypical human-computer inter-action task, and as such its study is likely to shed light on other human-computer interaction tasks. Furthermore, there is substantial variation in user performance, a result of both different editor designs and because of individual differences among users (Q7). The slowest editor designs require users to spend three times as long to make the same edits as do the fastest designs; the slowest users (long-time users, but not necessarily "experts") spend four times longer to make the same edits as the fastest users; the slowest experts spend 50% longer than the fastest experts. The faster-to-use editors are faster because they require users to type fewer keystrokes or perform fewer other actions than the slower-to-use editors. The faster users get their speed by possessing the following characteristics (each about equally effective in our studies): using the system regularly, technical background, and typing speed.

How can an information-processing analysis be constructed for text-editing (Q8)? The simplest model of user editing performance is the Constant Time per Modification Model. We investigated this model for the WYLBUR editor and showed, in a case study comparing text-editing and typewriting, that it could be used to predict tradeoffs between the two and could also be used in sensitivity analysis (Q9).

More detailed are models based on an information-processing analysis of the user's goals, operators, methods, and selection rules. We investigated a GOMS model for the POET editor and found that it was capable of predicting users' method selection about 90% of the time (using 2~4 rules) and of predicting editing time to a *RMS* error of about 35% (Q10). To discover how much the accuracy of the GOMS model depended on its level of detail (Q11), we constructed nine models of POET editing, with detail ranging from gross (12 sec/operator) to very fine (.5 sec/operator). These models fell at four levels: the Unit-Task Level (closely related to the Constant Time per Modification Model above, but with a stricter definition of what constitutes a task), the Functional Level (an operator for each major phase of an editing task), the Argument Level (an operator for each command and argument), and the Keystroke Level (an operator for each keystroke or other action). We found that accuracy improved in going from the Unit-Task Level to the Functional Level (*RMS* error 40%~30%, for one of the measures) and did not decline thereafter. It was shown that a GOMS analysis could be extended to the display-oriented BRAVO editor, to include stochastic elements, and to give a detailed account of the flow of task information (Q12).

We also considered a set of models for a component of the GOMS models, the operation of selecting a piece of text on a display (Q13). Here, experiments showed that the mouse is a faster, less error-prone device than step keys, text keys, or the rate-controlled isometric joystick measured. Models of each of these devices showed that pointing time for the analogue devices (mouse and joystick) is proportional to the log of the ratio of target distance and target size, as given by Fitts's Law, and that pointing time for the key devices is proportional to the number of keystrokes. Further analysis showed that the time to point using the mouse is not limited by the device itself, but by the information-processing rate of the human eye-hand coordination system.

ENGINEERING MODELS

How can the models above be adapted for use in system design? If the user's method is known, a simplification of the GOMS model at the Keystroke Level can be used to predict editing time, to calculate benchmarks, and to conduct parametric and sensitivity analyses (Q14). The Keystroke-Level Model is a compromise between simplicity (only five operators) and accuracy (much less error than for simpler models such as counting only keystrokes, adding a prorated mental overhead, or counting each operator at a constant cost) (Q15). The Keystroke-Level Model was shown to predict execution time, with an *RMS* error of 21% for text-editors, graphics programs, and various system utilities. This compares well with the accuracy of the GOMS models and is sufficient for practical work.

At the early stages of design, or when the user's method is not known, the Unit-Task-Level Model can be used to predict times by breaking the user's task into unit tasks and assigning each a constant time.

EXTENSIONS AND GENERALIZATIONS

Can the GOMS analysis be extended to more "creative," less routine tasks (Q16)? A case study on VLSI circuit design suggests that part of the behavior in such tasks is problem solving, to which the GOMS analysis is not applicable, and part of the behavior is skill, to which the GOMS/Keystroke-Level-Model analysis is appropriate.

It is important to ask how text-editing fits with other cognitive tasks (Q17). Computer text-editing is an example of a cognitive skill. Cognitive skill is a limiting case of problem solving in which the search for a solution has been greatly reduced through practice and experience. The transition from problem solving to cognitive skill can be seen by starting with a problem-solving task, such as the classic Tower of Hanoi puzzle, and observing how search through the problem space of the puzzle is reduced and finally eliminated as a consequence of practice, until solving the puzzle has become a skill. The same transition can be seen from the other end, from a task in which we know users exhibit cognitive skill, such as text-editing, and observing how the skill arose, with practice, out of earlier problem solving. Closer examination of the transition from problem solving to cognitive skill shows that the mechanisms whereby search is reduced are (1) accumulation of control knowledge and (2) the formation of new, larger-scale operators, which effectively partition the problem space into a reduced problem space and a skill space. Unit tasks are a consequence of constraints on the task resulting from the memory limitations of the user or lack of information. Other cognitive skills, whereas they share much of the above characterization with computer text-editing, also differ from text-editing along a number of dimensions: in the basic skill/problem-solving character, in their demands on Working Memory and Long-Term Memory, and in their external demands (such as pacing and accuracy).

Finally, there is the question of what role psychological research on human-computer interaction can play in design (Q18). We have argued that research should emphasize the development of performance models, enabling designers to predict the performance consequences of design alternatives. In addition to their use in considering design alternatives, performance models can also be used as an evaluation methodology, such as the one designed by Roberts (1979), and in the formulation of design principles, such as those listed in Chapter 12. These examples suggest how performance models may be made the link for transferring understanding to practice in a psychology of human-computer interaction.

Symbol Glossary

This glossary lists the principal symbols used in the book and the pages on which they are defined. It includes algebraic symbols and operator symbols from the Keystroke-Level Model; additional goal and operator symbols are listed in the Subject Index. The nominal value and range (in square brackets) is also given for each constant. An F after a page number indicates that the reference is to a figure on the page.

τ_P Cycle time of Perceptual Processor (= 100 [50~200] msec) . . 32

D Distance to target 52

H Entropy of a decision 72–76

H   Home hands (Keystroke-Level Model operator) 262–264

I_C Uncertainty Principle constant (= 150 msec/bit) 72, 74–76

I_M Fitts's Law constant (= 100 [50~120] msec/bit) 53–57

K   Keystroke (Keystroke-Level Model operator) 262, 264F

L_c Length crossover point 123

M   Mental preparation (Keystroke-Level Model operator) 262, 264F

P   Point with mouse (Keystroke-Level Model operator) . . . 262–264

R   Response by system (Keystroke-Level Model operator) . 262–264

S Size of target 52–55

t_H Time to home hands (= .40 sec) 263, 264F

t_K Time to type a keystroke (= [.08~1.20] sec) 262, 264F

t_M Time to make mental preparation (= 1.35 sec) 263, 264F

t_P Time to point with mouse (= 1.10 sec) 262, 264F

$T_{acquire}$ Total time to acquire a task 261

$T_{execute}$ Total time to execute a task 261

T_H Total homing time during task 262

T_K Total keystroke time during task 262

T_M Total mental preparation time during task 262

T_P Total pointing time during task 262

T_R Total system response time during task 262

Bibliographic Index

The boldface numbers to the right of each reference indicate the pages in book on which the reference is cited. An F after a page number indicates that the citation is in a figure, and an n after a page number indicates that it is in a footnote.

ABRUZZI, A. (1952). *Work Measurement.* New York: Columbia University Press. **182**

ABRUZZI, A. (1956). *Work, Workers, and Work Measurement.* New York: Columbia University Press. . . . **13, 161, 182, 183, 183F, 184**

AKIN, O., and CHASE, W. (1978). Quantification of three-dimensional structures. *Journal of Experimental Psychology 4*, 397–410. . . . . **43F**

ALDEN, D. G., DANIELS, R. W., and KANARICK, A. (1972). Keyboard design and operation: A review of the major issues. *Human Factors 14*, 275–293. . . . . **91**

ANANDAN, P., EMBLEY, D. W., and NAGY, G. (1980). An application of file-comparison algorithms to the study of program editors. *International Journal of Man-Machine Studies 13*, 201–211. . . . . **102**

ANDERSON, J. R. (1976). *Language Memory and Thought.* Hillsdale, New Jersey: Lawrence Erlbaum Associates. **96**

ANDERSON, J. R. (1980). *Cognitive Psychology and its Implications.* San Francisco: W. R. Freeman. **2n1, 91, 95, 36n, 359**

ANDERSON, J. R., ed. (1981*a*). *Cognitive Skills and their Acquisition.* Hillsdale, New Jersey: Lawrence Erlbaum Associates. **2n1, 359**

ANDERSON, J. R. (1981*b*). Acquisition of Cognitive Skills. Technical Report, Department of Psychology, Carnegie-Mellon University. . . **359**

ANDERSON, J. R., and BOWER, G. H. (1973). *Human Associative Memory.* Washington, D.C.: V. H. Winston and Sons. **2n4, 95**

ATKINSON, R. C., and SHIFFRIN, R. M. (1968). Human memory: A proposed system and its control processes. *The Psychology of Learning and Motivation 2*, 89–195. . . . . **91**

Atwood, M. E. *See* Ramsey and Atwood (1979); Ramsey, Atwood, and Kirshbaum (1978).

AVERBACH, E., and CORIELL, A. S. (1961). Short-term memory in vision. *Bell System Technical Journal 40*, 309–328. . . . . **29n, 30F**

BADDELEY, A. D. (1966). Short-term memory for word sequences as a function of acoustic, semantic and formal similarity. *Quarterly Journal of Experimental Psychology 18*, 362–365. . . . . **80F**

BADDELEY, A. D. (1976). *The Psychology of Memory*. New York: Basic Books. **2n4, 93**

BADDELEY, A. D. (1981). The concept of working memory: A view of its current state and probably future development. *Cognition 10*, 17 – 23. . . . . **93**

Baddeley, A. D. *See also* Long and Baddeley (1981).

Barnaby, J. R. *See* Myer and Barnaby (1973).

BARNARD, P. J., HAMMOND, N. V., MORTON, J., LONG, J. B., and CLARK, I. A. (1981). Consistency and compatibility in human-computer dialogue. *International Journal of Man-Machine Studies 15*, 87-134. . . . . **91**

Baron, S. *See* Pew, Baron, Feehrer, and Miller (1977).

BARTLETT, F. C. (1958). *Thinking*. London: Allen and Unwin. . . . **187**

BBN (1973). TENEX Text Editor and Corrector Manual. Cambridge, Massachusetts: Bolt, Beranek, and Newman. **109F, 415**

Bell, G. *See* Siewiorek, Bell, and Newell (1981).

BELMONT, L., and BIRCH, H. G. (1951). Re-individualizing the repression hypothesis. *Journal of Abnormal & Social Psychology 46*, 226–235. . . . . **81F**

BENNETT, J. (1972). The user interface in interactive systems. *Annual Review of Information Science and Technology 7*, 159–196. . . . . . **91**

Berman, M. L. *See* English, Englebart, and Berman (1967).

BERNBACH, H. A. (1970). A multiple-copy model for post perceptual memory. In D. A. Norman, ed., *Models of Human Memory*, 103–116, New York: Academic Press. **94**

Birch, H. G. *See* Belmont and Birch (1951).

Bisseret, A. *See* Sperandio and Bisseret (1974).

BLANKENSHIP, A. B. (1938). Memory span: A review of the literature. *Psychological Bulletin 35*, 1–25. **92**

BLUMENTHAL, A. L. (1977). *The Process of Cognition*. Englewood Cliffs, New Jersey: Prentice-Hall. **42n20**

BLOCH, A. M. (1885). Expérience sur la vision. *Comptes Rendus de Séances de la Société de Biologie (Paris) 37*, 493–495. 32

BOBROW, D., and RAPHAEL, B. (1974). New programming languages for artificial intelligence research. *Computing Surveys 6*, 155-174. . . 362

Bobrow, D. *See also* Norman and Bobrow (1975).

BOIES, S. J. (1974). User behavior on an interactive computer system. *IBM Systems Journal 13*, 2–18. . . . . 91, 101

Boies, S. J. *See also* Posner, Boies, Eichelman, and Taylor (1969).

BOOK, W. F. (1908). The psychology of skill with special reference to its acquisition in typewriting. *University of Montana Studies in Psychology 1*. Reprinted, New York: Gregg, 1925. 15

Bower, G. H. *See* Anderson and Bower (1973).

BREITMEYER, B. G., and GANZ, L. (1976). Implications of sustained and transient channels for theories of visual pattern masking, saccadic suppression, and information processing. *Psychological Review 83*, 1–36. 96

BROADBENT, D. E. (1958). *Perception and Communication*. London: Pergamon Press. 2n2

BROOKS, R. (1977). Towards a theory of the cognitive processes in computer programming. *International Journal of Man-Machine Studies 9*, 737–751. 91

BROWN, J. S., and VANLEHN, K. (1980). Repair theory: A generative theory of bugs in procedural skills. *Cognitive Science 4*, 379–426. . . . . 423n

BRYAN, W. L., and HARTER, N. (1898). Studies in the physiology and psychology of the telegraphic language. *Psychological Review 4*, 27–53. . . . . 15, 47

BRYAN, W. L., and HARTER, N. (1899). Studies on the telegraphic language, the acquisition of a hierarchy of habits. *Psychological Review 6*, 345–375. 15

Burr, B. J. *See* Card, English, and Burr (1978).

BUSSWELL, G. T. (1922). Fundamental reading habits: A study of their development. *Education Monographs (Supplement) 21*. 28n

CAKIR, A., HART, D. J., and STEWART, T. F. M. (1980). *Visual Display Terminals*. New York: Wiley. 91, 218

CALFEE, R. C. (1975). *Human Experimental Psychology*. New York: Holt, Rinehart, and Winston. 80F

CARD, S. K. (1978). Studies in the Psychology of Computer Text-editing Systems. Ph.D. Thesis, Department of Psychology, Carnegie-Mellon University. **102**

CARD, S. K., ENGLISH, W. K., and BURR, B. J. (1978). Evaluation of mouse, rate-controlled isometric joystick, step keys, and text keys for text selection on a CRT. *Ergonomics 21*, 601–613. . . . . **xiii**

CARD, S. K., MORAN, T. P., and NEWELL, A. (1976). The Manuscript Editing Task: A Routine Cognitive Skill. Palo Alto, California: Xerox Palo Alto Research Center, Technical Report SSL-76-8. . . . . . .
. . . . **102, 109F**

CARD, S. K., MORAN, T. P., and NEWELL, A. (1980*a*). Computer text-editing: An information-processing analysis of a routine cognitive skill. *Cognitive Psychology 12*, 32–74. . . . . **xiii, 102, 109F**

CARD, S. K., MORAN, T. P., and NEWELL, A. (1980*b*). The Keystroke-Level Model for user performance time with interactive systems. *Communications of the ACM 23*, 396–410. . . . . **xiii, 102, 109F**

Card, S. K. *See also* Moran and Card (1982).

CARLTON, L. G. (1980). Movement control characteristics of aiming responses. *Ergonomics 23*, 1019–1032. . . . . **53n24**

CAVANAUGH, J. P. (1972). Relation between the immediate memory span and the memory search rate. *Psychological Review 79*, 525–530.
. . . . **43F**

Chaffin, D. B. *See* Langolf, Chaffin, and Foulke (1976).

CHAPANIS, A., GARNER, W. R., and MORGAN, C. T. (1949). *Applied Experimental Psychology: Human Factors in Engineering Design.* New York: John Wiley and Sons. **34n**

CHASE, W. G., and ERICSSON, K. A. (1981). Skilled memory. In J. R. Anderson, ed., *Cognitive Skills and their Acquisition.* Hillsdale, New Jersey: Lawrence Erlbaum Associates. **37n**

CHASE, W. G., and ERICSSON, K. A. (1982). Skill and Working Memory. Pittsburgh, Pennsylvania: Department of Psychology, Carnegie-Mellon University, Technical Report No. 7; to appear in *The Psychology of Learning and Motivation 16*, in press. **93**

CHASE, W. G., and SIMON, H. A. (1973). Perception in chess. *Cognitive Psychology 4*, 55–81. . . . . **268**

Chase, W. G. *See also* Akin and Chase (1978); Ericsson, Chase, and Faloon (1980).

CHEATHAM, P. G., and WHITE, C. T. (1954). Temporal numerosity: III. Auditory perception of number. *Journal of Experimental Psychology 47*, 425–428. . . . . **33F**

CHENG, N. Y. (1929). Retroactive effect and degree of similarity. *Journal of Experimental Psychology 12*, 444–458. . . . . **81F**

CHI, M. T., and KLAHR, D. (1975). Span and rate of apprehension in children and adults. *Journal of Experimental Child Psychology 19*, 434–439. . . . . **43F**

Chiba, S. *See* Sakoe and Chiba (1978).

CLARK, H. H., and CLARK, E. V. (1976). *Psychology and Language: An Introduction to Psycholinguistics.* New York: Harcourt, Brace, Jovanovich. **3n7**

Clark, E. V. *See* Clark and Clark (1976).

Clark, I. A. *See* Barnard, Hammond, Morton, Long, and Clark (1981).

CLAUDE, J. (1972). A comparison of five variable weighting procedures. *Educational and Psychological Measurement 32*, 311–322. . . . . . . **296**

Colatla, V. *See* Tulving and Colatla (1970).

CONRAD, R. (1964). Acoustic confusions in immediate memory. *British Journal of Psychology 55*, 75–83. . . . . **79, 92**

Conway, L. *See* Mead and Conway (1980).

Coriell, A. S. *See* Averback and Coriell (1961).

CORNOG, J. R., and CRAIG, J. C. (1965). Keyboards and coding systems under consideration for use in the sorting of United States mail. *6th Annual Symposium of the IEEE G-HFE*, Boston, Massachusetts. **61F**

CORNSWEET, T. N. (1970). *Visual Perception.* New York: Academic Press. **2n2, 96**

Craig, J. C. *See* Cornog and Craig (1965).

CRAIK, K. J. W., and VINCE, MARGARET A. (1963). Psychological and physiological aspects of control mechanisms. *Ergonomics 6*, 419–440. . . . . **243, 254, 254F**

CROSSMAN, E. R. F. W. (1958). Discussion of Paper 7 in National Physical Laboratory Symposium. In *Mechanisation of Thought Processes* (Vol 2). London: H. M. Stationery Office. **31n5**

CROSSMAN, E. R. F. W., and GOODEVE, P. J. (1963). Feedback control of hand movements and Fitts' Law. Paper prepared for a meeting of the Experimental Psychology Society, Oxford, July 1963. . . . . . **51n22**

ELKIND, J. I., and SPRAGUE, L. T. (1961). Transmission of information in simple manual control systems. *IEEE Transaction on Human Factors in Electronics HFE-2*, 58–60. . . . . **55n**

EMBLEY, D. W., LAN, M. T., LEINBAUGH, D. W., and NAGY, G. (1978). A procedure for predicting program editor performance from the user's point of view. *International Journal of Man-Machine Studies 10*, 639–650. . . . . **102, 294**

EMBLEY, D. W., and NAGY, G. (1981). Behavioral aspects of text editors. *Computing Surveys 13*, 33–70. . . . . **103, 294**

Embley, D. W. *See also* Anandan, Embley, and Nagy (1980).

ENGELBART, D. C., and ENGLISH, W. K. (1968). A research center for augmenting human intellect. *Proceedings of the 1968 Fall Joint Computer Conference*, 395–410. Montvale, New Jersey: AFIPS Press.
. . . . **107, 109F, 415**

Engelbart, D. C. *See also* English, Engelbart, and Berman (1967).

ENGLISH, W. K., ENGELBART, D. C., and BERMAN, M. L. (1967). Display-selection techniques for text manipulation. *IEEE Transactions on Human Factors in Electronics HFE-8*, 5–15. . . . **230, 242**

English, W. K. *See also* Card, English, and Burr (1978); Engelbart and English (1968).

ERICKSEN, C. W., and SCHULTZ, D. W. (1978). Temporal factors in visual information processing: A tutorial review. In J. Requin, ed., *Attention and Performance VII*, Hillsdale, New Jersey: Lawrence Erlbaum Associates. **32n10**

ERICSSON, K. A., CHASE, W. G., and FALOON, S. (1980). Acquisition of memory skill. *Science 208*, 1181–1182. . . . . **37n**

Ericsson, K. A. *See also* Chase and Ericsson (1981); Chase and Ericsson (1982).

ESTES, W. K., ed. (1975–1978). *Handbook of Learning and Cognitive Processes* (6 vols). Hillsdale, New Jersey: Lawrence Erlbaum Associates. **2n1**

FAIRBAIRN, D. G., and ROWSON, J. H. (1978). ICARUS: An interactive integrated circuit layout program. *Proceedings of the 15th Annual Design Automation Conference, IEEE*, 188–192. . . . . **336**

Faloon, S. *See* Ericsson, Chase, and Faloon (1980).

Feehrer, C. E. *See* Pew, Baron, Feehrer, and Miller (1977).

GLANZER, M., and RAZEL. M. (1974). The size of the unit in short-term storage. *Journal of Verbal Learning and Verbal Behavior 13*, 114–131. . . . . **39n16**

Glenn, F. A. *See* Lane, Strieb, Glenn, and Wherry (1980).

Goodeve, P. J. *See* Crossman and Goodeve (1963).

GOODWIN, NANCY C. (1975). Cursor positioning on an electronic display using lightpen, lightgun, or keyboard for three basic tasks. *Human Factors 17*, 289–295. . . . . **230, 248**

GOULD, J. (1968). Visual factors in the design of computer-controlled CRT displays. *Human Factors 10*, 359–376. . . . . **91**

GREEN, D. M., and SWETS, J. A. (1966). *Signal Detection Theory and Psychophysics.* Huntington, New York: Robert E. Krieger Publishing Company. **2n2**

Green, T. R. G. *See* Sime, Fitter, and Green (1975); Sime and Green (1974) Fitter and Green (1979); Smith and Green (1980).

GUTTMAN, N., and JULESZ, B. (1963). Lower limits of auditory periodicity analysis. *Journal of the Acoustical Society of America 35*, 610. . . . . **31n5**

HAMMER, J. M. (1981). The Human as a Constrained Optimal Text Editor. Ph. D. thesis, Department of Computer Science, University of Illinois. Also Report T-105, Coordinated Science Laboratory, University of Illinois, Urbana, Illinois. **102**

HAMMER, J. M., and ROUSE, W. B. (1979). Analysis and modeling of freeform text editing behavior. *Proceedings of the 1979 International Conference on Cybernetics and Society,* Denver. **102**

Hammond, N. V. *See* Barnard, Hammond, Morton, Long, and Clark (1981).

HANSEN, W. J. (1971). User engineering principles for interactive systems. *Proceedings of the Fall Joint Computer Conference 39*, 523–532. . . . . **7, 418**

Hart, D. J. *See* Cakir, Hart, and Stewart (1980).

HARTER, M. R. (1967). Excitability and cortical scanning: A review of two hypotheses of central intermittency in perception. *Psychological Bulletin 68*, 47–58. . . . . **32n9**

Harter, N. *See* Bryan and Harter (1898); Bryan and Harter (1899).

Hatfield, S. A. *See* Mills and Hatfield (1974).

HERSHMAN, R. L., and HILLIX, W. A. (1965). Data processing in typing, typing rate as a function of kind of material and amount exposed. *Human Factors 7*, 483–492. . . . . **61F**

HICK, W. E. (1952). On the rate of gain of information. *Quarterly Journal of Experimental Psychology 4*, 11–26. . . . . **42n**

Hillix, W. A. *See* Hershman and Hillix (1965).

HIRSCHBERG, D. S. (1975). A linear space algorithm for computing maximal common subsequences. *Communications of the ACM 18*, 341–343. . . . . **157, 190**

HOCHBERG, J. (1976). Toward a speech-plan eye-movement model of reading. In R. A. Monty and J. W. Senders, eds., *Eye Movements and Psychological Processes*, 397–416, Hillsdale, New Jersey: Lawrence Erlbaum Associates. **51n21**

Hollan, J. D. *See* Williams and Hollan (1981).

HOVLAND, C. I. (1940). Experimental studies in rote learning theory. VI. Comparison of retention following learning to same criterion by massed and distributed practice. *Journal of Experimental Psychology 26*, 568–587. . . . . **81F**

HUNT, E. B., FROST, N. H., and LUNNEBORG, C. (1973). Individual differences in cognition: A new approach to intelligence. *The Psychology of Learning and Motivation 7*, 87–123. . . . . **3n9**

HYMAN, R. (1953). Stimulus information as a determinant of reaction time. *Journal of Experimental Psychology 45*, 188–196. . . . .
. . . . **43F, 75, 75F**

INGALLS. D. H. (1978). The Smalltalk-76 programming system: Design and implementation. *Conference Record of the Fifth Annual ACM Symposium on Principles of Programming Languages*, 9–16, Tucson, Arizona. **204**

JOHNSON, L. M. (1939). The relative effect of a time interval upon learning and retention. *Journal of Experimental Psychology 24*, 169–179. . . . . **81F**

JOHNSON, W. J. (1965). Analysis of Independence of Predetermined Time System Elements. M.S. Thesis, Department of Industrial Engineering, University of Miami. **223**

Julesz, B. *See* Guttman and Julesz (1963).

Kanarick, A. *See* Alden, Daniels, and Kanarick (1972).

Karlin, J. E. *See* Pierce and Karlin (1957).

KAPLAN, R. M., SHEIL, B. A., and SMITH, E. R. (1978). Interactive Data-Analysis Language Reference Manual. Palo Alto, California: Xerox Palo Alto Research Center, Technical Report SSL-78-4. xi

KAY, A. (1977). Microelectronics and the personal computer. *Scientific American*, September, 230–244. 204

KEELE, S. W. (1968). Movement control in skilled motor performance. *Psychological Bulletin 70*, 387–403. . . . . 51n22, 53

Kinkade, R. G. *See* Van Cott and Kinkade (1972).

KINKEAD, R. (1975). Typing speed, keying rates, and optimal keyboard layouts. *Proceedings of the 19th Annual Meeting of the Human Factors Society.* 34, 49, 60, 62F, 65, 167, 222, 222F

KINTSCH, W. (1974). *The Representation of Meaning in Memory.* Hillsdale, New Jersey: Lawrence Erlbaum Associates. 2n3

Kirshbaum, P. J. *See* Ramsey, Atwood, and Kirshbaum (1978).

KLAHR, D., and WALLACE, J. G. (1976). *Cognitive Development: An Information-Processing View.* Hillsdale, New Jersey: Lawrence Erlbaum Associates. 2n5

Klahr, D. *See also* Chi and Klahr (1975).

KLATZKY, R. L. (1980). *Human Memory: Structures and Processes,* (2nd ed.) San Francisco, California: W. H. Freeman and Company. . . . .
. . . . 31n5

KLEMMER, E. T. (1962). Communication and human performance. *Human Factors 4*, 75–79. . . . . 59F

KLEMMER, E. T., and LOCKHEAD, G. R. (1962). Productivity and errors in two keying tasks: A field study. *Journal of Applied Psychology 46*, 401–408. . . . . 61F

KOLESNIK, P. E., and TEEL, K. S. (1965). A comparison of three manual methods for inputting navigational data. *Human Factors 7*, 451–456.
. . . . 61F

KORNBLUM, S. (1973). *Attention and Performance IV.* New York: Academic Press. 2n1

KREUGER, W. C. (1929). The effect of overlearning on retention. *Journal of Experimental Psychology 12*, 71–78. . . . . 81F

Lampson, B. W. *See* Deutsch and Lampson (1967).

Lan, M. T. *See* Embley, Lan, Leinbaugh, and Nagy (1978).

LANE, N. E., STRIEB, M. I., GLENN, F. A., and WHERRY, R. J. (1980). The Human Operator Simulator: An overview. *NATO AGARD Conference on Manned Systems Design: New Methods and Equipment.* Frieburg, Federal Republic of Germany. **91**

LANDAUER, T. K. (1962). Rate of implicit speech. *Perception and Psychophysics 15,* 646. . . . . **42, 43F**

LANGOLF, G. D. (1973). Human Motor Performance in Precise Microscopic Work. Ph.D. Thesis, University of Michigan. Also published by the MTM Association, Fairlawn, New Jersey, 1973. . . . . . **53n24**

LANGOLF, G. D., CHAFFIN, D. B., and FOULKE, J. A. (1976). An investigation of Fitts's Law using a wide range of movement amplitudes. *Journal of Motor Behavior 8,* 113–128. . . . . **53n24**

Leinbaugh, D. W. *See* Embley, Lan, Leinbaugh, and Nagy (1978).

LESTER, O. P. (1932). Mental set in relation to retroactive inhibition. *Journal of Experimental Psychology 15,* 681–699. . . . . **81F**

LINDSAY, P. H., and NORMAN, D. A. (1977). *Human Information Processing: An Introduction to Psychology* (2nd ed.). New York: Academic Press. **2n1, 91, 95**

Lockhead, G. R. *See* Klemmer and Lockhead (1962).

LOFTUS, E. F. (1979). *Eyewitness Testimony.* Cambridge, Massachusetts: Harvard University Press. **3n8**

LONG, J. B. (1976). Visual feedback and skilled keying: Differential effects of masking the printed copy and the keyboard. *Ergonomics 19,* 93–110. . . . . **180, 215**

LONG, J. B., and BADDELEY, A., eds. (1981). *Attention and Performance IX.* Hillsdale, New Jersey: Lawrence Erlbaum Associates. **2n1**

Long, J. B. *See also* Barnard, Hammond, Morton, Long, and Clark (1981).

Love, T. *See* Shepard, Curtis, Milliman, and Love (1979).

LUH, C. W. (1922). The conditions of retention. *Psychological Monographs 31,* No. 3 (Whole No. 142). **81F**

Lunneborg, C. *See* Hunt, Frost, and Lunneborg (1973).

MARTIN, J. (1973). *Design of Man-Computer Dialogues.* Englewood Cliffs, New Jersey: Prentice-Hall. **91, 115**

Margolius, G. *See* Weiss, and Margolius (1954).

MASSARO, D. W. (1970). Preperceptual auditory images. *Journal of Experimental Psychology* 85, 411–417. . . . . **31n5**

MAYNARD, H. B. (1971). *Industrial Engineering Handbook, 3rd ed.* New York: McGraw-Hill. **161, 274, 298**

MAYZNER, M. S., and TRESSELT, M. E. (1965). Tables of single-letter and digram frequency counts for various word-length and letter-position combinations. *Psychonomic Monograph Supplements 1,* 13–32. . . . . **48F**

MCCORMICK, E. J. (1976). *Human Factors in Engineering and Design.* New York: McGraw-Hill. **3n10**

MEAD, C., and CONWAY, L. (1980). *Introduction to VLSI Systems.* Reading, Massachusetts: Addison-Wessley. **336**

MEISTER, D. (1976). *Behavioral Foundations of System Development.* New York: John Wiley and Sons. **91**

MELTON, A. (1963). Implications of short-term memory for a general theory of memory. *Journal of Verbal Learning and Verbal Behavior* 2, 1–21. . . . . **38n**

MICHOTTE, A. (1946/1963). *The Perception of Causality.* New York: Basic Books, 1963. Originally published as *La Perception de la Causalité.* Louvain: Publications Universitaires de Louvain, 1946. . .
. . . . **50F**

MICHON, J. A. (1978). The making of the present: A tutorial review. In J. Requin, ed., *Attention and Performance VII,* 89–111, Hillsdale, New Jersey: Lawrence Erlbaum Associates. **42n**

MILLER, G. A. (1956). The magical number seven plus or minus two: Some limits on our capacity for processing information. *Psychological Review 63,* 81–97. **39n17, 92, 93**

MILLER, L. A., and THOMAS, J. C., Jr. (1977). Behavioral issues in the use of interactive systems. *International Journal of Man-Machine Studies 9,* 509–536. . . . . **91**

Miller, D. C. *See* Pew, Baron, Feehrer, and Miller (1977).

Milliman, D. *See* Shepard, Curtis, Milliman, and Love (1979).

MILLS, R. G., and HATFIELD, S. A. (1974). Sequential task performance, task module relationships, reliabilities, and times. *Human Factors* 16, 117–128. . . . . **85F**

MINOR, F. J., and PITTMAN, G. G. (1965). Evaluation of variable format entry terminals for a hospital information system. Paper presented at the *Sixth Annual Symposium of the IEEE Professional Group on Human Factors in Electronics,* Boston. 61F

MINOR, F. J., and REVESMAN, S. L. (1962). Evaluation of input devices for a data setting task. *Journal of Applied Psychology 46,* 332–336. . . .
. . . . 61F

MORAN, T. P. (1980). Compiling Cognitive Skill. AIP Memo 150, Xerox Palo Alto Research Center. 279

MORAN, T. P. (1981a). The Command Language Grammar: A representation for the user interface of interactive computer systems. *International Journal of Man-Machine Systems 15,* 3–50. . . . .
. . . . 3, 91, 336, 404n, 421n

MORAN, T. P., ed. (1981b). Special Issue: The Psychology of Human-Computer Interaction. *Computing Surveys 13,* March. 252

MORAN, T. P., and CARD, S. K. (1982). Applying Cognitive Psychology to Computer Systems. *Proceedings of the Conference on Human Factors in Computer Systems,* Gaithersburg, Maryland. viii

Moran, T. P. *See also* Card, Moran, and Newell (1976, 1980a, 1980b); Roberts and Moran (1982).

Morton, J. *See* Barnard, Hammond, Morton, Long, and Clark (1981).

MUNGER, S. J., SMITH, R. W., and PAYNE, D. (1962). *An Index of Electronic Equipment Operability.* Pittsburgh, Pennsylvania: American Institute for Research, Report AIR-C43-1/62-RP(1). 61F

MURDOCK, B. B., JR. (1960b). The immediate retention of unrelated words. *Journal of Experimental Psychology 60,* 222–234. . . . . 39n16

MURDOCK, B. B., JR. (1961). Short-term retention of single paired-associates. *Psychological Reports 8,* 280. . . . . 38F, 38n

MURDOCK, B. B., JR. (1963). Short-term retention of single paired associates. *Journal of Experimental Psychology 65,* 433–443. . . . 82n

MURDOCK, B. B., JR. (1967). Recent developments in short-term memory. *British Journal of Psychology 58,* 421–433. . . . . 39n16

MURDOCK, B. B., JR. (1974). *Human Memory: Theory and Data.* Hillsdale, New Jersey: Lawrence Erlbaum Associates. 2n4

MYER, T. H., and BARNABY, J. R. (1973). TENEX executive language manual for users. Cambridge, Massachusetts: Bolt, Beranek, and Newman, Inc. 271F

Nagy, G. *See* Embley, Lan, Leinbaugh, and Nagy (1978); Anandan, Embley, and Nagy (1980); Embley and Nagy (1981).

NANDA, R. (1968). The additivity of elemental times. *Journal of Industrial Engineering 19*(5), 235–242. 223

NEAL, A. S. (1977). Time intervals between keystrokes, records, and fields in data entry with skilled operators. *Human Factors 19*, 163–170. . . . 61F

NEISSER, U. (1967). *Cognitive Psychology.* New York: Appleton-Century-Crofts. . . . 2n2, 91

NEWELL, A. (1973). Production systems: Models of control structures. In W. G. Chase, ed., *Visual Information Processing,* 283–308, New York: Academic Press. . . . 96

NEWELL, A. (1980). Reasoning, problem solving, and decision processes: The problem space as a fundamental category. In R. Nickerson, ed., *Attention and Performance VIII,* Hillsdale, New Jersey: Lawrence Erlbaum Associates. . . . 361, 361n, 364, 368F

NEWELL, A., and ROSENBLOOM, P. S. (1981). Mechanisms of skill acquisition and the law of practice. In J. R. Anderson, ed., *Cognitive Skills and their Acquisition,* 1–51, Hillsdale, New Jersey: Lawrence Erlbaum Associates. . . . 59n, 363

NEWELL, A., and SIMON, H. A. (1972). *Human Problem Solving.* Englewood Cliffs, New Jersey: Prentice-Hall. . . . 2n6, 41n, 86n, 88F, 89F, 91, 96, 147

Newell, A. *See also* Card, Moran, and Newell (1976, 1980a, 1980b); Siewiorek, Bell, and Newell (1981).

NEWMAN, W., and SPROULL, R. (1979). *Principles of Interactive Computer Graphics, 2nd ed.* New York: McGraw-Hill. . 271F, 404n

NILSSON, N. (1971). *Problem-Solving Methods in Artificial Intelligence.* New York: McGraw-Hill. . . . 364

NORMAN, D. A. (1980). Cognitive engineering and education. In D. T. Tuma, and F. Reif, eds., *Problem Solving in Education: Issues in Teaching and Research,* 97-107, Hillsdale, New Jersey: Lawrence Erlbaum Associates. . . . 3n16

NORMAN, D. A. (1981). Categorization of action slips. *Psychological Review 88*, 1–15. . . . 423n

NORMAN. D. A., and BOBROW, D. (1975). On data-limited and resource-limited processes. *Cognitive Psychology 7*, 44–64. . . . . **94, 387**

NORMAN, D. A., and RUMELHART, D. E. (1975). *Explorations in Cognition.* San Francisco: W. H. Freeman. **95**

Norman, D. A. *See also* Waugh and Norman (1965); Lindsay and Norman (1977).

OREN, S. S. (1972). A mathematical model for computer-assisted document creation. *Proceedings of the Fourth International Symposium on Computer and Information Sciences,* Miami Beach, Florida. . . **102**

OREN, S. S. (1974). A mathematical theory of man-machine text editing. *IEEE Transactions on Systems, Man, and Cybernetics SMC-4,* 256–267. . . . . **102**

OREN, S. S. (1975). A mathematical theory of man-machine document assembly. *IEEE Transactions on Systems, Man, and Cybernetics SMC-5,* 256–267. . . . . **102**

PARSONS, H. M. (1972). *Man-Machine Systems Experiments.* Baltimore, Maryland: Johns Hopkins University Press. **91**

Payne, D. *See* Munger, Smith, and Payne (1962).

Peterson, J. R. *See* Fitts and Peterson (1964).

PETERSON, L. R., and PETERSON, M. J. (1959). Short-term retention of individual verbal items. *Journal of Experimental Psychology 58,* 193–198. . . . . **38F, 38n**

Peterson, M. J. *See* Peterson and Peterson (1959).

PEW, R. W., BARON, S., FEEHRER, C. E., and MILLER, D. C. (1977). *Critical Review and Analysis of Performance Models Applicable to Man-Machine Systems Evaluation.* Cambridge, Massachusetts: Bolt, Beranek, and Newman, Inc., Report 3446. . . . . **91**

PIERCE, J. R., and KARLIN, J. E. (1957). Reading rates and the information rate of the human channel. *Bell System Technical Journal 36,* 497–516. . . . . **55n**

Pittman, G. G. *See* Minor and Pittman (1965).

POLLACK, W. T., and GILDNER, G. G. (1963). Study of Computer Manual Input Devices. Hanscom Field, Bedford, Massachusetts: Air Force Systems Command, Electronic Systems Division, September, Report ESD-TDR-63-545. . . . . **61F**

POSNER, M. I. (1978). *Chronometric Explorations of Mind.* Hillsdale, New Jersey: Lawrence Erlbaum Associates. **2n3**

POSNER, M. I., BOIES, S. J., EICHELMAN, W. H., and TAYLOR, R. L. (1969). Retention of visual and name codes of single letters. *Journal of Experimental Psychology 79*, 1–16. **72F**

Posner, M. I. *See also* Fitts and Posner (1967).

POULTON, E. C. (1974). *Tracking Skill and Manual Control.* New York: Academic Press. **243, 248, 254, 254F**

QUICK, J. H. (1962). *Work Factor Time Standards.* New York: McGraw-Hill. **91**

RABBITT, P. M. A., and DORNIČ, S. (1975). *Attention and Performance V.* London: Academic Press. **2n1**

Radford, B. *See* Fitts and Radford (1966).

RAMSEY, H. R., and ATWOOD, M. E. (1979). Human Factors in Computer Systems: A Review of the Literature. Englewood, Colorado: Science Applications, Inc., Technical Report SAI-79-111-DEN, NTIS AD A075679. . . . . **91, 404n**

RAMSEY, H. R., ATWOOD, M. E., and KIRSHBAUM, P. J. (1978). A Critically Annotated Bibliography of the Literature on Human Factors in Computer Systems. Englewood, Colorado: Science Applications, Inc., Technical Report SAI-78-070-DEN, NTIS AD-A057081. . . . . **91**

Raphael, B. *See* Bobrow and Raphael (1974).

RAYMOND, B. (1969). Short-term storage and long-term storage in free recall. *Journal of Verbal Learning and Verbal Behavior 8*, 567–574. . . . . **39n16**

Razel, M. *See* Glanzer and Razel (1974).

Reed, S. K. *See* Simon and Reed (1976).

REISNER, PHYLLIS (1981). Using a formal grammar in human factors design of an interactive graphics system. *IEEE Transactions on Software Engineering SE-7*, 229–240. . . . . **91, 421n**

REQUIN, J. (1978). *Attention and Performance VII.* Hillsdale, New Jersey: Lawrence Erlbaum Associates. **2n1**

Revesman, S. L. *See* Minor and Revesman (1962).

Rice, D. E. *See* Van Dam and Rice (1971).

Richardson, J. *See* Underwood and Richardson (1956).

RIDDLE, ELIZABETH A. (1976). *Comparative Study of Various Text Editors and Formatting Systems.* Washington, D.C.: Air Force Data Services Center, The Pentagon, AD-A029 050. 102

ROBERTS, TERESA L. (1979). Evaluation of Computer Text Editors. Ph.D. Thesis, Department of Computer Science, Stanford University. Reprinted as Xerox Palo Alto Research Center Technical Report SSL-79-9. . . . . 102, 413, 416F, 424, 431

ROBERTS, TERESA L., and MORAN, T. P. (1982). A methodology for evaluating text editors. *Proceedings of the Conference on Human Factors in Computer Systems,* Gaithersburg, Maryland. 413

Rosenbloom, P. S. *See* Newell and Rosenbloom (1981).

ROUSE, W. B. (1977). Human-computer interaction in multi-task situations. *IEEE Transactions on Systems, Man, and Cybernetics SMC-7,* 384–392. 91

ROUSE, W. B. (1980). *Systems Engineering Models of Human-Machine Interaction.* New York: North Holland. 91

Rouse, W. B. *See also* Hammer and Rouse (1979).

Rowson, J. H. *See* Fairbairn and Rowson (1978).

Rumelhart, D. E. *See* Norman and Rumelhart (1975).

RUSSELL, D. S. (1973). POET: A Page Oriented Editor for TENEX. Computer Science Division, University of Utah. 109F

RUSSO, J. E. (1978). Adaptation of cognitive processes to the eye-movement system. In J. W. Senders, D. F. Fisher, and R. A. Monty, eds., *Eye Movements and the Higher Psychological Functions,* 89–109. Hillsdale, New Jersey: Lawrence Erlbaum Associates. 25n, 28n

SACKMAN, H. (1970). Experimental analysis of man-computer problem-solving. *Human Factors 12,* 187–201. 119

SAKOE, H., and CHIBA, S. (1978). Dynamic programming algorithm optimization for spoken word recognition. *IEEE Transactions on Acoustics, Speech, and Signal Processing ASSP-26,* 43–49. . 157, 190

SAVITSKY, S. (1969). Son of STOPGAP. Stanford, California: Stanford University Artificial Intelligence Laboratory, Operating Note 50.1. . . . . 109F, 271F

SCHMIDTKE, H., and STIER, F. (1961). An experimental evaluation of the validity of predetermined elemental time systems. *The Journal of Industrial Engineering XII* (3), 182–204. 13

Schneider, W. *See* Shiffrin and Schneider (1977).

Schultz, D. W. *See* Ericksen and Schultz (1978).

Schulz, R. W. *See* Underwood and Schulz (1960).

SEIBEL, R. (1964). Data entry through chord, parallel entry devices. *Human Factors 6*, 189–192. . . . . 61F

SEIBEL, R. (1972). Data entry devices and procedures. In H. P. Van Cott and R. G. Kinkade, eds., *Human Engineering Guide to Equipment Design*, Washington, D. C.: U. S. Government Printing Office. 91

SHEIL, B. (1981). The psychological study of programming. *Computing Surveys 13*, 101–120. 91

Sheil, B. A. *See also* Kaplan, Sheil, and Smith (1968).

SHEPARD, S., CURTIS, B., MILLIMAN, P., and LOVE, T. (1979). Modern coding practices and programmer performance. *Computer 12*, 41–49.
. . . . 91

SHERIDAN, T. B., and FERRELL, W. R. (1974). *Man-machine systems: Information, control, and decision models of human performance.* Cambridge, Massachusetts: M.I.T. Press. 3n10, 91

SHIFFRIN, R. M. and DUMAIS, SUSAN T. (1981). The development of automatism. In J. R. Anderson, ed., *Cognitive Skills and their Acquisition*, 111–140, Hillsdale, New Jersey: Lawrence Erlbaum Associates. 363

SHIFFRIN, R. M., and SCHNEIDER, W. (1977). Controlled and automatic human information processing: II. Perceptual learning, automatic attending, and a general theory. *Psychological Review 84*, 127–190. .
. . . . 93

Shiffrin, R. M. *See also* Atkinson and Shiffrin (1968).

SHNEIDERMAN, B. (1980). *Software Psychology.* Cambridge, Massachusetts: Winthrop. 3n14, 91

SHURTLEFF, D. A. (1980). *How to Make Displays Legible.* La Mirada, California: Human Interface Design. 91

SIEGAL, A. I., and WOLF, J. J. (1969). *Man-machine simulation models.* New York: John Wiley and Sons. 91

SIEWIOREK, D., BELL, G., AND NEWELL, A. (1981). *Computer Structures.* New York: McGraw-Hill. **24n**

SIME, M. E., FITTER, M., and GREEN, T. R. G. (1975). Why is programming computers so hard? *New Behaviour* (September 4). **3n13**

SIME, M. E., and GREEN, T. R. G. (1974). Psychology and the Syntax of Programming. Medical Research Council, Social and Applied Psychology Unit, Department of Psychology, The University, Sheffield, MRC Memo No. 52. . . . . **3n12**

SIMON, H. A. (1947). *Administrative Behavior.* New York: Macmillan. **86n**

SIMON, H. A. (1969). *The Sciences of the Artificial.* Cambridge, Massachusetts: M.I.T. Press. **86n**

SIMON, H. A. (1974). How big is a chunk? *Science 183,* 482–488. . . **36n**

SIMON, H. A. (1975). The functional equivalence of problem solving skills. *Cognitive Psychology 7,* 268–288. . . . . **364, 370**

Simon, H. A. *See also* Newell and Simon (1972); Chase and Simon (1973).

Smith, E. R. *See also* Kaplan, Sheil, and Smith (1978).

SMITH, G. A. (1977). Studies of compatibility and a new model of choice reaction time. In S. Dornič, ed., *Attention and Performance VI,* Hillsdale, New Jersey: Lawrence Erlbaum Associates. **71n**

SMITH, H. T., and GREEN, T. R. G., eds. (1980). *Human Interaction with Computers.* London: Academic Press. **91**

Smith, R. W. *See also* Munger, Smith, and Payne (1962).

SNODDY, G. S. (1926). Learning and stability. *Journal of Applied Psychology 10,* 1–36. . . . . **57**

SPERANDIO, J. C., and BISSERET, A. (1974). Human Factors in the Study of Information Input Devices. Royal Aircraft Establishment Library Translation No. 1728. Originally published as Facteurs humains dans l'étude des dispositifs d'entrée d'informations, *Bulletin du CERP 17,* 4, 269–294 (1968). **91**

SPERLING, G. (1960). The information available in brief visual presentations. *Psychological Monographs 74* (11, Whole No. 498). **29n, 30F, 31n6, 92**

SPERLING, G. (1963). A model for visual memory tasks. *Human Factors 5*, 19–31. . . . . **31n6**

Sprague, L. T. *See* Elkind and Sprague (1961).

Sproull, R. *See* Newman and Sproull (1979).

STALLMAN, R. M. (1981). EMACS—The extensible, customizable self-documenting display editor. *Proceedings of the ACM SIGPLAN SIGOA Symposium on Text Manipulation,* Portland, Oregon, 147–156. . . . . **415**

STANFORD CENTER FOR INFORMATION PROCESSING (1975). Wylbur/ 370—the Stanford Timesharing System—Reference Manual, 3rd ed. Stanford, California: Stanford University. **121, 415**

STERNBERG, S. (1975). Memory scanning: New findings and current controversies. *Quarterly Journal of Experimental Psychology 27,* 1–32. . . . . **42n**

Stewart, T. F. M. *See* Cakir, Hart, and Stewart (1980).

Stier, F. *See* Schmidtke and Stier (1961).

Strieb, M. I. *See* Lane, Strieb, Glenn, and Wherry (1980).

Swets, J. A. *See* Green and Swets (1966).

Taylor, R. C. *See* Posner, Boies, Eichelman, and Taylor (1969).

Teel, K. S. *See* Kolesnik and Teel (1965).

TEITELMAN, W. (1978). INTERLISP Reference Manual. Palo Alto, California: Xerox Palo Alto Research Center. . . . . **xi**

Thompson, D. M. *See* Tulving and Thompson (1973).

Thomas, J. C., Jr. *See* Miller and Thomas (1977).

Tresselt, M. E. *See* Mayzner and Tresselt (1965).

TULVING, E., and COLATLA, V. (1970). Free recall of trilingual lists. *Cognitive Psychology 1,* 86–98. . . . . **39n16**

TULVING, E., and THOMPSON, D. M. (1973). Encoding specificity and retrieval processes in episodic memory. *Psychological Review 80,* 352–373. . . . . **40n**

Turvey, M. T. *See* Darwin, Turvey, and Crowder (1972).

TVERSKY, A. (1977). Features of similarity. *Psychological Review 84,* 327–352. . . . . **2n3**

UNDERWOOD, B. J. (1952). Studies of distributed practice: VII. Learning and retention of serial nonsense lists as a function of intralist similarity. *Journal of Experimental Psychology 44*, 80–87. . . . . . **81F**

UNDERWOOD, B. J. (1953a). Studies of distributed practice: VIII. Learning and retention of paired nonsense syllables as a function of intralist similarity. *Journal of Experimental Psychology 45*, 133–142. . . . . **81F**

UNDERWOOD, B. J. (1953b). Studies of distributed practice: IX. Learning and retention of paired adjectives as a function of intralist similarity. *Journal of Experimental Psychology 45*, 143–149. . . . . **81F**

UNDERWOOD, B. J. (1953c). Studies of distributed practice: X. The influence of intralist similarity on learning and retention of serial adjective lists. *Journal of Experimental Psychology 45*, 253–259. . . . . **81F**

UNDERWOOD, B. J. (1957). Interference and forgetting. *Psychological Review 64*, 49–60. . . . . **81, 81F**

UNDERWOOD, B. J., and RICHARDSON, J. (1956). The influence of meaningfulness, intralist similarity, and serial position on retention. *Journal of Experimental Psychology 52*, 119–126. . . . . **81F**

UNDERWOOD, B. J., and SCHULZ, R. W. (1960). *Meaningfulness and Verbal Learning*. Philadelphia: Lippincott. **64F**

VALLEE, J. (1976). There ain't no user science: a tongue-in-cheek discussion of interactive systems. *Proceedings of the American Society for Information Science Annual Meeting 13*, San Francisco. **3n11**

VAN COTT, H. P., and KINKADE, R. G. (1972). *Human Engineering Guide to Equipment Design* (revised ed.). Washington, D. C.: U. S. Government Printing Office. **218**

VAN DAM, A., and RICE, D. E. (1971). On-line text editing: A survey. *Computing Surveys 3*, 93–114. . . . . **102**

VanLehn, K. *See* Brown and VanLehn (1980).

VINCE, MARGARET A. (1948). Corrective movements in a pursuit task. *Quarterly Journal of Experimental Psychology 1*, 85–103. . . . **53, 253**

Vince, Margaret A. *See also* Craik and Vince (1963).

WAINER, H. (1976). Estimating coefficients in linear models: It don't make no nevermind. *Psychological Bulletin 83*, 213–217. . . . . . . **296**

Wallace, J. G. *See* Klahr and Wallace (1976).

Subject Index

An F after a page number indicates that the reference is located in a figure on the page, and an n indicates that the reference is located in a footnote.